MOHAMED HE

D1797159

Tons. Men. For

Sunderland   18

VIRTUTE ET LABORE

Lambert Sc. Newcastle on Steel

I have shipped on board the

of

Master:

Tons &   Cwt of

Hetton Wallsend Coals.

WROUGHT & GOTTEN OUT OF HETTON COLLIERY.

At

Shillings & Sixpence pr. 53 Cwt loaded by drop.

FOR THE Hetton Coal Co.

Custom House. SUNDERLAND.

Ent.d No

Collector

Tons | Men   For,

Sold and delivered on board the

NETHERTON COAL COMPANY.

Blyth

18

Tons and   Cwt of

Lamberts on Steel

HOWARD'S WEST HARTLEY NETHERTON MAIN COALS.

AT   PER TON.

Custom House. **BLYTH.**

No   Entd

TO  THE
PUBLIC.

The Ship-builders of the Tyne having stated in a Hand-bill that they had taken it into Consideration to Employ 150 Ship-wrights, (out of 700 unemployed,) at 21s. per Week in consequence of the great Distress prevailing in the Neighbourhood, (and Sunderland in particular,) we, the Ship-wrights of the Tyne, are ready to serve any Ship-builder at the aforesaid Wages, either for six, eight or twelve Month but we would wish to intimate to the Public that we have been accosted by Mr. Mc' Leod, the Agent for the Middle Dock Company, and Mr. James Edwards to work for 21s. per Week, but instead of 21s. per Week it proves to be 3s. 6d. per Day; now for the last twelve Months we have been working for 4s. per Day, and our average Wages per Week for the best employed Men in the River, has not exceeded 14s. 7d. per Week, and for the second best employed, not more than 9s. 5d. per Week, and for the worst part not more than 6s. 4d. per Week; now we would leave it to the better Judgment of the Public, to ascertain what might be our average Wages at 3s. 6d. per Day, when we, the Ship-wrights of the Tyne have given our average Wages per Week, at 4s. per Day, and we would wish to inform the Public that Mr. Metcalfe intimated to us for a Reduction of Wages, and he told us that if the Men would go to work for 1s. per Day, Mr. Forsyth either would not or could not employ one Man; yet the time may come when we will have an opportunity to raise our Wages, then the Ship-owners of the Tyne may thank Mr Mc' Leod, Mr. Smith, and Mr. Edwards, for the high rate of Wages paid in the Tyne.

South Shields, 4th August, 1842

Market Place Printing Offices; R. M. Kelly, South Shields.

EMIS Datareviews Series No. 11

Series Advisor: Dr. B. L. Weiss

PROPERTIES OF
Group Ⅲ Nitrides

ELECTRONIC MATERIALS INFORMATION SERVICE

Other books in the EMIS Datareviews Series from INSPEC:

PROPERTIES OF
Group III Nitrides

Edited by

JAMES H. EDGAR
Kansas State University, USA

Published by: INSPEC, the Institution of Electrical Engineers, London, United Kingdom

© 1994: INSPEC, the Institution of Electrical Engineers

Apart from any fair dealing for the purposes of research or private study, or criticism or review, as permitted under the Copyright, Designs and Patents Act, 1988, this publication may be reproduced, stored or transmitted, in any forms or by any means, only with the prior permission in writing of the publishers, or in the case of reprographic reproduction in accordance with the terms of licences issued by the Copyright Licensing Agency. Inquiries concerning reproduction outside those terms should be sent to the publishers at the undermentioned address:

Institution of Electrical Engineers
Michael Faraday House,
Six Hills Way, Stevenage,
Herts. SG1 2AY, United Kingdom

While the editor and the publishers believe that the information and guidance given in this work is correct, all parties must rely upon their own skill and judgment when making use of it. Neither the editor nor the publishers assume any liability to anyone for any loss or damage caused by any error or omission in the work, whether such error or omission is the result of negligence or any other cause. Any and all such liability is disclaimed.

The moral right of the authors to be identified as authors of this work has been asserted by them in accordance with the Copyright, Designs and Patents Act 1988.

British Library Cataloguing in Publication Data

A CIP catalogue record for this book
is available from the British Library

ISBN 0 85296 818 3

Printed in England by Short Run Press Ltd., Exeter

Contents

10   MATERIAL INTERFACES WITH GROUP III NITRIDES

At the editor's request, all royalties on the sale of this book are to be paid to the international relief charity CARE.

Foreword

Group III nitrides encompass a range of materials with bandgaps corresponding to wavelengths from red to vacuum UV. The mechanical and thermal properties of these nitrides with bandgaps well into the UV are extremely robust. A case in point is BN, which has many applications in tool coating and packaging for operation in adverse environments. Other nitrides, such as GaN and its alloys, are exploited for light emitting diodes (LED) in the UV, blue and green regions with applications to full colour displays, traffic lights and automobile and aircraft lighting. Moreover, the wide bandgaps afforded by group III nitrides, coupled with favourable transport properties, pave the way for high temperature and/or high power electronic applications.

Group III nitrides have captured the limelight recently with the demonstration of bright InGaN blue LEDs having intensities of several cd, ample for displays. Moreover, demonstration in the laboratory of InGaN green LEDs with intensities of about 1 cd (unmatched by GaP), together with the available AlGaAs red counterparts, is certain to impact not only the display area but also large volume markets such as traffic lights, automobile and aircraft lights and colour-on-demand illumination for home and office use.

GaN, InN and AlN when used together in appropriate combinations are conducive for the generation of coherent light sources in the visible and UV part of the spectrum. With the short wavelengths afforded by these potential coherent sources, digital information storage density can be enhanced quadratically with decreasing wavelength. It is estimated that information at a density approaching $1 \, \mathrm{Gbit \, cm^{-2}}$ should be possible. With parallel advances in the recording media, write and read operation by the consumer will be possible.

Some of these wide gap nitrides, such as AlN and high Al mole fraction AlGaN, are purported to have negative electron affinity surfaces. If so, mono-energetic cold cathodes for vacuum electronics, flat panel displays, and electron microscopes could be attained. These cathodes would have longer lifetimes, operate in poor vacuum and, in the case of displays, consume much less power as the deflection voltages required would be small.

Large bandgaps and heterojunction capability afforded by at least some of the group III nitrides pave the way for electronic devices operative at high temperatures and/or high power levels. For example, AlN with its excellent insulating properties can be used as the gate dielectric, in conjunction with pseudomorphic InGaN on GaN, to achieve FETs with large transconductances, large current capabilities, and large gate swings in the forward direction. These features are essential for linear high power devices. Emerging advances in the materials technology suggest that these high performance devices will be produced in due time.

Among all the nitrides, GaN and its allied alloys have received by far the most attention. As early as mid-1970, blue LEDs based on the MIS concept were achieved with very good longevity. A metal layer was used to inject electrons into Zn-doped GaN where they radiatively drop to acceptor levels, giving off the blue emission. This diode was temporarily marketed, but it suffered from a large forward voltage. With advances in the buffer layers and achievement of active Mg doping, high quality p-n junctions became possible in the past few years. A flurry of activity then followed leading to blue LEDs with over 1 cd brightness being available from Nichia Chemical Industries, Inc., and soon to be available from Toyoda

Gosei. High vacuum deposition techniques, reminiscent of molecular beam epitaxy, have also seen a boost with active nitrogen species generated with ion sources and Electron Cyclotron Resonance (ECR) sources. At the present time, many researchers are immersed in the investigation of the nitrogen species most conducive for the growth of nitrides. These sources aided greatly the achievement of good GaN films by vacuum deposition.

Boron nitride films are also grown under non-equilibrium conditions using the active nitrogen sources mentioned above. Although not yet as successful as for GaN and its allied alloys, great strides have been made in the growth of the cubic phase of BN. In addition, high temperature/high pressure conversion of industrial grade BN to cubic BN has led to UV p-n junction electroluminescent devices. Hardness and thermal conductivity of BN, second only to diamond, are of importance for its applications to tools and the coating of other devices and circuits.

With many exciting developments, all of the group III nitride family have gained considerable interest in the last few years. However, the data on the properties of these emerging materials are scattered over many decades in time and many journals. One must spend an inordinate amount of time and effort to gather the information needed. This book, with contributions from experts in the field, is very timely and certainly is an ideal and compact forum for the dissemination of the most recent data. It will certainly be one of the most referenced books in my office.

Hadis Morkoc
University of Illinois
September 1994

Introduction

The group III nitrides have been potential candidates for semiconductor devices for many years, but only recently has control of the material quality improved sufficiently to develop devices containing p-n junctions. As described in the Foreword, the group III nitrides have great potential for device applications, primarily as short wavelength emitters and detectors and in high temperature electronics. Thus, as devices are beginning to be produced, it is timely to collect and evaluate the properties of these materials.

The chapters are organised by property, and contain summaries of each property of each nitride. The physical, thermal, and structural properties of the group III nitrides, which are significantly different from the traditional semiconductors, are presented in Chapter 1. The phase diagrams of the group III nitrides presented in Chapter 2 help to explain why the group III nitrides are so difficult to produce under equilibrium conditions. The collected electrical properties described in Chapter 3 suggest that the nitrides are excellent candidates for devices operating under extreme conditions. The application of theory to group III nitrides and their alloys in Chapters 4 and 5 enables prediction of properties not yet measured, gives help in interpreting experimental results, and provides guidance for future research by predicting property values to be expected. The important fundamental optical functions necessary for designing many devices are presented in Chapter 6. The photoluminescence, cathodoluminescence and Raman spectroscopy of Chapters 7 and 8 provide standards for characterizing the quality of material. Chapter 9 discusses common structural defects in the group III nitrides and their impact on the materials' properties. The properties of interfaces formed with group III nitrides, both metallic and semiconducting heterojunctions, are detailed in Chapter 10.

The purpose of this book is to serve as a resource for those studying the group III nitrides. There are many opportunities for measuring fundamental physical properties, synthesizing new materials and developing new devices, and this book should help to guide such research.

James H. Edgar
Kansas State University
September 1994

Contributing Authors

I. Akasaki	Meijo University, Faculty of Science & Technology, Department of Electrical Engineering, Tempaku-Ku, Nagoya 468, Japan	1.4, 7.2
H. Amano	Meijo University, Faculty of Science & Technology, Department of Electrical Engineering, Tempaku-Ku, Nagoya 468, Japan	1.4, 7.2
W.A. Bryden	The Johns Hopkins University, Applied Physics Laboratory, Laurel, MD 20723, USA	3.3
R.F. Davis	North Carolina State University, Department of Materials Science & Engineering, Raleigh, NC 27695, USA	10.2, 10.3
G.L. Doll	General Motors Research Laboratory, Physics Department, 30500 Mound Road, Warren, PO Box 9055, MI 48090-9055, USA	6.1, 8.2
K. Doverspike	The Naval Research Laboratories, Code 6861, 4555 Overlook Avenue, Washington, DC 20375, USA	3.2
J.H. Edgar	Kansas State University, Department of Chemical Engineering, Durland Hall, Manhattan, KS 66506-5102, USA	1.1, 1.2
R.H. French	E.I. du Pont de Nemours & Company, Central Research & Development, Wilmington, DE 19880, USA	6.2
D.K. Gaskill	The Naval Research Laboratories, Code 6861, 4555 Overlook Avenue, Washington, DC 20375, USA	3.2
D.R. Gilbert	The University of Florida, Department of Materials Science & Engineering, Gainsville, FL 32611, USA	10.1
I. Grzegory	Polish Academy of Sciences, High Pressure Research Center, Ul. Sokolowska 29/37, Warsaw 01-142, Poland	2.2-2.4
J.H. Harris	BP Research, Warrensville Research Center, Cleveland, OH 44128, USA	7.1
D. Jenkins	Amoco Research Center, MS-F5, PO Box 3011, Naperville, IL 60566, USA	5.1

T.J. Kistenmacher	The Johns Hopkins University, Applied Physics Laboratory, Laurel, MD 20723, USA	3.3
W.R.L. Lambrecht	Case Western Reserve University, Department of Physics, Cleveland, OH 44106, USA	4.1-4.5, 5.2
V.V. Lopatin	High Voltage Research Institute, Polytechnical University, Lenin Street 2a, Tomsk 634050, Russia	3.1
S. Loughin	General Electric, 29B12, PO Box 8555, Philadelphia PA 19101-8555, USA	6.2
T. Matsuoka	Nippon Telegraph and Telephone Corporation, NTT Opto-electronics Laboratories, 3-1, Morinosato Wakamiya, Atsugi-shi, Kanagawa Pref., 243-01 Japan	7.3
L.E. McNeil	University of North Carolina, Department of Physics & Astronomy, Phillips Hall CB #, Chapel Hill, NC 27599-3255, USA	8.1, 8.3-8.5
W.J. Meng	General Motors Research Laboratories, 30500 Mound Road, Warren, MI 48090-9055, USA	1.3
J.A. Miragliotta	The Johns Hopkins University, Applied Physics Department, John Hopkins Road, Laurel, MD 20723, USA	6.3, 6.4
S. Porowski	Polish Academy of Sciences, High Pressure Research Center, Ul. Sokolowska 29/37, Warsaw 01-142, Poland	2.2-2.4
L.B. Rowland	The Naval Research Laboratories, Code 6861, 4555 Overlook Avenue, Washington, DC 20375, USA	3.2
B. Segall	Case Western Reserve University, Department of Physics, Cleveland, OH 44106, USA	4.1-4.5, 5.2
R.K. Singh	The University of Florida, Department of Materials Science & Engineering, Gainsville, FL 32611, USA	10.1
L.L. Smith	North Carolina State University, Department of Materials Science & Engineering, Raleigh, NC 27695, USA	10.2, 10.3
V.L. Solozhenko	Institute for Superhard Materials, Ukrainian Academy of Sciences, 2, Avtozavodskaya Street, Kiev, 254153, Ukraine	2.1

S.C. Strite   IBM Research Division, Saumerstrasse 4,   9.1-9.5
                     Ruschlikon, CH 8803, Switzerland

T.L. Tansley   MacQuarie University, School of Mathematics,   1.5
                     Physics, Computing & Electronics, Sydney,
                     NSW 2109, Australia

R.A. Youngman   BP Research, Warrensville Research Center,   7.1
                     Cleveland, OH 44128, USA

Acknowledgements

The editor is grateful for the helpful comments provided by these reviewers of the manuscripts incorporated in this book.

C.R. Abernathy   AT&T Bell Laboratories, Murray Hill, NJ 07974, USA

F.P. Bundy   Lebanon, OH 45036, USA

R.J. Caveney   De Beers Diamond Research Laboratory, Johannesburg 2000, South Africa

W.Y. Ching   University of Missouri, Kansas City, MO 64110, USA

W.J. Choyke   University of Pittsburgh, Pittsburgh, PA 15260, USA

N.E. Christensen   Aarhus University, Aarhus, Denmark

D. Debowska   Jagellonian University, Krakow, Poland

R. DeVries   Burnt Hills, NY 12027, USA

T.J. Drummond   Sandia National Laboratories, Albuquerque, NM 87185, USA

J.A. Freitas   Naval Research Laboratory, Washington, DC 20375, USA

U. Kaufman   Fraunhofer-Institute fur Angewandte Festkorperphysik, Tullastraße 72, D-79108 Freiburg i Br., Germany

S. Krishnankutty   APA Optics, Blaine, MN 55449, USA

Z.H. Levine   Ohio State University, Columbus, OH 43210, USA

A.W. Moore   Union Carbide Coating Services Corporation, Cleveland, OH 44101, USA

H. Morkoc   University of Illinois, Urbana, IL 61801, USA

N. Newman   Lawrence Berkeley Laboratories, Berkeley, CA 94720, USA

R.M. Park   University of Florida, Gainsville, FL 32611, USA

A. Yoshida   Toyohashi University of Technology, Toyohashi, Japan

S. Yoshida   Electrotechnology Laboratory, Ibaraki 305, Japan

Abbreviations

The following abbreviations are used throughout the book.

ASA	atomic sphere approximation
BG	blue-green (band)
BN	boron nitride
β-BN	cubic boron nitride
c-BN	cubic boron nitride
CL	cathodoluminescence
CP	critical point
C-V	capacitance-voltage
CVD	chemical vapour deposition
DC	direct current
2DEG	two-dimensional electron gas
DFT	density functional theory
DSC	differential scanning calorimetry
DTA	differential thermal analysis
ECR	electron cyclotron resonance
EELS	electon energy loss spectroscopy
EP	empirical pseudopotential
EPR	electron paramagnetic resonance
FCC	face centred cubic
FET	field effect transistor
FLAPW	full-potential linearized augmented plane-wave
FP	full-potential
γ-BN	wurtzite BN
GW	Approximation to the self energy based on the one-electron Green's function G and screened Coulomb interaction W (Hedlin & Lundquist, 1969)
h-BN	hexagonal boron nitride
HCP	hexagonal close packed
HEMT	high electron mobility transistor
HOPBN	highly oriented pyrolytic boron nitride
HVPE	hydride vapour phase epitaxy
IC	integrated circuit
IDB	inversion domain boundary
IR	infrared
I-V	current-voltage
LA	longitudinal acoustic
LA-CVD	laser assisted chemical vapour deposition

LDA	local density approximation
LED	light emitting diode
LEEBI	low energy electron beam irradiation
LMTO	linear muffin-tin orbital
LO	longitudinal optical
LVB	lower valence band
MBE	molecular beam epitaxy
MESFET	metal semiconductor field effect transistor
MIS	metal-insulator-semiconductor
MOCVD	metal-organic chemical vapour deposition
MOVPE	metal-organic vapour phase epitaxy
MOMBE	metal-organic molecular beam epitaxy
NBE	near band edge
OLCAO	orthogonalised linear combination of atomic orbitals
OMVPE	organometallic vapour phase epitaxy
PBN	pyrolytic boron nitride
PCVD	plasma chemical vapour deposition
PF	ab initio pseudofunction method
PL	photoluminescence
PP-PW	norm-conserving pseudopotential plane-wave
r-BN	rhombohedral boron nitride
R	red (band)
RF	radio frequency
RHEED	reflection high energy electron diffraction
RT	room temperature
SCF	self-consistent field
SEM	scanning electron microscopy
SIMS	secondary ion mass spectrometry
SIS	semiconductor-insulator-semiconductor
TA	transverse acoustic
TB	semi-empirical tight binding method
t-BN	turbostratic boron nitride
TEM	transmission electron microscopy
TLM	transfer-length measurements
TMA	thermomechanical analysis
TO	transverse ptical
TSC	thermally stimulated conductivity
TSL	thermally stimulated luminescence
UPS	ultraviolet photoelectron spectroscopy
UV	ultraviolet
VB	violet-blue (band)

VPE	vapour phase epitaxy
VUV	vacuum ultraviolet
w-BN	wurtzite boron nitride
WZ	wurtzite
XRD	X-ray diffraction
YG	yellow-green (band)
ZB	zinc blende

CHAPTER 1

BASIC PHYSICAL PROPERTIES

1.1 Common crystal structures of group III nitrides

J.H. Edgar

February 1994

There are three common crystal structures shared by the group III nitrides: the wurtzite, zinc blende (also known as sphalerite, or the β-polytype), and rocksalt structures. At ambient conditions, the thermodynamically stable structures are wurtzite for bulk AlN, GaN and InN, and zinc blende for BN. The zinc blende structure for GaN and InN has been stabilized by epitaxial growth of thin films on the (001) crystal planes of cubic substrates such as Si [1], MgO [2] and GaAs [3]. In these cases, the intrinsic tendency to form the wurtzite structure is overcome by topological compatibility. Metastable wurtzite type BN can be synthesized by shock compression [4]. The rocksalt or NaCl structure can be induced in AlN, GaN and InN at very high pressures (further details are presented in Chapter 2 [5]). The rocksalt structure has not been observed for BN. Boron nitride does have unique, hexagonal structures similar to graphite; these are presented in Datareview 1.2 Section B. The wurtzite structure has a hexagonal unit cell and thus two lattice constants, c and a (FIGURE 1). It contains 6 atoms of each type. The space grouping for the wurtzite structure is P6₃mc (C_{6v}^4). The wurtzite structure consists of two interpenetrating hexagonal close packed (HCP) sublattices, each with one type of atom, offset along the c-axis by 5/8 of the cell height (5/8 c).

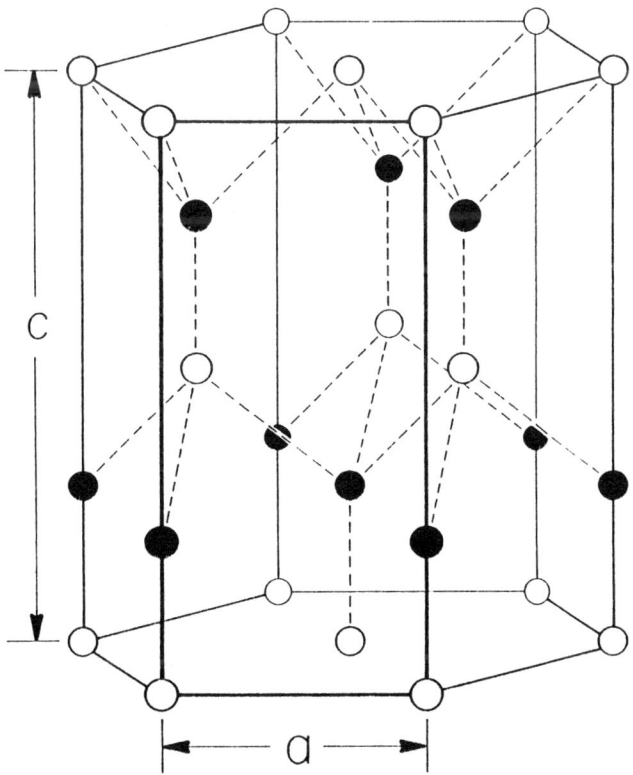

FIGURE 1 The wurtzite crystal structure.

The zinc blende structure has a cubic unit cell, containing four group III elements and four nitrogen elements (FIGURE 2). The space grouping for the zinc blende structure is F43m (T_d^2). The position of the atoms within the unit cell is identical to the diamond crystal structure; both structures consist of two interpenetrating face centred cubic (FCC) sublattices, offset by one quarter of the distance along a body diagonal. Each atom in the structure may be viewed as positioned at the centre of a tetrahedron, with its four nearest atomic neighbours defining the four corners of the tetrahedron.

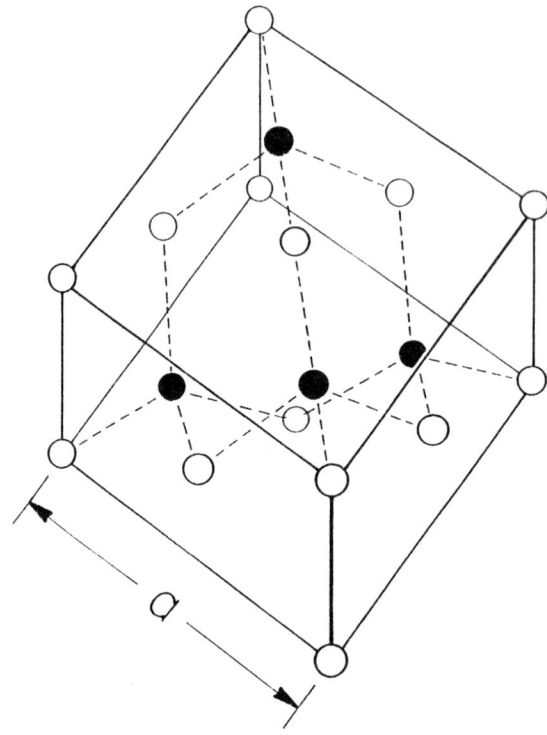

FIGURE 2 The zinc blende crystal structure.

The rocksalt crystal structure is cubic, and each atom has six nearest neighbours located at the corners of an octahedron (FIGURE 3). The space grouping for rocksalt is Fm3M (O_h^5). The rocksalt structure also consists of two interpenetrating FCC sublattices each with one type of atom, but offset along the cube edge by half the edge length (a/2). There are four atoms of each type in the unit cell.

The zinc blende and wurtzite structures are similar. In both cases, each group III atom is coordinated by four nitrogen atoms; conversely, each nitrogen atom is coordinated by four group III atoms. The main difference between these two structures is in the stacking sequence of closest packed diatomic planes. For the wurtzite structure, the stacking sequence of (0001) planes is ABABAB in the <0001> direction; for the zinc blende structure, the stacking sequence of (111) planes is ABCABC in the <111> direction. The difference in the zinc blende and wurtzite crystal structures can be seen by viewing along a chemical bond in the <111> or <0001> (c-axis) directions: the second nearest neighbours are staggered in the zinc blende crystal structure but are eclipsed in the wurtzite structure (FIGURE 4). All group III - nitrogen bond lengths are equivalent in the zinc blende structure but there are two slightly different bond lengths in the wurtzite structure. One bond length R is equal to uc, while the

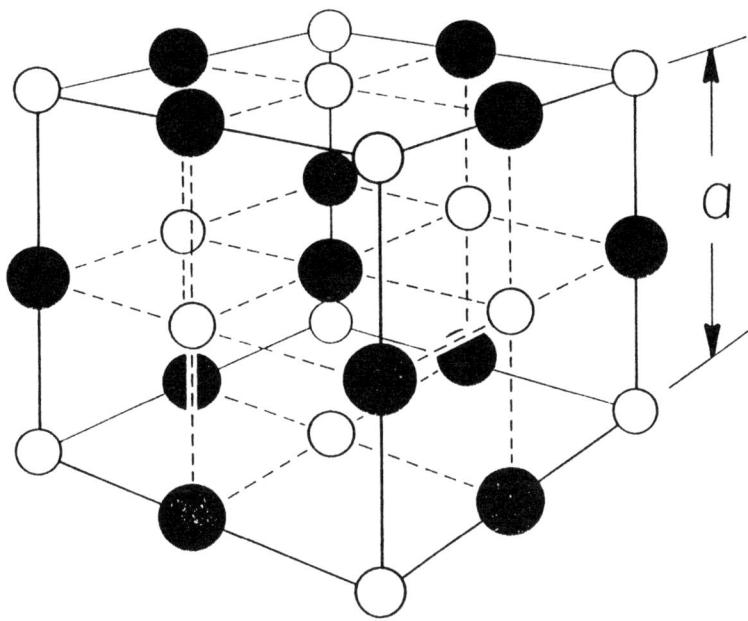

FIGURE 3 The rocksalt crystal structure.

other three bonds are equal to

$$a \left[\frac{1}{3} + \left(\frac{1}{2} - u \right)^2 \left(\frac{c}{a} \right)^2 \right]^{\frac{1}{2}}$$

where u is the dimensionless cell internal structure parameter. For an ideal wurtzite structure, u = 8/3 and c/a = √(8/3). In real crystals, including the group III nitrides, u and c/a deviate from these values; for AlN, GaN and InN, c/a is slightly less, while for BN it is slightly more. This is in agreement with the general observation that wurtzite structures with c/a greater than √(8/3) are unstable while those with c/a less than √(8/3) are stable [6].

All three structures contain polar crystal planes: planes terminating with either all group III atoms or all nitrogen atoms. For the zinc blende and rocksalt structures, the (111) plane is the primary polar plane, while the basal or (0001) plane is the primary polar plane for the wurtzite structure. In cubic crystal structures, group III terminated planes are denoted as (111) or (111)A and group V terminated planes are denoted as ($\bar{1}\bar{1}\bar{1}$) or (111)B. In general, physical properties such as piezoelectric effects, etching characteristics, oxidation rates and photoemission characteristics can vary with the surface polarity, although these effects have not been studied for the group III nitrides.

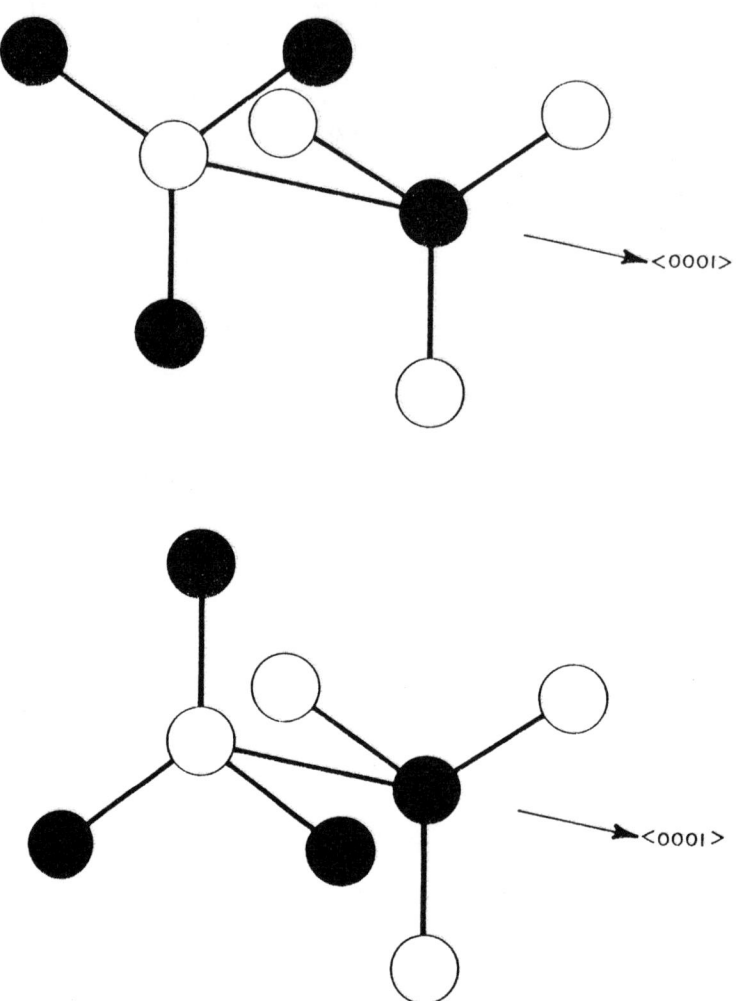

FIGURE 4 A comparison of bonds between close packed planes for the wurtzite (top) and zinc blende (bottom) crystal structures. In the wurtzite structure there is an eclipsed bond configuration while in the zinc blende structure the bonding configuration is staggered.

REFERENCES

[1]   T. Lei, T.D. Moustakas, R.J. Graham, Y. He, S.J. Berkowitz [*J. Appl. Phys. (USA)* vol.71 (1992) p.4933-4]

[2]   R.C. Powell, N.-E. Lee, Y.-W Kim, J.E. Greene [*J. Appl. Phys. (USA)* vol.73 (1993) p.189-204]

[3]   S. Strite et al [*J. Vac. Sci. Technol. B (USA)* vol.9 (1991) p.1924-9]

[4]   T. Soma, A. Sawaoka, S. Saito [*Mater. Res. Bull. (USA)* vol.9 (1974) p.755-62]

[5]   S. Porowski, I. Grzegory [Datareviews in this book: 2.2 Phase diagram of AlN; 2.3 Phase diagram of GaN; 2.4 Phase diagram of InN]

[6]   C.-Y. Yeh, Z.W. Lu, S. Froyen, A. Zunger [*Phys. Rev. B (USA)* vol.46 (1992) p.10086-97]

1.2 Crystal structure, mechanical properties and thermal properties of BN

J.H. Edgar

March 1994

A INTRODUCTION

In this Datareview, the basic physical properties of all modifications of boron nitride are presented. The similar properties, structures, processing and applications of boron nitride and carbon naturally lead to comparisons of these materials [1]. Cubic boron nitride (also known as sphalerite boron nitride and abbreviated as z-BN, c-BN or β-BN) and diamond have similar crystal structures and lattice constants, and physical properties such as extreme hardness, wide energy bandgaps, and high thermal conductivities. Both are generally produced at high temperatures (T > 1800 K) and pressures (P > 4.0 GPa). Hexagonal boron nitride (h-BN or α-BN) and graphite also have similar crystal structures and lattice constants, and physical properties such as thermal conductivity and mechanical strength. Both are produced at high temperatures but low pressure ($\leq$ 100 kPa). The relatively rare wurtzitic boron nitride (w-BN or γ-BN) is structurally similar to lonsdaleite, the hexagonal form of diamond, and both are metastable crystal structures.

There are also significant differences between carbon and boron nitride, primarily due to the differences in chemical bonding: BN has mixed covalent-ionic bonding while bonding in carbon is completely covalent. As a result, both h-BN and c-BN have lower mechanical strength, thermal conductivity and Debye temperatures than their carbon counterparts, but larger lattice constants and energy bandgaps. Consequently, h-BN is an electrical insulator while graphite is a conductive semi-metal. There are practical differences between BN and carbon as well. Cubic BN can be doped either n- or p-type, but diamond is readily doped only p-type [2,3]. Both types of boron nitride are more resistant to oxidation than their carbon analogues, due to the formation of a nonvolatile boron oxide.

Measurements of the physical properties of boron nitride have been hampered by the small size of single crystals available (the maximum size of single crystal c-BN reported to date has been 3 mm in diameter [4], and single crystal h-BN is only rarely produced [5]) and the wide variation in material quality. In comparison to other semiconductors, the crystal quality of c-BN is quite poor, with high dislocation densities (5×10^7 to $10^9 \, cm^{-2}$ [6]), and typically high residual impurity concentrations (see TABLE 1) [7]. Point defects are quite common in c-BN including both B and N vacancies, vacancy clusters, self and impurity interstitials, and antisite defects [8]. As a result, the physical properties vary from crystal to crystal and are non-uniform within a crystal. The structure of the material on which the measurements were taken is noted when presented by the original authors.

Because of its strong bonding in two dimensions and weak bonding in the third, many of the properties of h-BN are highly anisotropic. How the h-BN is processed strongly affects its crystal size, crystal orientation, and density, which in turn affects its other properties. The direction in which a property was measured and how the h-BN was processed has been stated when presented in the original work.

TABLE 1 Mass % of impurities in c-BN (a) synthesized from a system of Mg-B-N, and (b) direct conversion of h-BN to c-BN [7].

	C	O	Si	Ca	Al	Mg	Fe
(a)	0.1	0.1	0.05	0.01	0.01	0.2	0.002
(b)	0.04	0.08	0.02	0.05	0.002	0.004	0.002

Several reviews have been written covering other aspects of boron nitride such as its processing and applications. The reviews by Meller [9,10] cover the properties, processing, and applications of all modifications of boron nitride. The properties and methods of direct and catalytic synthesis of bulk cubic boron nitride have been reviewed by Vel et al [11]. Paine and Narula [12] discuss chemical precursors and the chemistry of h-BN synthesis. Karim et al [13] recently published a thorough review of methods of depositing BN thin films.

B   CRYSTAL STRUCTURES OF BORON NITRIDE

Boron nitride exists in both high density (tetrahedral diamond-like) and low density (graphite-like) forms or modifications. The high density forms include both the cubic and wurtzite crystal structures. The low density forms include the hexagonal and rhombohedral structures, and their variations such as the turbostratic and pyrolytic forms. Mixed crystal structures and amorphous boron nitride are also possible.

All three major variations of the graphite-like form consist of layers of hexagonal rings of sp^2 bonded B and N atoms. These hexagonal rings are composed of three B and three N atoms at alternating positions around the ring. In the most common graphite-like structure, hexagonal boron nitride, designated as h-BN or α-BN, the layers are stacked in a sequence of AA'AA', with the hexagonal rings in all layers coinciding, but alternating from N to B at the same position from layer to layer (FIGURE 1). This stacking sequence is different from graphite: the hexagonal rings in graphite do not coincide but are translated with each layer producing a stacking sequence ABAB. The space group for h-BN is $P6_3/mmc$. The intralayer bonds between N and B are strong covalent bonds while the interlayer bonds are weak Van der Waals bonds. As a result, the B-N bond length within the layers is much shorter (0.1446 nm) than the B-N bond length between layers (0.33306 nm).

Hexagonal boron nitride formed by chemical vapour deposition is referred to as pyrolytic boron nitride (PBN). PBN is deposited with a preferred orientation of the c-axis perpendicular to the deposited surface. Stacking disorder is typical in PBN resulting in slightly higher (2-4%) interlayer spacing. Typically, the maximum grain size in PBN is on the order of 500 nm.

A less common form of graphitic BN is the rhombohedral structure, designated r-BN, with a stacking sequence of ABCABC [14]. For r-BN, the hexagonal rings no longer coincide but are translated parallel to each other from one layer to the next. The sequence of N to B to N between layers is maintained (FIGURE 2). The space group for this structure is R3m.

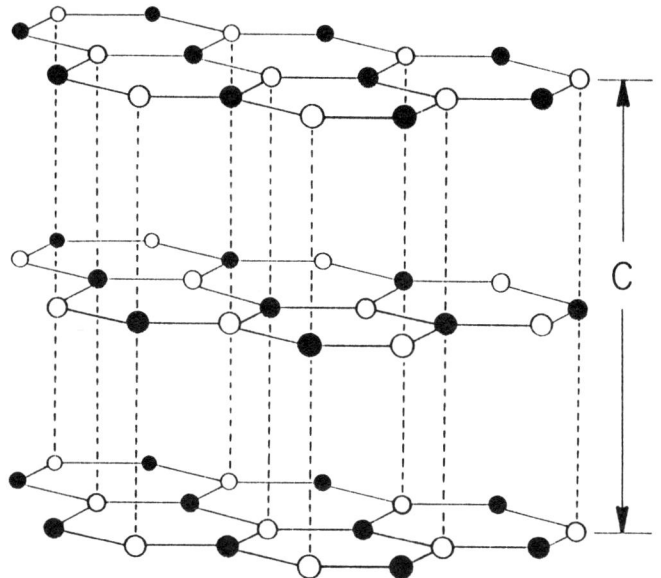

FIGURE 1   Crystal structure of hexagonal boron nitride.

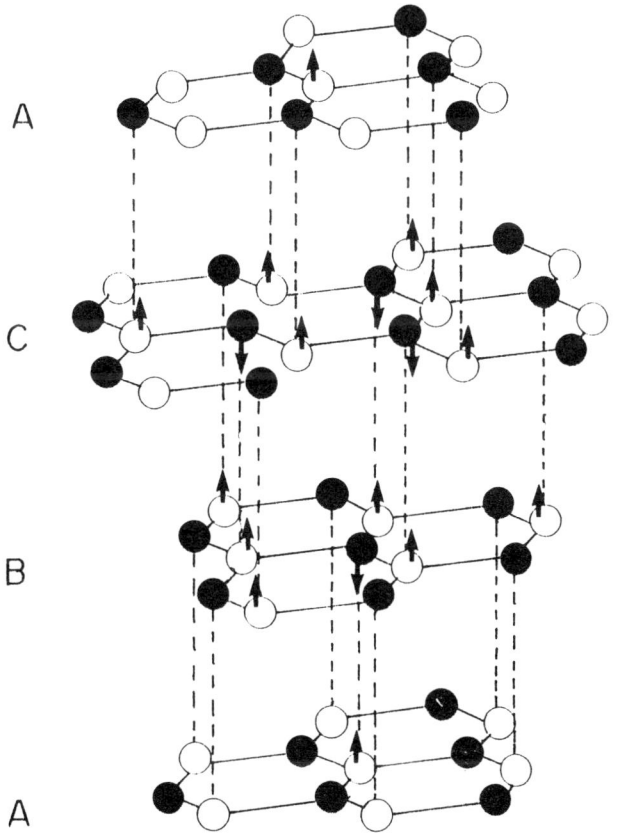

FIGURE 2   Crystal structure of rhombohedral boron nitride.

Turbostratic boron nitride (t-BN) is another graphite-like form of BN in which the two-dimensional atomic layers are roughly parallel, but each layer is randomly translated and rotated to the layer above and below it [15]. There is limited three-dimensional order, and as a result the distance between layers is 3 - 4 % larger than in a perfectly ordered structure. Typically, the angular spread of the X-ray diffraction from the (002) plane indicating the degree of orientation of turbostratic h-BN is 50° or greater. The crystalline orientation can be improved by high temperature compression annealing to produce highly oriented pyrolytic boron nitride (HOPBN), reducing the angular spread to as low as 2° [16].

Of the high density forms, the zinc blende or sphalerite structure (c-BN or β-BN) is the most stable and best characterized. The wurtzite structure (w-BN or γ-BN) is metastable, usually highly defective and is impure, contaminated with other modifications. Boron nitride is unique among the group III nitrides in that its zinc blende structure is the most stable form. Figures, detailed descriptions, and comparisons of the wurtzite and zinc blende structures for all group III nitrides are presented in Datareview 1.1.

The lattice constants for the four BN forms and their densities are listed in TABLES 2 and 3. The density of the wurtzite structure is slightly less (1 %) than the zinc blende form. The B-N distance in the c-BN structure is uniformly 0.1565 nm. Kurdyumov et al [17] observed a reduction in the c lattice constant and improved thermal stability for w-BN after annealing which they attributed to the annihilation of point defects.

TABLE 2 Lattice constants of hexagonal, zinc blende and wurtzite boron nitride.

	a (nm)	c (nm)	Comments	Ref
h-BN	0.250399 ± 0.000005	0.66612 ± 0.00005 0.686 ± 0.001	HOPBN Turbostratic*	[18] [16]
r-BN	0.2507 ± 0.0003	1.000 ± 0.001		[19]
c-BN	0.3615 ± 0.0001 0.36160			[1] [20]
w-BN	0.2556 0.255 ± 0.001 0.2553 ± 0.0003 0.25502 ± 0.00003 0.25502 ± 0.00003	0.4175 0.420 ± 0.001 0.4228 ± 0.0004 0.42190 ± 0.00006 0.42131 ± 0.00006	Theoretical Shock compression Initial After annealing	[1] [21] [22] [17] [17]

* The c lattice constant is not actually defined for t-BN: reported is twice the distance between layers for comparison with HOPBN.

C MECHANICAL PROPERTIES OF HEXAGONAL BORON NITRIDE

The mechanical properties of hexagonal boron nitride are dominated by the significant difference in the chemical bond strengths in and between layers. As noted previously, there is strong sp^2 bonding within layers and weak Van der Waals bonding between layers. As a

TABLE 3 Densities of hexagonal, zinc blende and wurtzite boron nitride.

Structure	Density (g/cm³)	Comments	Ref
h-BN	2.28 ± 0.01 2.28 ± 0.01 $2.0 - 2.2$	Theoretical HOPBN PBN as deposited	[18] [16] [16]
r-BN	2.276		[23]
c-BN	3.48 3.45	Calculated from X-ray data Measured	[1] [20]
w-BN	3.49 ± 0.03 3.454 ± 0.009 3.470	Calculated from X-ray data Calculated from X-ray data Calculated from X-ray data	[21] [22] [23]

TABLE 4 Mechanical properties of h-BN.

Physical property	Numerical value	Comments	Ref
c_{11} (GPa)	830 750	HOPBN Pyrolytic	[25] [26]
c_{12} (GPa)	130 150	 Pyrolytic	[25] [26]
c_{33} (GPa)	31.2 32.4 ± 3 35.6 18.7	Theoretical Measured Pyrolytic	[27] [27] [25] [26]
c_{44} (GPa)	0.5 2.52 3.0 6.2	Pyrolytic Theoretical Determined from Raman spectroscopy	[28] [26] [27] [29]
Tensile strength (MPa)	41 ∥ a 103 ∥ a 2.8 ∥ c	300 K 2500 K 300 K	[30] [30] [30]
Flexural strength (MPa)	83	300 K	[30]
E (GPa)	22 ∥ a 19.6 ∥ a 6.9 ∥ c	300 K 300 K 300 K	[30] [31] [31]
B (GPa)	20 ± 2 335	Calculated Theoretical	[32] [33]

result, many of the mechanical properties of h-BN are highly anisotropic. The interlayer shear modulus (c_{44}) is much lower than the other elastic constants, suggesting that shear would be the predominant failure mechanism. The ultimate tensile strength of pyrolytic boron nitride

(PBN) increases with temperature due to a reduction of stress concentration and an increased ductility [24]. The ultimate strength of PBN is roughly 1/3 to 1/2 that of pyrolytic graphite [24].

Somewhat surprisingly, the bulk modulus (B) has not been reported for h-BN. An estimate (20 GPa) can be made from data reported by Lynch and Drickamer [32] but this is more than a factor of 10 less than that determined theoretically by Xu and Ching (335 GPa) [33], by a method usually considered to be reasonably accurate. This discrepancy indicates the difficulty in measuring the intrinsic properties of h-BN due to the effects of processing on its properties.

D   MECHANICAL PROPERTIES OF CUBIC BORON NITRIDE

Hardness is the most extensively characterized mechanical property of c-BN since it can be readily measured on even small crystals. The hardness is anisotropic (dependent on the crystal plane and direction) and is affected by the crystallinity and impurity content of the sample. Novikov et al [34] found the value of hardness was dependent on the load applied, increasing as the load was reduced. Values of Vickers hardness were consistently higher than Knoop hardness measurements on the same crystals. Knoop hardness is more accurate than other hardness measurements since it does not produce surface cracks [34]. From hardness measurements, Brookes [35] identified the {111} crystal plane and the <110> direction as the predominant slip system for single crystal c-BN. Adding up to 2.5 wt% of transition metals (Mn, Cr and Fe) to sintered c-BN compacts increased the hardness due to the formation of metal borides [36]. On the Mohs comparative scale, c-BN has a hardness of 9.5 (second only to diamond) [37] while h-BN has a hardness of 1.5 [38]. The hardness of c-BN is attributed to the large stresses necessary to initiate dislocations movement.

Few of the other mechanical properties of c-BN have been measured since these generally require larger single crystals than are readily available. A general consensus has been established for the bulk modulus (B) of c-BN between calculated and measured values. As with diamond (B = 435 GPa), the small distance between nearest neighbour atoms in c-BN imparts it with a high bulk modulus. Values for the components of elastic stiffness (c_{ij}), modulus of elasticity (E) and critical stress intensity factor (K_{IC}) have been reported by only a few researchers, and should be regarded as preliminary. The modulus of elasticity for sintered polycrystalline c-BN as a function of temperature is presented in FIGURE 3 [48].

E   MECHANICAL PROPERTIES OF WURTZITIC BORON NITRIDE

Wurtzitic boron nitride is generally produced in very small crystal sizes, with high dislocation densities, and the final product typically contains retained h-BN or c-BN. Thus, the measurement of its mechanical properties is very difficult, and there have been only two reports of the elastic constants (c_{ij}), Young's modulus (E) and shear modulus (μ) for w-BN [51,52]. Pesin [51] calculated the elastic constants for w-BN based on the measured values for c-BN. These values are significantly different from those calculated by Sheleg and Savastenko [52] from measurements of thermal expansion. The bulk modulus has been measured [51] and calculated theoretically [51,53] and is slightly higher than B measured and calculated for c-BN.

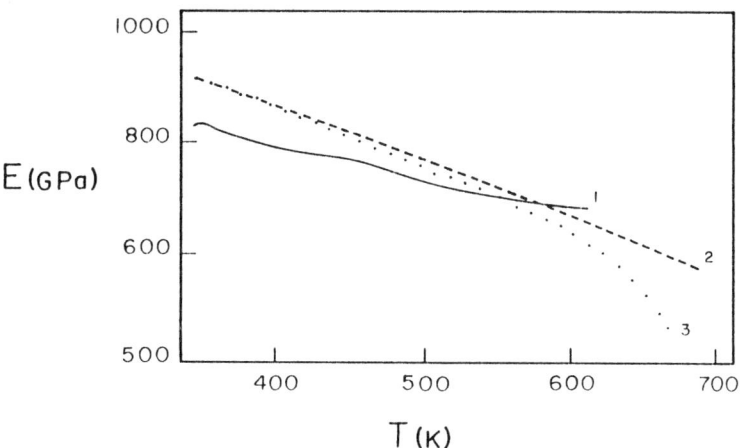

FIGURE 3 Young's modulus (E) for a sintered (at 2800 K and 7.5 - 8 GPa) c-BN compact composed of 20 - 50 µm particles and annealed at (1) 790, (2) 870 and (3) 1010 K [48].

TABLE 5 Mechanical properties of c-BN.

Physical property	Numerical value	Comments	Ref
c_{11} (GPa)	712 442 831		[1] [39] [40]
c_{12} (GPa)	80 107 420		[1] [39] [40]
c_{44} (GPa)	334 191 450		[1] [39] [40]
B (GPa)	290 369 ± 14 367 353 465 382 ± 3	Measured Calculated Calculated Single crystal	[1] [41] [42,43] [44] [45] [46]
E (GPa)	794 900	Polycrystalline Polycrystalline See also FIGURE 3	[47] [48] [48]
H (GPa)	43.12 29.89 41.5 39.45 46.8 ± 3.5 50.2 ± 1.4 90 70	Knoop, (001) [100] Knoop, (001) [110] Knoop, (111) [11$\underline{2}$] Knoop, (111) [1$\underline{1}$0] Knoop, (111) <110> Knoop, (111) <112> Vickers, (1$\underline{1}$1) face Vickers, (1$\overline{1}$1) face	[35] [35] [35] [35] [34] [34] [49] [49]
K_{IC} (MN/m$^{3/2}$)	3.5 2.8	 {111} plane	[34,50] [35]

TABLE 6 Mechanical properties of w-BN.

Physical property	Numerical value	Comments	Ref
c_{11} (GPa)	876 520 ± 21	Calculated	[51] [52]
c_{12} (GPa)	119 431 ± 15	Calculated	[51] [52]
c_{13} (GPa)	80 370 ± 13	Calculated	[51] [52]
c_{33} (GPa)	916 424 ± 15	Calculated	[51] [52]
c_{44} (GPa)	346 65 ± 6	Calculated	[51] [52]
E (GPa)	790 ± 80 831	Measured Calculated	[51] [51]
B (GPa)	410 ± 80 358 426 392	Measured Calculated Calculated	[51] [51] [52] [53]
μ (GPa)	330 ± 40 373	Measured Calculated	[51] [51]

F   THERMAL PROPERTIES OF BORON NITRIDE

Solozhenko [54] has determined the volume coefficient β

$$\beta = \frac{1}{V} \left[\frac{\partial V}{\partial T} \right]$$

of thermal expansion for all modifications of boron nitride by thermomechanical analysis (TMA) and X-ray diffraction. β for each modification is given by the polynomial equation:

$$\beta = A + BT + CT^2 + DT^3 + ET^4 + FT^5$$

with the temperature in kelvin and the constants for each modification given in TABLE 7. Above 1300 K, β becomes temperature independent, and is 4.29×10^{-5} and 1.76×10^{-5} for h-BN and c-BN respectively.

The heat capacities for all modifications of BN over a wide temperature range are presented in Datareview 2.1 by Solozhenko. Representative values are presented in the following sections.

TABLE 7 Coefficients of equations approximating temperature dependences of volume coefficients of thermal expansion for BN polymorphous modifications [54].

	$A \times 10^6$	$B \times 10^8$	$C \times 10^{11}$	$D \times 10^{15}$	$E \times 10^{17}$	$F \times 10^{21}$
h-BN.r-BN	35.259	0.2095	1.2360	-7.1994	0	0
c-BN	-4.2586	4.4762	-3.0020	6.6287	0	0
w-BN	3.5763	2.5366	-5.1101	72.988	-4.4872	9.5179

G   THERMAL PROPERTIES OF HEXAGONAL BORON NITRIDE

The vector properties of hexagonal boron nitride (h-BN) (thermal conductivity and thermal expansion coefficient) are highly anisotropic due to its crystal structure. The thermal conductivity in the plane of the atomic layers (a-direction) is an order of magnitude higher than perpendicular to the layers (c-direction) due to the difference in distance between nearest neighbour atoms. The thermal conductivity of h-BN is dependent on the crystalline order, and is significantly increased by annealing under pressure to produce highly oriented pyrolytic boron nitride (HOPBN). Thermal conductivity in h-BN is principally by lattice vibrations due to the lack of free electrons in this wide bandgap insulator.

While the c-axis lattice for h-BN expands on heating, the a lattice constant decreases, resulting in a negative thermal expansion coefficient for this direction over large temperature ranges [18]. The thermal decomposition of h-BN is slow; typically the nitrogen vapour pressure is below predicted thermodynamic equilibrium values even at very high temperatures. No evidence of PBN decomposition was observed in PBN vacuum (5×10^{-9} torr) baked at 1873 K for 1 hr [55].

H   THERMAL PROPERTIES OF CUBIC BORON NITRIDE

Slack [60] has estimated the thermal conductivity (k) of single crystal cubic boron nitride to be second only to that of diamond, with a room temperature value of 1300 W m^{-1} K^{-1}. Measured thermal conductivities of single crystal c-BN by Novikov et al [61] ranged from 416 to 742 W m^{-1} K^{-1}, depending on the concentration of colour centres. Thermal conductivities of compact polycrystalline c-BN have been as high as 700 W m^{-1} K^{-1} [62,63].

Linear thermal expansion coefficients

$$\alpha \left[1^{-1} \left[\frac{dl}{dT} \right] \right]$$

for c-BN have been measured by Slack and Bartram [64] and Kolupayeva et al [65] who also report values for wurtzite BN (FIGURE 4). The room temperature thermal expansion coefficient of c-BN is smaller than that of many other semiconductors (AlN, BP, GaP, Si and Ge); only diamond has a slightly smaller value [64].

TABLE 8 Thermal properties of h-BN and r-BN.

Physical property	Numerical value		Comments	Ref
	a-direction	c-direction		
$\alpha \times 10^6$ (K^{-1})	-2.9	40.5	293 K	[18]
			273 - 800 K	[18]
	-2.9		230 K	[56]
	-1.4		600 K	[56]
k (W m^{-1} K^{-1})	50 at 300 K		HIP-BN	[57]
	20 at 1273 K		HIP-BN	[57]
	63 at 300 K	1.5	p-BN	[30]
	63 at 1073 K	2.9	p-BN	[30]
	400 at 25 °C		HOPBN	[58]
θ_D (K)	323			[18]
	410			[58]
C_p (J mol^{-1} K^{-1})	19.9		h-BN	[59]
	19.85 ± 0.06		h-BN	[23]
	21.72		t-BN	[23]
	20.63 ± 0.06		r-BN	[23]
			See also Datareview 2.1	

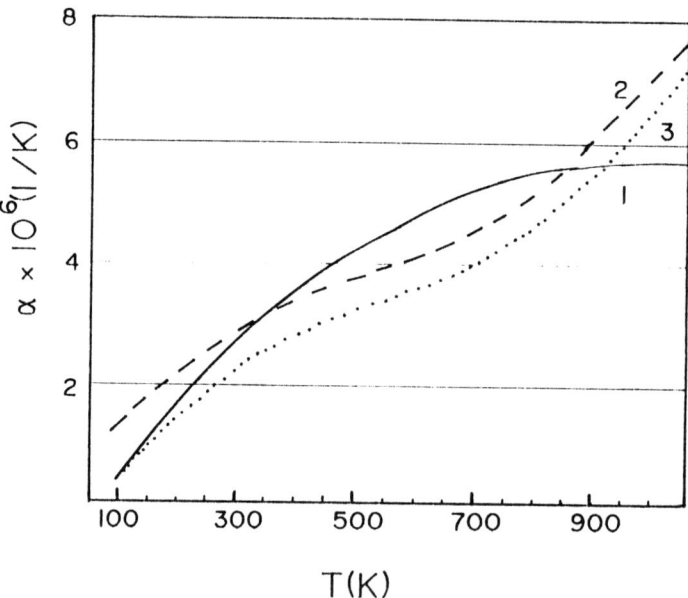

FIGURE 4 Thermal expansion coefficients for (1) c-BN, (2) c-axis w-BN and (3) a-axis w-BN as a function of temperature [65].

The rate of decomposition of c-BN to h-BN is dependent on its initial structure including the crystal quality, distribution of crystal sizes, and impurities present. Solozhenko and Turkevich

[7] observed the beginning of the c-BN to h-BN transformation at 1670 ± 5 K. For metastable w-BN, the transformation to h-BN begins at temperatures as low as 873 K [66].

TABLE 9 Thermal properties of c-BN.

Physical property	Numerical value	Comments	Ref
$\alpha \times 10^6$ (K^{-1})	1.15 at 300 K 1.9 at 300 K	See also TABLE 10 and FIGURE 4	[64] [48]
k (W m^{-1} K^{-1})	416 - 742 at 300 K 200 at 300 K 1300 at 300 K 700 at 300 K	Various single crystals Polycrystalline compact Theoretical Polycrystalline compact	[61] [1] [60] [62,63]
Debye temperature (K)	1900 1700 1610 1700 1730 ± 70	Theoretical From IR data From C_p measurements From X-ray data	[69] [70] [68] [71] [65]
C_p (J mol^{-1} K^{-1})	15.97 at 300 K 12.5 at 300 K 36.8 at 600 K 46.1 at 1000 K 46 - 48 at 1300 - 1600 K 16.28 at 300 K	See also Datareview 2.1	[67] [1]

TABLE 10 Thermal expansion coefficient of cubic boron nitride as a function of temperature [64].

T (K)	$\alpha \times 10^6$ (K^{-1})
0	0.00
100	?
200	0.50
300	1.15
400	1.80
500	2.50
600	3.23
800	4.70
1000	5.96
1200	6.45

I   THERMAL PROPERTIES OF WURTZITE BORON NITRIDE

There have been very few measurements of the thermal properties of the wurtzite modification of BN. As with most wurtzite structure materials, thermal expansion parallel to the c-axis is larger than that parallel to the a-axis [65]. Fedoseev et al [72] observed that the thermal conductivity of mixtures of w-BN and c-BN increased with the w-BN content. The heat capacity (C_p) for w-BN was measured by Sirota and Kofman [73], Gorbunov et al [59], and Gavrichev et al [23]. The results of Sirota and Kofman [73] deviate considerably from the other investigators, possibly due to the low purity of the sample (92% w-BN) [23].

TABLE 11   Thermal properties of w-BN.

Physical property	Numerical value	Comments	Ref
$\alpha \times 10^6$ (K^{-1})	2.0 ∥ a 2.6 ∥ c	300K 300K See also FIGURE 4	[65] [65]
k (W m^{-1} K^{-1})	26 at 1150 K 60 at 1150 K	75% c-BN, 25% w-BN 15% c-BN, 85% w-BN	[72] [72]
θ_D (K)	1400 1460 ± 70		[73] [65]
C_p (J mol^{-1} K^{-1})	16.63 16.45 ± 0.05	298 K See also Datareview 2.1	[59] [23]

J   CONCLUSION

Many of the fundamental properties of all modifications of boron nitride have been measured, but only a few can be considered to be definitively determined. The measurements of density, lattice constants, heat capacities and thermal expansion coefficients are reliably established, but values for the elastic constants, tensile strengths and thermal conductivities should be considered tentative. Definitive measurements of many of these properties await methods of producing better quality material, improved processing, and measurement techniques requiring less material. It is clear that the mechanical and thermal properties of boron nitride have the highest values of any of the group III nitrides, and in many cases are second only to carbon. Thus developments in producing high quality boron nitride are worthwhile from both basic science and technological viewpoints.

REFERENCES

[1]   R.C. DeVries [*Cubic Boron Nitride: Handbook of Properties*, General Electric Report No. 72CRD178 (1972)]

[2]   R.H. Wentroff [*J. Chem. Phys. (USA)* vol.36 (1962) p.1990-1]

[3]   O. Mishima, J. Tanaka, S. Yamaoka, O. Fukunaga [*Science (USA)* vol.238 (1987) p.181-3]

[4] M. Kagamida, H. Kanda, M. Akaishi, A. Nukui, T. Osawa, S. Yamaoka [*J. Cryst. Growth (Netherlands)* vol.94 (1989) p.261-9]

[5] T. Ishii, T. Sato [*J. Cryst. Growth (Netherlands)* vol.61 (1983) p.689-90]

[6] S.I. Futergendler, L.I. Fel'dgun, V.I. Labes, V.M. Davidenko [*Sint. Almazy (USSR)* no.1 (1979) p.20-2]

[7] V.L. Solozhenko, V.Z. Turkevich [*J. Therm. Anal. (UK)* vol.38 (1992) p.1181-8]

[8] V.B. Shipilo, A.E. Rud', N.G. Anichenko, L.M. Gameza, A.I. Vrublevskii [*Inorg. Mater. (USA)* vol.26 (1991) p.1397-1400]

[9] A. Meller [*Gmelin Handbuch der Anorganische Chemie, Boron Compounds, 2nd Supplement*, vol.1 (Springer-Verlag, Berlin, 1983) p.304-60]

[10] A. Meller [*Gmelin Handbuch der Anorganische Chemie, Boron Compounds, 3rd Supplement*, vol.3 (Springer-Verlag, Berlin, 1988) p.1-91]

[11] L. Vel, G. Demazeau, J. Etourneau [*Mater. Sci. Eng. B (Switzerland)* vol.10 (1991) p.149-64]

[12] R.T. Paine, C.K. Narula [*Chem. Rev. (USA)* vol.90 (1990) p.73-91]

[13] M.Z. Karim, D.C. Cameron, M.S.J. Hashmi [*Mater. Des. (UK)* vol.13 (1992) p.207-14]

[14] T. Ishii, T. Sato, Y. Sekikawa, M. Iwata [*J. Cryst. Growth (Netherlands)* vol.52 (1981) p.285-9]

[15] J. Thomas, N.E. Weston, T.E. O'Connor [*J. Am. Chem. Soc. (USA)* vol.84 (1963) p.4619-22]

[16] A.W. Moore [*Nature (UK)* vol.221 (1969) p.1133-4]

[17] A.V. Kurdyumov, V.L. Solozhenko, W.B. Zelyavsky, I.A. Petrusha [*J. Phys. Chem. Solids (UK)* vol.54 (1993) p.1051-3]

[18] R.S. Pease [*Acta Crystallogr. (Denmark)* vol.5 (1952) p.356-61]

[19] T. Sado [*Proc. Jpn. Acad. B. (Japan)* vol.61 (1985) p.459-63]

[20] V.L. Solozhenko, V.V. Chernyshev, G.V. Fetisov, V.B. Rybakov, I.A. Petrusha [*J. Phys. Chem. Solids (UK)* vol.51 (1990) p.1011-12]

[21] F.P. Bundy, R.H. Wentroff [*J. Chem. Phys. (USA)* vol.38 (1963) p.1144-9]

[22] T. Sôma, A. Sawaoka, S. Saito [*Proc. 4th Int. Conf. on High Pressure*, Kyoto, 1974 (Physico-Chemical Society of Japan, Kyoto, Japan, 1975) p.446-53]

[23] K.S. Gavrichev, V.L. Solozhenko, V.E. Gorbunov, L.N. Golushina, G.A. Totrova, V.B. Lazarev [*Thermochim. Acta (Netherlands)* vol.217 (1993) p.77-89]

[24] W.V. Kotlensky, H.E. Martens [*Nature (UK)* vol.196 (1962) p.1091-2]

[25] B.T. Kelly [*J. Nucl. Mater. (Netherlands)* vol.68 (1977) p.9-12]

[26] L. Duclaux, B. Nysten, J-P. Issi, A.W. Moore [*Phys. Rev. B (USA)* vol.46 (1992) p.3362-7]

[27] J.F. Green, T.K. Bolland, J.W. Bolland [*J. Chem. Phys. (USA)* vol.64 (1976) p.656-62]

[28] B.T. Kelly [*Philos. Mag. (UK)* vol.32 (1975) p.859-67]

[29] T. Kuzuba, Y. Sato, S. Yamaoka, K. Era [*Phys. Rev. B (USA)* vol.18 (1978) p.4440-3]

[30] A. W. Moore [*J. Cryst. Growth (Netherlands)* vol.106 (1990) p.6-15]

[31] N.J. Archer [High Temp. Chem. Inorg. Ceram. Mater. Proc. Conf. (UK) vol. 30 Spec. Publ. - Chem. Soc. (1977) p. 167-80]

[32] R.W. Lynch, H.G. Drickamer [*J. Chem. Phys. (USA)* vol.44 (1966) p.181-4]

[33] Y.-N. Xu, W.Y. Ching [*Phys. Rev. B (USA)* vol.44 (1991) p.7787-98]

[34] N.V. Novikov, S.N. Dub, V.I. Mal'nev [*Sov. J. Superhard Mater. (USSR)* vol.5 (1983) p.16-21]

[35] C.A. Brookes [*Proc. Int. Conf. on Science of Hard Materials*, Rhodes, Greece, 23-28 September 1984, Eds E.A. Almond, C.A. Brookes, R. Warren (Adam Hilger, Bristol, England, 1986) p. 207-20]

[36] V.B. Shipilo, A.E. Rud', N.G. Anichenko, V.S. Kuzmin, I.I. Ugolev, S.E. Bogushevich [*Inorg. Mater. (USA)* vol.27 (1991) p.1212-6]

[37] C.F. Gardinier [*Ceram. Bull. (USA)* vol.67 (1988) p.1006-9]

[38] D.A. Lelonis [Union Carbide Sales Brochure (USA) (1989)]

[39] N.N. Sirota, A. F. Revinskii [*Vesti Akad, Nauk BSSR, Ser. Fiz.-Mat. Navuk (USSR)* no.6 (1981) p.64-7]

[40] T.D. Sokolovskii [*Inorg. Mater. (USA)* vol.19 (1983) p.1311-4]

[41] E. Knittle, R.M. Wentzcovitch, R. Jeanloz, M.L. Cohen [*Nature (UK)* vol.337 (1989) p.349-52]

[42] R.M. Wentzcovitch, K.J. Chang, M.L. Cohen [*Phys. Rev. B (USA)* vol.34 (1986) p.1071-9]

[43] P.E. Van Camp, V.E. Van Doren, J.T. Devreese [*Phys. Status Solidi B (Germany)* vol.146 (1988) p.573-87]

[44] K.T. Park, K. Terakura, N. Hamada [*J. Phys. C (UK)* vol.20 (1987) p.1241-51]

[45] J.A. Sanjurjo, E. López-Cruz, P. Vogl, M. Cardona [*Phys. Rev. B (USA)* vol.28 (1983) p.4579-84]

[46] E.V. Yakovenko, I.V. Aleksandrov, A.F. Goncharov, S.M. Stishov [*Sov. Phys.-JETP (USA)* vol.68 (1989) p.1213-5]

[47] V. Kh. Nurmukhamedov, G.A. Adler, V.I. Veprintsev, N.A. Luchsheva, V.A. Botsulyak [*Inorg. Mater. (USA)* vol.14 (1978) p.1497-8]

[48] V.B. Shipilo, N.A. Shishonok, A.V. Mazovko [*Inorg. Mater. (USA)* vol.26 (1990) p.1401-3]

[49] V.B. Shipilo, E.M. Shishonok, E.M. Zaitsev, N.G. Anichenko, L.S. Unyarkha [*Inorg. Mater. (USA)* vol.26 (1990) p.1404-8]

[50] S.N. Dub, S.N. Sokolov [*Proizvod. Primen. Sverkhtverd. Mater. (USSR)* (1983) p.72-5]

[51] V.A. Pesin [*Sverktverd. Mater. (USSR)* no.6 (1980) p.5-7]

[52] A.U. Sheleg, V.A. Savastenko [*Inorg. Mater. (USA)* vol.15 (1979) p.1257-61]

[53] Y.-N. Xu, W.Y. Ching [*Phys. Rev. B. (USA)* vol.48 (1993) p.4335-51]

[54] V.L. Solozhenko [PhD Dissertation, Moscow State University, Moscow, Russia (1993) p.23-4]

[55] F.A. Chambers, G.W. Zajac, T.H. Fleisch [*J. Vac. Sci. Technol. B (USA)* vol.4 (1986) p.1310-5]

[56] B. Yates, M.J. Overy, O. Pirgon [*Philos. Mag. (UK)* vol.32 (1975) p.847-57]

[57] A. Lipp, K.A. Schwetz, K. Hunold [*J. Eur. Ceram. Soc. (UK)* vol.5 (1989) p.3-9]

[58] E.K. Sichel, R.E. Miller, M.S. Abrahams, C.J. Buiocchi [*Phys. Rev. B (USA)* vol.13 (1976) p.4607-11]

[59] V.E. Gorbunov, K.S. Gavrichev, G.A. Totrova, A.V. Bochko, V.B. Lazarev [*Russ. J. Phys. Chem. (UK)* vol.62 (1988) p.9-12]

[60] G.A. Slack [*J. Phys. Chem. Solids (UK)* vol.34 (1973) p.321-35]

[61] N.V. Novikov, T.D. Osetinskaya, A.A. Shul'zhenko, A.P. Podoba, A.N. Sokolov, I.A. Petrusha [*Dopov. Akad. Nauk. Ukr. RSR, Ser. A Fiz.-Mat. Tekh Nauki (USSR)* no.10 (1983) p.72-5]

[62] H. Sumiya, K. Tsuji, S. Yazu [*Proc. 2nd Int. Conf. on New Diamond Sci. Technol.*, Washington, DC, USA, 23-27 Sept 1990 (MRS, Pittsburgh, USA, 1991) p.1063-8]

[63] F.R. Corrigan [*High-Pressure Science and Technology, 6th AIRAPT Conf.*, Boulder, CO, 25-29 July 1977, Eds K.D. Timmerhaus, M.S. Barber (Plenum, New York, 1979) p. 994-9]

[64] G.A. Slack, S.F. Bartram [*J. Appl. Phys. (USA)* vol.46 (1975) p.89-98]

[65] Z.I. Kolupayeva, M. Ya Fuks, L.I. Gladkikh, A.V. Arinkin, S.V. Malikhin [*J. Less-Common Met. (Switzerland)* vol.117 (1986) p.259-63]

[66] A.N. Pilyankevich, A.V. Kurdyumov, N.F. Ostrovskaya [*Phys. Status Solidi A (Germany)* vol.116 (1989) p.K1-K5]

[67] V.L. Solozhenko, V.E. Yachmenev, V.A. Vil'kovskii, A.N. Sokolov, A.A. Shul'zhenko [*Russ. J. Phys. Chem. (UK)* vol.61 (1987) p.1480-2]

[68] T. Atake, S. Takai, A. Honda, Y. Saito, K. Saito [*Rep. Res. Lab. Eng. Mater. Tokyo Inst. Technol. (Japan)* no.16 (1991) p.15-25]

[69] E.F. Steigmeier [*Appl. Phys. Lett. (USA)* vol.3 (1963) p.6-8]

[70] P.J. Gielisse et al [*Phys. Rev. (USA)* vol.155 (1967) p.1039-46]

[71] V.A. Manzher [*Sint. Almazy (USSR)* no.3 (1978) p.6-9]

[72] D.V. Fedoseev, A.V. Lavrent'ev, I.G. Varshavskaya, A.V. Bochko, G.G. Karyuk [*Poroshk. Metall. (Ukraine)* no.3 (1978) p.92-4]

[73] N.N. Sirota, N.A. Kofman [*Sov. Phys.-Dokl. (USA)* vol.21 (1976) p.516-7]

1.3 Crystal structure, mechanical properties, thermal properties and refractive index of AlN

W.J. Meng

December 1993

A INTRODUCTION

Non-metallic adamantine (diamond-like) compounds, such as silicon carbide (SiC), beryllium oxide (BeO), cubic boron nitride (c-BN), and aluminium nitride (AlN), have a combination of interesting physical properties such as high temperature stability, high thermal conductivity, high elastic stiffness, and varying electrical properties from semiconducting to insulating. AlN, together with the other group III nitrides, is being intensely investigated for possible electronic, opto-electronic, and electronic packaging applications [1-4]. We describe here the basic physical properties of AlN.

B CRYSTAL STRUCTURE OF ALUMINIUM NITRIDE

AlN crystallizes in the hexagonal wurtzite structure, with two formula units per unit cell (4 atoms per cell) and molar mass $20.495 \, \text{g mol}^{-1}$. The ideal wurtzite structure is tetrahedrally coordinated with Al(N) in the centre of an N(Al) regular tetrahedron, with a c/a ratio of $\sqrt{(8/3)} = 1.633$. The space group symmetry is C_{6v}^4 ($P6_3mc$) and the point group symmetry is C_{6v} (6mm) [5]. Reported lattice parameters for AlN range from $3.110 - 3.113 \, \text{Å}$ for a and $4.978 - 4.982 \, \text{Å}$ for c, while the c/a ratio varies from 1.600 to 1.602 [6-9]. The deviation of the c/a ratio from that of the ideal wurtzite structure has been related to lattice stability and ionicity [10,11]. The corresponding X-ray densities range from $3.279 - 3.228 \, \text{g cm}^{-3}$. Oxygen impurities are known to influence the lattice parameters [12].

AlN is generally reported to be non-polymorphous [5]. However, several reports suggested the occurrence of a metastable zinc blende polytype of AlN [13-15], with a lattice parameter $a = 4.38 \, \text{Å}$ [13]. One theoretical estimate of the lattice parameter of the AlN zinc blende polytype also suggested $a = 4.38 \, \text{Å}$ [16]. No other independent experimental verification of the occurrence of zinc blende AlN is known to date. A pressure-induced rocksalt (space group O_h^5) phase of AlN has been observed experimentally [17,18]. The equilibrium transformation pressure was reported to be less than $14 \, \text{GPa}$ [17]. The rocksalt structure is retained at room temperature with a lattice parameter $a = 4.043 - 4.045 \, \text{Å}$ [17,18].

C MECHANICAL PROPERTIES OF ALUMINIUM NITRIDE

Early investigations of the elastic properties of AlN were carried out on sintered polycrystalline specimens, due to the unavailability of large single crystals [19]. The measured bulk modulus B and Young's modulus E are shown in TABLE 1. The entire single crystal elastic constant matrix has been obtained by fitting results of surface acoustic wave measurements made on epitaxial AlN films [20-22], and also by Brillouin scattering measurements made on an AlN single crystal [23]. The elastic constants so obtained are

TABLE 1 Mechanical properties of AlN: c_{ij}'s are elastic constants; elastic moduli denoted calc. are calculated from single crystal elastic constants. μ_{TO} and μ_{LO} are zone centre optical phonon wavenumbers obtained from IR and Raman scattering experiments.

Physical property	Value	Refs	Measurement method
c_{11}	345 (GPa)	[20-22]	Surface acoustic wave velocity
	411	[23]	Brillouin scattering
c_{12}	125	[20-22]	Surface acoustic wave velocity
	149	[23]	Brillouin scattering
c_{13}	120	[20-22]	Surface acoustic wave velocity
	99	[23]	Brillouin scattering
c_{33}	395	[20-22]	Surface acoustic wave velocity
	389	[23]	Brillouin scattering
	394	[24]	Longitudinal sound velocity
c_{44}	118	[20-22]	Surface acoustic wave velocity
	125	[23]	Brillouin scattering
B	160	[19]	Sound velocity
	201	calc.	
E	308	[19]	Sound velocity
	293	calc.	
μ_{TO}	664 (cm^{-1})	[26]	IR reflectance
	667	[27]	
μ_{TO} (E$_1$)	657	[28]	
μ_{LO}	897	[26]	
	916	[27]	
μ_{LO} (A$_1$)	895	[28]	
	886	[28]	
μ_{TO} (A$_1$)	614	[23]	Raman scattering
	667	[29]	
	659	[30]	
	659	[31]	
	607	[32]	
μ_{LO} (A$_1$)	893	[23]	
	897	[30]	
	888	[31]	
μ_{TO} (E$_1$)	673	[23]	
	667	[29]	
	672	[30]	
	672	[31]	
μ_{LO} (E$_1$)	916	[23]	
	910	[29]	
	912	[30]	
	895	[31]	
	924	[32]	
μ (E$_2$)	252	[23]	
	241	[32]	
μ (E$_2$)	660	[23,32]	
	655	[29]	

shown in TABLE 1; only five constants c_{11}, c_{12}, c_{13}, c_{33} and c_{44} are distinct due to the hexagonal symmetry. The value of c_{33} has been obtained by measurement of longitudinal sound velocity on epitaxial AlN films [24]. From the relations between the elastic moduli (B and E) and the elastic compliance tensor S_{ijkl}, $B = 1/S_{iikk}$ and $E = 1/S_{1111}$, both B and the averaged E can be obtained from the single crystal elastic constants [25]. Such calculated moduli are also shown in TABLE 1.

The phonon dispersion spectrum of AlN has 12 branches, 3 acoustic and 9 optical. Transverse optical (TO) and longitudinal optical (LO) phonon energies have been obtained from fits to infrared reflectivity measurements, the results of which are listed in TABLE 1 [26-28]. Raman active optical phonon modes belong to the A_1, E_1 and E_2 group representations. Several Raman scattering studies on AlN have been conducted and the measured phonon energies are listed in TABLE 1 [23,29-32].

The hardness of AlN has been measured to be ~12 GPa on the basal plane (0001), using a Knoop diamond indenter [33]. Some anisotropy in Knoop hardness has been observed with the indent direction perpendicular to the c axis, with measured values ranging from 10 - 14 GPa [34].

D   THERMAL PROPERTIES OF ALUMINIUM NITRIDE

Several measurements of thermal expansion of AlN have been made [34-28]. Representative data are shown in TABLE 2. A summary of previous measurements has been presented, where the recommended thermal expansion values are given (293 - 1700 K) [39]:

$$\Delta a/a_0 = -8.679 \times 10^{-2} + 1.929 \times 10^{-4} T + 3.400 \times 10^{-7} T^2 - 7.969 \times 10^{-11} T^3 \qquad (1)$$

$$\Delta c/c_0 = -7.006 \times 10^{-2} + 1.583 \times 10^{-4} T + 2.719 \times 10^{-7} T^2 - 5.834 \times 10^{-11} T^3 \qquad (2)$$

TABLE 2 Thermal expansion of AlN.

T (K)	$(\Delta a/a_0) \times 10^3$	$(\Delta c/c_0) \times 10^3$	Ref
291	0.000	0.000	[37]
427	0.611	0.261	
611	1.511	1.145	
767	2.314	1.787	
939	3.342	2.390	
1073	4.146	3.253	
297	0.000	0.000	[38]
503	0.80	0.64	
628	1.38	1.14	
683	1.70	1.53	
901	3.02	2.69	
1099	4.37	3.74	
1269	5.56	4.64	

The heat capacity of AlN has been measured independently on powder samples down to 53 K [40,41] and 5 K [42], respectively. Good agreement exists between different measurements. The measured constant pressure specific heats (C_p) are shown in TABLE 3 [40-42]. Assuming a Debye expression for the low temperature specific heat, $C_v = (12\pi^4/5)R(T/\theta_D)^3$

[43], and ignoring differences between C_p and C_v, a Debye temperature θ_D of 825 K is obtained by fitting the combined data from 5 - 75 K. For hexagonal crystals, θ_D can be estimated from the single crystal elastic constants [44]. Using the elastic constants shown in TABLE 1, a θ_D of 903 K is obtained. Using polycrystalline elastic constants, a θ_D of 950 K has been suggested [45].

TABLE 3 Constant pressure specific heat of AlN: measurements were carried out on polycrystalline specimens, see [40-42].

T (K)	C_p (cal g^{-1} K^{-1})	T (K)	C_p (cal g^{-1} K^{-1})
5	0.0000055	245.7	0.1453
50	0.005107	256.33	0.152
52.91	0.00598	266.12	0.1579
57.44	0.007683	276.14	0.1638
62.13	0.009783	286.38	0.1693
67.15	0.01223	296.24	0.1748
72.49	0.01509	350	0.2034
77.19	0.01786	400	0.2211
81.59	0.02057	450	0.2335
86.79	0.02391	500	0.2427
95.08	0.02962	550	0.2497
104.95	0.03684	600	0.2553
114.74	0.04421	650	0.2599
124.54	0.05194	700	0.2637
135.81	0.06102	750	0.267
145.91	0.06941	800	0.2698
156.3	0.07802	850	0.2724
165.8	0.08573	900	0.2747
176.29	0.09417	950	0.2767
185.94	0.1019	1000	0.2787
195.99	0.1095	1050	0.2805
206.46	0.1176	1100	0.2821
216.42	0.125	1150	0.2837
236.17	0.1388	1200	0.2852

Thermal conductivity κ of AlN at room temperature has been estimated to be $\sim 3.2\,\mathrm{W\,cm^{-1}\,K^{-1}}$, approaching that of copper [12,45]. Measurement of a single crystal has yielded κ (300 K) = $2.5\,\mathrm{W\,cm^{-1}\,K^{-1}}$ [46], while measurements made on polycrystalline specimens generally yielded lower values [47,48]. The highest value of room temperature thermal conductivity κ (300 K) = $2.85\,\mathrm{W\,cm^{-1}\,K^{-1}}$ has been obtained from a single crystal specimen [45]. The measured thermal conductivities as a function of temperature are shown in TABLE 4 [45].

Thermodynamic potentials for AlN have been tabulated [49]. The heat of formation of AlN from the reaction Al(c) + 1/2 N_2(g) = AlN(c) was determined to be ΔH_{298K} = - 75.6 kcal mol^{-1} [40]. Together with the known thermodynamic potentials for Al and N_2 [49], the equilibrium N_2 vapour pressure has been calculated [50]. The vapour pressures of Al and N_2 in equilibrium with solid AlN and liquid Al are shown in FIGURE 1. The calculated temperatures at which the equilibrium N_2 pressure reaches 1, 10 and 100 atmospheres are 2836 K, 3088 K and 3390 K, respectively [50]. Note solid AlN would decompose into liquid Al if the N_2 overpressure falls below the N_2 equilibrium vapour pressure, thus explaining the widely varying values of the melting temperature of AlN reported [34,50]. Under 100 atmospheres of N_2, the melting temperature of AlN has been reported to be 3073 K [51].

TABLE 4 Thermal conductivity of AlN: measurements were carried out on a single crystal, see [45].

T (K)	κ (W cm^{-1} K^{-1})
0.4	0.0038
0.6	0.0097
1	0.032
2	0.21
4	1.08
6	2.4
10	5.7
15	10.2
20	15
30	20
45	23
60	22
100	17.5
150	11
200	6.5
300	2.85
400	1.8
600	0.96
1000	0.48
1800	0.24

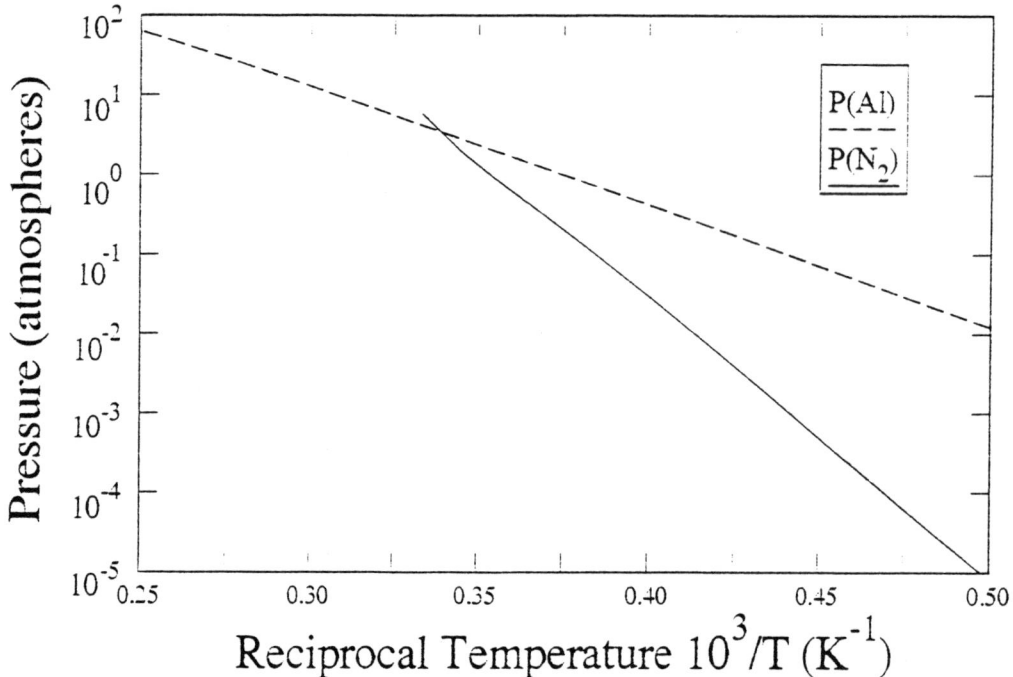

FIGURE 1 Vapour pressures of Al and N_2 in equilibrium with solid AlN and liquid Al, calculated from data in [49]; see also [50].

E   REFRACTIVE INDEX OF ALUMINIUM NITRIDE

The refractive index of AlN has been measured on amorphous, polycrystalline and epitaxial thin films as well as on single crystals [52-61]. The refractive index is observed to increase with increasing structural order, ranging from 1.8-1.9 for amorphous films [52], 1.9-2.1 for polycrystalline films [53-56], and 2.1-2.2 for epitaxial films [57,58]. The refractive index for single crystals is reported around 2.2 [59-61]. The spectral dependence of the refractive index has been measured, which showed near constant refractive index in the wavelength range of 400-600 nm [60,61]. A small birefringence of AlN (~ 2-4%) has been reported [60-63]. The reported refractive index is in general agreement with low and high frequency dielectric constants (ε_0 and ε_∞) determined from infrared measurements, with ε_∞ ranging from 4.6-4.8 and ε_0 from 8.5-9.1 [26,27]. Low frequency electrical measurements also yielded ε_0 around 8.5 [34,64].

F   CONCLUSION

A survey of the basic physical properties of AlN has been made. It is evident from the reference section that experimental work devoted to the basic physical properties of AlN is rather sparse. As research activities on the group III nitrides continue, it is hoped that more definitive work will become available.

REFERENCES

[1]   T.D. Moustakas, J.I. Pankove, Y. Hamakawa (Eds) [*Mater. Res. Soc. Symp. Proc. (USA)* vol.242 (1992)]

[2]   S. Strite, H. Morkoc [*J. Vac. Sci. Technol. B (USA)* vol.10 (1992) p.1237-66]

[3]   R.F. Davis [*Proc. IEEE (USA)* vol.79 (1991) p.702-12]

[4]   J.H. Edgar [*J. Mater. Res. (USA)* vol.7 (1992) p.235-52]

[5]   H.A. Wriedt [*Bull. Alloy Phase Diagrams (USA)* vol.7 no.4 (1986) p.329-33]

[6]   G.A. Jeffrey, G.S. Parry [*J. Chem. Phys. (USA)* vol.23 (1955) p.406]

[7]   G.A. Jeffrey, G.S. Parry, R.L. Mozzi [*J. Chem. Phys. (USA)* vol.25 (1956) p.1024-31]

[8]   H. Schulz, K.H. Thiemann [*Solid State Commun. (USA)* vol.23 (1977) p.815-9]

[9]   I.C. Huseby [*J. Am. Ceram. Soc. (USA)* vol.66 (1983) p.217-20]

[10]   P. Lawaetz [*Phys. Rev. B (USA)* vol.5 (1972) p.4039-45]

[11]   C.Y. Yeh, Z.W. Lu, S. Froyen, A. Zunger [*Phys. Rev. B (USA)* vol.46 (1992) p.10086-97]

[12]   G.A. Slack [*J. Phys. Chem. Solids (UK)* vol.34 (1973) p.321-35]

[13]   I. Petrov, E. Mojab, R.C. Powell, J.E. Greene, L. Hultman, J.E. Sundgren [*Appl. Phys. Lett. (USA)* vol.60 (1992) p.2491-3]

[14]   M.J. Paisley, R.F. Davis [*J. Cryst. Growth (Netherlands)* vol.127 (1993) p.136-42]

[15]   S. Strite et al [*J. Cryst. Growth (Netherlands)* vol.127 (1993) p.204-8]

[16]   M.E. Sherwin, T.J. Drummond [*J. Appl. Phys. (USA)* vol.69 (1991) p.8423-5]

[17]   Q. Xia, H. Xia, A.L. Ruoff [*J. Appl. Phys. (USA)* vol.73 (1993) p.8198-200]

[18]   H. Vollstadt, E. Ito, M. Akaishi, S. Akimoto, O. Fukunaga [*Proc. Jpn. Acad. B (Japan)* vol.66 (1990) p.7-9]

[19]   D. Gerlich, S.L. Dole, G.A. Slack [*J. Phys. Chem. Solids (UK)* vol.47 (1986) p.437-41]

[20]   K. Tsubouchi, K. Sugai, N. Mikoshiba [*1981 Ultrasonics Symp. Proc.* (IEEE, New York, 1981) p.375-80]

[21]   K. Tsubouchi, N. Mikoshiba [*1983 Ultrasonics Symp. Proc.* (IEEE, New York, 1983) p.299-310]

[22]   K. Tsubouchi, N. Mikoshiba [*IEEE Trans. Sonics Ultrason. (USA)* vol.SU-32 (1985) p.634-44]

[23]   L.E. McNeil, M. Grimsditch, R.H. French [*J. Am. Ceram. Soc. (USA)* vol.76 (1993) p.1132-6]

[24]   W.J. Meng, J.A. Sell, T.A. Perry, G.L. Eesley [*J. Vac. Sci. Technol. A (USA)* vol.11 (1993) p.1377-82]

[25]   J.F. Nye [*Physical Properties and Crystals* (Oxford University Press, Glasgow, 1957)]

[26]   I. Akasaki, M. Hashimoto [*Solid State Commun. (USA)* vol.5 (1967) p.851-3]

[27]   A.T. Collins, E.C. Lightowlers, P.J. Dean [*Phys. Rev. (USA)* vol.158 (1967) p.833-8]

[28]   M.F. MacMillan, R.P. Devaty, W.J. Choyke [*Appl. Phys. Lett. (USA)* vol.62 (1993) p.750-2]

[29]   O. Brafman, G. Lengyel, S.S. Mitra, P.J. Gielisse, J.N. Plendl, L.C. Mansur [*Solid State Commun. (USA)* vol.6 (1968) p.523-6]

[30]   R. Tsu, R.F. Rutz [*3rd Int. Conf. on Light Scattering in Solids* Eds M. Balkanski, R.C.C. Leite, S.P.S. Porto (Flammarion, Paris, 1976) p.393-5]

[31]   J.A. Sanjurjo, E. Lopez-Cruz, P. Vogl, M. Cardona [*Phys. Rev. B (USA)* vol.28 (1983) p.4579-84]

[32]   P. Perlin, A. Polian, T. Suski [*Phys. Rev. B (USA)* vol.47 (1993) p.2874-7]

[33]   C.F. Cline, J.S. Kalm [*J. Electrochem. Soc. (USA)* vol.110 (1963) p.773-5]

[34]   K.M. Taylor, C. Lenie [*J. Electrochem. Soc. (USA)* vol.107 (1960) p.308-14]

[35]   G. Long, M. Foster [*J. Am. Ceram. Soc. (USA)* vol.42 (1959) p.53-9]

[36]   T.J. Davies, P.E. Evans [*J. Nucl. Mater. (Netherlands)* vol.13 (1964) p.152]

[37]   W.M. Yim, R.J. Paff [*J. Appl. Phys. (USA)* vol.45 (1974) p.1456-7]

[38]   G.A. Slack, S.F. Bartram [*J. Appl. Phys. (USA)* vol.46 (1975) p.89-98]

[39]   Y.S. Touloukian, R.K. Kirby, R.E. Taylor, T.Y.R. Lee (Eds) [*Thermophysical Properties of Matter* vol.13 (Plenum Press, New York, 1977) p.1127-30]

[40]   A.D. Mah, E.G. King, W.W. Weller, A.U. Christensen [*Bur. Mines Rep. Invest. (USA)* RI-5716 (1961)]

[41]   Y.S. Touloukian, E.H. Buyco (Eds) [*Thermophysical Properties of Matter* vol.5 (Plenum Press, New York, 1970) p.1075-7]

[42]   V.I. Koshchendo, Y.K. Grinberg, A.F. Demidenko [*Inorg. Mater. (USA)* vol.20 (1985) p.1550-3]

[43]   M. Born, K. Huang [*Dynamical Theory of Crystal Lattices* (Oxford University Press, Oxford, 1985)]

[44]   N.M. Wolcott [*J. Chem. Phys. (USA)* vol.31 (1959) p.536-40]

[45]   G.A. Slack, R.A. Tanzilli, R.O. Pohl, J.W. Vandersande [*J. Phys. Chem. Solids (UK)* vol.48 (1987) p.641-7]

[46]   G.A. Slack, T.F. McNelly [*J. Cryst. Growth (Netherlands)* vol.42 (1977) p.560-3]

[47]   N. Kuramoto, H. Taniguchi [*J. Mater. Sci. Lett. (UK)* vol.3 (1984) p.471-4]

[48]   R.B. Dinwiddie, D.G. Onn [*Thermal Conductivity 21* Eds C.J. Cremers, H.A. Fine (Plenum Press, New York, 1990) p.499-508]

[49] D.R. Stull, H. Prophet (Eds) [*JANAF Thermochemical Tables*, 2nd edition (NSRDS-NBS37, US National Bureau of Standards, 1971)]

[50] G.A. Slack, T.F. McNelly [*J. Cryst. Growth (Netherlands)* vol.34 (1976) p.263-79]

[51] W. Class [NASA Contract Rep. NASA CR-1171 (1968)]

[52] R.G. Gordon, D.M. Hoffman, U. Riaz [*J. Mater. Res. (USA)* vol.6 (1991) p.5-7]

[53] T.L. Chu, R.W. Kelm [*J. Electrochem. Soc. (USA)* vol.122 (1975) p.995-1000]

[54] H. Demiryont, L.R. Thompson, G.J. Collins [*Appl. Opt. (USA)* vol.25 (1986) p.1311-8]

[55] F.S. Ohuchi, R.H. French [*J. Vac. Sci. Technol. A (USA)* vol.6 (1988) p.1695-6]

[56] E. Rille, R. Zarwasch, H.K. Pulker [*Thin Solid Films (Switzerland)* vol.228 (1993) p.215-7]

[57] J.A. Sell, W.J. Meng, T.A. Perry [*J. Vac. Sci. Technol. A (USA)* vol.10 (1992) p.1804-8]

[58] W.J. Meng, J.A. Sell, G.L. Eesley, T.A. Perry [*J. Appl. Phys. (USA)* vol.74 (1993) p.2411-4]

[59] G.A. Cox, D.O. Cummins, K. Kawabe, R.H. Tredgold [*J. Phys. Chem. Solids (UK)* vol.28 (1967) p.543-8]

[60] J. Pastrnak, L. Roskovcova [*Phys. Status Solidi (Germany)* vol.14 (1966) p.K5-8]

[61] L. Roskovcova, J. Pastrnak, R. Babuskova [*Phys. Status Solidi (Germany)* vol.20 (1967) p.K29-32]

[62] F. Keffer [*J. Chem. Phys. (USA)* vol.32 (1960) p.62-6]

[63] A.N. Pikhtin, A.D. Yaskov [*Sov. Phys.-Semicond. (USA)* vol.15 (1981) p.8-12]

[64] A.J. Noreika, M.H. Francombe, S.A. Zeitman [*J. Vac. Sci. Technol. (USA)* vol.6 (1969) p.194-7]

1.4 Crystal structure, mechanical properties and thermal properties of GaN

I. Akasaki and H. Amano

March 1994

A INTRODUCTION

Gallium nitride (GaN) is a promising semiconductor for optical, electronic and opto-electronic applications because of its advantageous properties such as large energy bandgap, good thermal and chemical stability, and physical hardness.

B CRYSTAL STRUCTURE

GaN usually crystallizes in the hexagonal wurtzite (WZ) structure, with two formula units per unit cell (4 atoms per cell) and a molecular weight of 83.728 g/mol. The space group symmetry is C_{6v}^4 (P6$_3$mc) and the point group symmetry is C_{6v}(6mm). Strain and defects may distort the lattice constants from their intrinsic values, and thus there is a wide dispersion in reported values. For WZ-GaN, at room temperature, lattice parameters of $a_0 = 3.1892 \pm 0.0009$ Å and $c_0 = 5.1850 \pm 0.0005$ Å [1,2] are generally accepted. For the zinc blende (ZB) polytype of GaN, the calculated lattice constant based on the measured Ga-N bond distance in WZ-GaN is $a = 4.503$ Å, while the measured values range from 4.49 to 4.55 Å [3-7]. A high pressure phase transition from wurtzite to either rocksalt structure or NiAs structure was reported both theoretically [8,9] and experimentally [10]. Perlin et al [10] found the phase transition occurred near 50 GPa. Munoz and Kunc [8] estimated the lattice constant of GaN in the rocksalt phase to be $a_0 = 4.22$ Å, and that in the NiAs phase to be $a_0 = 3.02$ Å. Camp et al [9] calculated the lattice constant of GaN in the rocksalt phase to be $a_0 = 4.098$ Å using first-principles non-local pseudopotentials, which is in agreement with the value calculated by Munoz and Kunc (see Datareview 2.2, Section E).

C MECHANICAL PROPERTIES

Investigations of the elastic constants of WZ-GaN were carried out by Savastenko and Sheleg on powdered GaN crystals using X-ray methods [11]. The measured elastic constants are shown in TABLE 1. Krishnankutty et al [12] reported similar elastic constants. The Poisson's ratio $\nu_{<0001>} = (\Delta a/a_0)/(\Delta c/c_0)$ is estimated to be 0.372 [11] using the elastic constants shown in TABLE 1. The Poisson's ratio $\nu_{<0001>}$ was experimentally characterized using Bond's X-ray method [2] on heteroepitaxially grown GaN having different thicknesses. The measured Poisson's ratio is $\nu_{<0001>} = 0.38$, which is in good agreement with the calculated one from the elastic constants shown in TABLE 1. The Young's modulus $E_{<0001>}$ is estimated to be 150 GPa [11] using the elastic constants shown in TABLE 1. Sherwin and Drummond [13] estimated the elastic properties of ZB-GaN using the values of those of WZ-GaN [11]. Results are also shown in TABLE 1.

TABLE 1 Mechanical properties of WZ-GaN and ZB-GaN: c_{ij}'s of WZ-GaN are the elastic constants obtained from X-ray diffraction measurements; those of ZB-GaN are calculated from those of WZ-GaN; ν is Poisson's ratio; E is Young's modulus; B is the bulk modulus. c_{ij}, E and B are written in units of GPa. ν and E are calculated from elastic constants.

	WZ		ZB	
	Numerical value	Refs	Numerical value	Refs
c_{11}	296 ± 18	[11]	264	[13] calc. from [11]
c_{12}	130 ± 11		153	
c_{13}	158 ± 6			
c_{33}	267 ± 18			
c_{44}	24.1 ± 2		68	
$\nu_{<0001>}$	0.372 0.38	calc. from [11] [2] X-ray	0.366	[13]
$E_{<0001>}$	150	[11]		
B	195 203 190 194.6 245	calc. [14] calc. [15] calc. [9] [11,16] [10] X-ray	173 185	calc. [9] [13]

Miwa and Fukumoto [14] performed first-principles calculations to get the bulk modulus of WZ-GaN to be 195 GPa. Xu and Ching [15] also calculated the bulk modulus of WZ-GaN to be 203 GPa using the first-principles orthogonalized linear combination of atomic orbitals method. Camp et al [9] also calculated the bulk modulus of WZ-GaN to be 190 GPa. These values are in good agreement with the value of 194.6 GPa [16] estimated from elastic constants shown in [11]. Perlin et al [10] reported a slightly larger value of 245 GPa.

As for the optical phonon modes of WZ-GaN, group theory predicts one A_1 and one E_1 mode, two E_2 modes and two B_1 modes. Wavenumbers of the Raman active optical phonon modes, that is, E_1-TO, A_1-TO, E_1-LO, A_1-LO and two E_2 phonons measured by Raman scattering, are listed in TABLE 2 [17-19]. Results of the first-principles pseudopotential calculation of the wavenumbers of these optical phonons are also listed in TABLE 2 [14]. The wavenumbers calculated in [14] coincide very well with the experimentally measured ones.

The optical phonon wavenumber of ZB-GaN has also been studied using Raman scattering [20]. The result is also listed in TABLE 2.

TABLE 2 Zone centre optical phonon wavenumbers of GaN obtained from Raman scattering measurements at 300 K. Numerical values are written in units of cm^{-1}. Calc. means the calculated values obtained using the first-principles pseudopotential calculation method [14].

WZ			ZB		
Mode	Numerical value	Refs	Mode	Numerical value	Refs
E_1-TO	559 558 556	[17] [19] calc. [14]	TO	558	calc. [14]
A_1-TO	533 534	[17,19] calc. [14]			
E_1-LO	741	[18]	LO	730	[20]
A_1-LO	710	[18]			
E_2 (Low)	145 144 146	[17] [19] calc. [14]			
E_2 (High)	568 560	[17,19] calc. [14]			

D   THERMAL PROPERTIES

Thermal expansion of single crystal WZ-GaN has been studied [1] in the temperature range from 300 K to 900 K, or from 80 K to 820 K [21]. Maruska and Tietjen [1] reported that the lattice constant a changes linearly with temperature with a coefficient of thermal expansion of $+5.59 \times 10^{-6}$ K^{-1}. Meanwhile, the expansion of lattice constant c shows a superlinear dependence on temperature. The mean coefficient of thermal expansion parallel to the c-axis is $+3.17 \times 10^{-6}$ K^{-1} for T = 300 - 700 K, and $+7.75 \times 10^{-6}$ K^{-1} for T = 700 - 900 K. Sheleg and Savastenko [21] reported the thermal expansion coefficient at 600 K perpendicular and parallel to the c-axis to be $+4.52 \pm 0.05 \times 10^{-6}$ and $+5.25 \pm 0.05 \times 10^{-6}$, respectively.

Other thermal properties of WZ-GaN have been studied by a number of authors. The specific heat of WZ-GaN at constant pressure (C_p) is given by [22]:

$$C_p(T) = 9.1 + (2.15 \times 10^{-3} \times T) \ (cal\,mol^{-1}\,K^{-1}) \tag{1}$$

The Debye temperature (θ_D) of GaN at 0 K was calculated to be $\theta_D \sim 600$ K [23]. Thermal conductivity (κ) of single crystal WZ-GaN was measured by Sichel and Pankove in the temperature range from 25 K to 360 K [24]. Thermal conductivity at 300 K is $\kappa = 1.3$ W cm^{-1} K^{-1}, which is slightly smaller than the predicted value of 1.7 W cm^{-1} K^{-1} [23].

Thermodynamic properties for WZ-GaN have been tabulated by Elwell and Elwell in their review paper [25]. The heat of formation of WZ-GaN from the reaction

$$Ga(s) + \tfrac{1}{2}N_2(g) = GaN(s) \tag{2}$$

was calculated to be $\Delta H_{298 K} = -26.4 \, \text{kcal mol}^{-1}$ [26], or as the standard heat of formation $\Delta H = -37.7 \, \text{kcal mol}^{-1}$ [27]. The equilibrium vapour pressure of N_2 over solid GaN has been found to be 10 MPa at 1368 K and 1 GPa at 1803 K [28]. A thorough description of the GaN phase diagram including the equilibrium vapour pressure of N_2 over GaN is given in Datareview 2.2.

E   CONCLUSION

A survey of the crystal structure, mechanical properties and thermal properties of GaN has been given. Compared to basic physical properties of GaN in the wurtzite phase, data on those of GaN in the zinc blende phase exist rather sparsely. Further advances in the growth of high quality single crystals will surely lead to a more clear understanding of the basic physical properties of GaN in both the zinc blende and wurtzite structures.

REFERENCES

[1]   H.P. Maruska, J.J. Tietjen [*Appl. Phys. Lett. (USA)* vol.15 (1969) p.327-9]
[2]   T. Detchprohm, K. Hiramatsu, K. Itoh, I. Akasaki [*Jpn. J. Appl. Phys. (Japan)* vol.31 (1992) p.L1454-6]
[3]   M. Mizuta, S. Fujieda, Y. Matsumoto, T. Kawamura [*Jpn. J. Appl. Phys. (Japan)* vol.25 (1986) p.L945-8]
[4]   S. Strite et al [*J. Vac. Sci. Technol. B (USA)* vol.9 (1991) p.1924-9]
[5]   M.J. Paisley, Z. Sitar, J.B. Posthill, R.F. Davis [*J. Vac. Sci. Technol. A (USA)* vol.7 (1989) p.701-5]
[6]   T. Lei, M. Fanciulli, R.J. Molnar, T.J. Moustakas, R.J. Graham, J. Scanlon [*Appl. Phys. Lett. (USA)* vol.59 (1991) p.944-6]
[7]   R.C. Powell, N.E. Lee, Y.W. Kim, J.E. Greene [*J. Appl. Phys. (USA)* vol.73 (1993) p.189-204]
[8]   A. Munoz, K. Kunc [*Phys. Rev. B (USA)* vol.44 (1991) p.10372-3]
[9]   P.E. Van Camp, V.E. Van Doren, J.T. Devreese [*Solid State Commun. (USA)* vol.81 (1992) p.23-6]
[10]   P. Perlin, C.J. Carillon, J.P. Itie, A.S. Miguel, I. Grzegory, A. Polian [*Phys. Rev. B (USA)* vol.45 (1992) p.83-9]
[11]   V.A. Savastenko, A.U. Sheleg [*Phys. Status Solidi A (Germany)* vol.48 (1978) p.K135-9]
[12]   S. Krishnankutty, R.M. Kolbas, M.A. Khan, J.N. Kuznia, J.M. Van Hove, D.T. Olson [*J. Electron. Mater. (USA)* vol.21 (1992) p.609-12]
[13]   M.E. Sherwin, T.J. Drummond [*J. Appl. Phys. (USA)* vol.69 (1991) p.8423-5]
[14]   K. Miwa, A. Fukumoto [*Phys. Rev. B (USA)* vol.48 (1993) p.7897-902]
[15]   Y.N. Xu, W.Y. Ching [*Phys. Rev. B (USA)* vol.48 (1993) p.4335-51]
[16]   D. Gerlich, S.L. Dole, G.A. Slack [*J. Phys. Chem. Solids (UK)* vol.47 (1986) p.437-41]
[17]   D.D. Manchon Jr., A.S. Barker Jr., P.J. Dean, R.B. Zetterstrom [*Solid State Commun. (USA)* vol.8 (1970) p.1227-31]
[18]   A. Cingolani, M. Ferrara, M. Lugara, G. Scamarcio [*Solid State Commun. (USA)* vol.58 (1986) p.823-4]

[19]   J. Nakahara, T. Kuroda, H. Amano, I. Akasaki, S. Minomura, I. Grzegory [*9th Symposium Record of Alloy Semiconductor Physics and Electronics (Japan)* vol.9 (1990) p.391-7]

[20]   S. Miyoshi et al [*J. Cryst. Growth (Netherlands)* vol.124 (1992) p.439-42]

[21]   A.U. Sheleg, V.A. Savastenko [*Vestsi Akad. Navuk BSSR Ser. Fiz.-Mat. Navuk (USSR)* (1977) p.126-8]

[22]   [*Thermochemical Properties of Inorganic Substances* (Springer, Berlin-Heidelberg-New York, USA, 1977)]

[23]   G.A. Slack [*J. Phys. Chem. Solids (UK)* vol.34 (1973) p.321-35]

[24]   E.K. Sichel, J.I. Pankove [*J. Phys. Chem. Solids (UK)* vol.38 (1977) p.330]

[25]   D. Elwell, M.M. Elwell [*J. Cryst. Growth (Netherlands)* vol.17 (1988) p.53-78]

[26]   C.D. Thurmond, R.A. Logan [*J. Electrochem. Soc. (USA)* vol.119 (1972) p.622-6]

[27]   J. Karpinski, S. Porowski [*J. Cryst. Growth (Netherlands)* vol.66 (1984) p.11-20]

[28]   J. Karpinski, J. Jun, S. Porowski [*J. Cryst. Growth (Netherlands)* vol.66 (1984) p.1-10]

1.5 Crystal structure, mechanical properties, thermal properties and refractive index of InN

T.L. Tansley

December 1993

A INTRODUCTION

In the absence of good quality single crystal samples, the physical properties of indium nitride have been measured on non ideal thin films, typically ordered polycrystalline with crystallites in the 50 nm to 500 nm range. Structural, mechanical and thermal properties have not been reported for epitaxial films on lattice-matched substrates.

Dissociation temperature is low, and nonstoichiometry ensues at growth temperatures high enough for adatom mobility to promote large-scale crystallization. The resulting high density of nitrogen vacancies affects many properties including lattice parameters.

Material not grown under ultra high vacuum conditions is susceptible to high levels of oxygen incorporation which may affect physical properties; few workers report measured concentrations of this impurity when describing other properties.

Basic data on the physical properties of indium nitride are sparse, and the following sections include instances of values uncorroborated by repeated measurement. In cases where data are absent, the chemical trends across the nitride group have made it possible for first-order estimates to be made and these are included where thought appropriate. The reader is invited to consider these extrapolations by consulting firmer values for AlN and GaN elsewhere in this volume. More convincing data await the advent of better quality material.

B CRYSTAL STRUCTURE OF INDIUM NITRIDE

Indium nitride normally crystallizes in the wurtzite (hexagonal) structure [1] with C_{6v}^4 (P6$_3$mc) space group and C_{6v} (6mm) point group symmetries. The zinc blende (cubic) form has been reported [2] to occur in films containing both polytypes.

TABLE 1 lists values [1-8] of basal and perpendicular axes a_0 and c_0. Note that refs [3-8] offer consistent ratios of c_0/a_0 of about 1.615 ± 0.008 and that the ratio approaches the ideal value of 1.633 only in the case of growth with precautions taken to reduce N vacancies [6]. It is also noteworthy that, amongst data of reasonable agreement in c_0/a_0 [3-8], values of a_0 also agree well (within about $\pm 0.1\%$) while uncertainties in c_0 are about ten times greater. This may be a function of nitrogen deficiency, since N atoms are close packed in (0001) planes and high vacancy densities may preferentially shrink the lattice in the perpendicular direction, parallel to c_0.

The single reported measurement of a_0 in the cubic polytype [2] yields a molecular cell volume ($a_0^3/4$) of 30.9 Å^3, compared to 31.2 ± 0.2 Å^3 [3-8] for the cell volume of the hexagonal polytype ($\sqrt{3}\, a_0^2 c_0/4$).

TABLE 1 Lattice parameters of indium nitride.

a_0	c_0	c_0/a_0	Ref (year)	Comment
0.353	0.596	1.69	[1] (38)	Thermally reacted at high pressure
0.3533	0.5963	1.611	[3] (66)	Powder
0.3544	0.5718	1.613	[4] (75)	
0.35446	0.57034	1.609	[5] (78)	c-axis oriented polycrystal
0.35480	0.576	1.6234	[6] (86)	Powder, low N vacancy
-	0.569	-	[7] (89)	
0.3540	0.5705	1.612	[8] (89)	c-axis oriented on sapphire
0.36	0.574	1.59	[2] (93)	Plasma on (100) GaAs
0.498	- cubic	-	[2] (93)	c-axis normal to GaAs (111)

Theoretical approaches to total energy as a function of volume predict a phase transition to the rocksalt structure under high pressure [9,10]. The transition is calculated to occur at a pressure of about 245 GPa and an experimental value of 230 GPa [11] has been quoted [9]. The critical volume ratio [9] is $V/V_0 = 0.83$, equivalent to a molecular volume reduction from 31 to 27 Å^3.

C   MECHANICAL PROPERTIES OF INDIUM NITRIDE

Directly measured density (by Archimedean displacement [12]) is $6.89 \times 10^3 \, \text{kg m}^{-3}$ at 25 °C. A comparable value of $6.81 \times 10^3 \, \text{kg m}^{-3}$ has been estimated from X-ray data [13]. The cell volumes cited in Section B, taken in conjunction with a molar mass of 128.827 g mol^{-1}, yield densities of $(6.81 \pm 0.05) \times 10^3 \, \text{kg m}^{-3}$ and $6.97 \times 10^3 \, \text{kg m}^{-3}$ for the wurtzite and zinc blende polytypes respectively.

Bulk modulus has been calculated from first principles by a local-density approximation [14] and by a linear muffin-tin orbital method [8], suggesting a value B = 165 GPa.

The five distinguishable second-order elastic moduli in a hexagonal crystal are c_{11}, c_{12}, c_{13}, c_{33} and c_{44}. There are reports of neither measured nor calculated values but, since each depends principally on the lattice constants [15] which vary by only about 10% across the nitrides, values for AlN (q.v.) may be used as a first approximation. The comparability of bulk moduli in indium, gallium and aluminium nitrides supports this approach. Estimates of the principal transverse and longitudinal elastic constants c_t and c_l are given in TABLE 2.

The piezoelectric constant has not been reported, but its dependence on the dielectric constants ε_r and e_{14} [16] allows values of about 50% of those found in AlN to be inferred [17].

Indium nitride has twelve phonon modes at the zone centre (symmetry group C_{6v}), three acoustic and nine optical with the acoustic branches essentially zero at k = 0. The IR active modes are $E_1(\text{LO})$, $E_1(\text{TO})$, $A_1(\text{LO})$ and $A_1(\text{TO})$. A transverse optical mode has been identified at 478 cm^{-1} (59.3 meV) by reflectance [4] and 460 cm^{-1} (57.1 meV) by transmission

TABLE 2 Mechanical properties of indium nitride.

Property	Value	Ref	Comment
Density (hex.)	$6.89 \times 10^3 \, \text{kg m}^{-3}$	[12]	Meas. by displacement
	$(6.81 \pm 0.05) \times 10^3 \, \text{kg m}^{-3}$	-	Various X-ray data
Density (cubic)	$6.97 \times 10^3 \, \text{kg m}^{-3}$	[2]	X-ray data
Molar mass	$128.827 \, \text{g mol}^{-1}$		
Mol. vol. (hex.)	$31.2 \, \text{Å}^3$		From lattice constants
Mol. vol. (cubic)	$30.9 \, \text{Å}^3$		From lattice constants
c_t	$4.42 \times 10^{11} \, \text{dyn cm}^{-2}$	[17]	Estimate
c_l	$2.65 \times 10^{12} \, \text{dyn cm}^{-2}$	[17]	Estimate
$h_{14}{}^2(4c_l + 3c_t)/12$	$7.78 \times 10^3 \, \text{V}^2 \text{dyn}^{-1}$	[17]	Estimate
Deformation potential	$7.1 \, \text{eV}$	[17]	Estimate
$\hbar\omega_{TO}$	$59.3 \, \text{meV} \, (478 \, \text{cm}^{-1})$	[4]	Reflectance meas.
	$57.1 \, \text{meV} \, (460 \, \text{cm}^{-1})$	[18]	Transmission meas.
$\hbar\omega_{LO}$	$86.2 \, \text{meV} \, (694 \, \text{cm}^{-1})$	[4]	Est. - Brout sum rule
	$89.2 \, \text{meV} \, (719 \, \text{cm}^{-1})$	[18]	Est. - Brout sum rule

[18]. In both reports the location of a longitudinal optical mode is inferred from the Brout sum rule, giving respective values of $694 \, \text{cm}^{-1}$ (86.1 meV) and $719 \, \text{cm}^{-1}$ (89.2 meV).

D THERMAL PROPERTIES OF INDIUM NITRIDE

The linear thermal expansion coefficients (perpendicular and parallel to the c-axis) have been measured at five temperatures between 190 K and 560 K [19] and are given in TABLE 3. These are the only data extant and, given the disagreement between these authors and others on thermal expansion in GaN [20], should be treated cautiously.

TABLE 3 Coefficient of linear thermal expansion of indium nitride ($\times 10^{-6} \, \text{K}^{-1}$) [19].

T (K)	α (perp)	α (parallel)
190	3.40	2.70
260	3.75	2.85
360	4.20	3.15
460	4.80	3.45
560	5.70	3.70

Thermal conductivity has not been reported but may be inferred from AlN and GaN data by using the Liebfried-Schloman scaling parameter [21]. A value of about $(80 \pm 20) \, \text{W m}^{-1} \text{K}^{-1}$ is estimated on the assumption that thermal conductivity is limited by intrinsic phonon-phonon scattering, but may be reduced by oxygen contamination and phonon scattering at defects or increased by very high electronic concentrations.

Heat capacity is $(9.1 + 2.9 \times 10^{-3} \, \text{T}) \, \text{cal mol}^{-1} \text{K}^{-1}$ between 298 and 1273 K, while entropy is $10.4 \, \text{cal mol}^{-1} \text{K}^{-1}$ at 298.15 K [22]. Enthalpies, entropies and Gibbs functions of fusion and formation are given in TABLE 4. Thermochemical data should also be regarded with circumspection, in particular the early value of $\Delta H_f^0 = -5 \, \text{kcal mol}^{-1}$ [1] which is often quoted but should now be discarded in favour of more recent results [23,29] found to lie close to those for the other nitrides.

TABLE 4 Thermal and thermodynamic properties of indium nitride.

(i)	General properties	Value	Ref	Comment
	Thermal conductivity, k	(80 ± 20) W m^{-1} K^{-1}	-	Estimate
	Heat capacity, C_p	$(9.1 + 2.9 \times 10^{-3}$ T) cal mol^{-1} K^{-1}	[21]	298 to 1273 K
	Entropy, S^0	10.4 cal mol^{-1} K^{-1}	[21]	298.15 K
	N_2 equilib. vap. press.	1 atm	[22]	800 K
	"	10^5 atm	[22]	1100K
(ii)	Thermodynamic state-function changes at formation and fusion			
	Formation: ΔH_f^0	- 34.3 kcal mol^{-1}	[22]	298.15 K
		- 30.5 kcal mol^{-1}	[28]	298.15 K
	ΔS_f^0	- 25.3 cal mol^{-1} K^{-1}	[22]	298.15 K
	ΔG_f^0	- 22.96 kcal mol^{-1}	[29]	298.15 K
	Fusion: ΔH_m	14 kcal mol^{-1}	[29]	Theoretical
	ΔS_m	10.19 cal mol^{-1} K^{-1}	[29]	Theoretical

The equilibrium partial pressure of N_2 over indium nitride is 1 atm at 800 K, increasing exponentially with 1/T to 10^5 atm at 1100 K [23]. Note that In metal is highly reactive with atomic N while largely inert in the molecular species, so the equilibrium partial pressure of the former is more important in growth kinetic studies. Low temperature growth is possible in the presence of N_1 and plasma induced or assisted methods are common as are those which liberate N_1 from a precursor molecule. The phase diagram of InN is discussed critically in Datareview 2.3 of this volume [S. Porowski and I. Grzegory].

E   REFRACTIVE INDEX OF INDIUM NITRIDE

Several single-wavelength measurements have been made on samples with $n < 10^{20}$ cm^{-3}. The long wavelength limit calculated from the imaginary part of the theoretical dielectric function [24] is 2.88 ± 0.15, very close to the measured values of $2.9 \pm 10\%$ [25], 2.9 [26] and $3.05 + 0.1\%$ [27]. Measurements in the range 2 to 20 eV have been made by specular reflection of synchrotron radiation and a value of n = 2.5 at 2.0 eV has been confirmed by ellipsometry [28]. The imaginary part of the dielectric function in this range agrees well with that derived from the calculated band structure [24].

The long-wavelength refractive index suggests a high frequency dielectric constant $\varepsilon_\infty = n^2 = 8.4$. Ionic polarizability makes a significant contribution at low frequencies where, although measurement is presently inhibited by high conductivity, $\varepsilon_0 \sim 15$ is recommended.

TABLE 5 Refractive index and dielectric constants of indium nitride.

(i)	Refractive index	Wavelength (nm)	Ref	Comment
	2.88 ± 0.15	Long limit	[23]	Theoretical
	2.9 ± 0.3	600 - 800	[24]	-
	2.9	900 - 1200	[25]	Transmission interference
	3.05 ± 0.03	-	[26]	-
	2.65	620	[27]	Normal incidence reflectance
	1.0	120	[27]	of synchrotron radiation
	2.93	820	[30]	$n > 10^{20} \, cm^{-3}$
(ii)	Dielectric constants			
	$\varepsilon_\infty = 8.4$ (from n^2 in long wavelength limit) $\varepsilon_0 = 15$ (estimated)			

F   CONCLUSION

Firm data on the basic physical properties of InN await improvement in growth techniques for monocrystalline, stoichiometric samples. Normal growth habit is wurtzite, but nitrogen vacancy levels verging on nonstoichiometry as well as susceptibility to oxygen inclusion are held responsible for uncertainties in unit-cell dimensions (TABLE 1).

No elastic properties have been reported, but legitimate inferences may be drawn from well-established AlN data and two principal, optical-mode lattice vibrations are identified (TABLE 2).

Thermal and thermodynamic properties are rather less dependent on crystalline perfection but present data (TABLES 3 and 4) should nevertheless be regarded circumspectly.

Optical properties appear to have been most widely studied, with consensus values reached for refractive indices. The symmetry suggests the presence of birefringence, but better samples are required for confirmation. High frequency permittivity may be deduced from optical results, but its counterpart at low frequencies remains speculative (TABLE 5).

REFERENCES

[1]   R. Juza, H. Hahn [*Z. Anorg. Allg. Chem. (Germany)* vol.239 (1938) p.282-8]

[2]   S. Strite, D. Chandrasekhar, D.J. Smith, J. Sariel, H. Chen, N. Teraguchi [*J. Cryst. Growth (Netherlands)* vol.127 (1993) p.204-6]

[3]   G. Gieseke [*Semicond. Semimet. (USA)* vol.2 (1966) p.63]

[4]   K. Osamura, S. Naka, Y. Murakami [*J. Appl. Phys. (USA)*, vol.46 (1975) p. 3432-7]

[5]   I.G. Pichugin, M. Tlachala [*Izv. Akad. Nauk SSSR Neorg. Mater. (USSR)* vol.14 (1978) p.175-7]

[6]   T.L. Tansley, C.P. Foley [*J. Appl. Phys. (USA)* vol.59 (1986) p.3241-7]

[7]   A. Wakahara, A. Yoshida [*Appl. Phys. Lett. (USA)* vol.54 (1989) p.2984-7]

[8]   K. Kubota, Y. Kobayashi, K. Fujimoto [*J. Appl. Phys. (USA)*, vol.66 (1989) p. 2984-7]

[9]   I. Gorczyca, N.E. Christensen [*Physica B (Netherlands)* vol.185 (1992) p.410-4]

[10]   A. Munoz [*Physica B (Netherlands)* vol.185 (1992) p.422-5]

[11]   P. Perlin [unpublished]

[12]   H. Hahn, R. Juza [*Z. Anorg. Allg. Chem. (Germany)* vol.244 (1940) p.111-2]

[13]   W.B. Pearson [*A Handbook of Lattice Spacings and Structures of Metals and Alloys* (Pergamon Press, Oxford, 1967)]

[14]   P.E. van Camp, V.E. van Doren, J.T. Devreese [*Phys. Rev. B (USA)* vol.41 (1990) p.1598-604]

[15]   S. Adachi [*J. Appl. Phys. (USA)* vol.58 (1985) p.R1-4]

[16]   C.M. Wolfe, N. Holonyak, G.E. Stillman [*Physical Properties of Semiconductors* (Prentice Hall, Eaglewood Cliffs, 1989)]

[17]   V.W.L. Chin, T.L. Tansley, T. Osotchan [to be published]

[18]   T.L. Tansley, R.J. Egan, E.C. Horrigan [*Thin Solid Films (Switzerland)* vol.164 (1988) p.441-8]

[19]   A.V. Sheleg, V.A. Savastenko [*Vestsi Akad. Nauk BSSR Ser. Fiz. Mat. Nauk (USSR)* vol.3 (1976) p.126-8]

[20]   H.P. Maruska, J.J. Tietjen [*Appl. Phys. Lett. (USA)* vol.15 (1969) p.327-8]

[21]   G.A. Slack, R.A. Tanzilli, R.O. Pohl, J.W. Vandersande [*J. Phys. Chem. Solids (UK)* vol.48 (1987) p. 641-7]

[22]   I. Barin, O. Knacke, O. Kubaschewski [*Thermodynamical properties of inorganic substances* (Springer Verlag, Berlin, 1977)]

[23]   J.B. MacChesney, P.M. Bridenbaugh, P.B. O'Connor [*Mater. Res. Bull. (USA)* vol. 5 (1970) p.783-90]

[24]   C.P. Foley, T.L. Tansley [*Phys. Rev. B. (USA)* vol.33 (1986) p.1430-3]

[25]   J. Misek, F. Srobar [*Elektrotech. Cas. (Czechoslovakia)* vol.30 (1979) p.690-2]

[26]   H. Hovel, J.J. Cuomo [*Appl. Phys. Lett. (USA)* vol.20 (1972) p.71-3]

[27]   M. Ilegems [*J. Cryst. Growth (Netherlands)* vol.13/14 (1972) p.360-4]

[28]   Q. Guo, O. Kato, M. Fujisawa, A. Yoshida [*Solid State Commun. (USA)* vol.83 (1992) p. 721-3]

[29]   S.P. Gordienko, B.V. Fenochka [*Zh. Fiz. Khim. (USSR)* vol.5 (1977) p.530-4]

[30]   J.A. van Vechten [*Phys. Rev. B. (USA)* vol.7 (1973) p.1479-86]

[31]   V.A. Tyagai, A.M. Evstigneev, A.N. Krasiko, A.F. Andreeva, V.Y. Malakhov [*Sov. Phys.-Semicond. (USA)* vol.11 (1977) p.1257-8]

CHAPTER 2

PHASE DIAGRAMS

2.1 Phase diagram of BN

V.L. Solozhenko

July 1994

A INTRODUCTION

Boron nitride and materials based on it are high on the list of the most important inorganic materials and provide a basis for many advanced technologies. The wide spectrum of properties of the four BN polymorphic modifications is responsible. These modifications are two graphite-like, hexagonal (h-BN) and rhombohedral (r-BN), and two dense, cubic (c-BN) and, wurtzitic (w-BN), structures. Materials for electronics and nuclear energy, superhard materials, lubricants and refractories are an incomplete list of applications of the different modifications of boron nitride. In recent years, a trend has been observed of an increase in commercial production of cubic boron nitride. Materials with a high thermal conductivity, semiconducting crystals and other high-tech products are in the greatest demand. In connection with this, the development of the fundamentals for producing materials with a preset complex of physico-chemical and exploitation properties has assumed great importance. This fact, in turn, suggests that reliable data on the boron nitride phase diagram needs to be collected.

In the past, the boron nitride phase diagram was based on Wentorf's experimental data on the h-BN→c-BN catalytic transition [1-3] and h-BN melting [4] under high pressures and temperatures. Based on the above data and on evaluation of the position of the h-BN - c-BN - liquid triple point and c-BN melting curve, carried out by analogy with the carbon phase diagram [5], in 1963 Bundy and Wentorf proposed the boron nitride phase p,T-diagram [6] (FIGURE 1 - here and subsequently pressure values are presented according to the revised NBS pressure scale [7]), which has been generally accepted up to now. In 1975, Corrigan and Bundy [8] extrapolated the h-BN↔c-BN boundary line to the low-temperature region by analogy with the graphite↔diamond equilibrium curve.

Subsequent attempts to refine and supplement the BN phase diagram pertained only to the h-BN↔c-BN boundary line. Various publications dealing with c-BN and h-BN mutual transformations at high pressures and temperatures [6,8-20] show a considerable disagreement between experimental data (FIGURE 1), the result of the presence of many different 'catalysts' chemically interacting with boron nitride. This makes us doubt the validity of identification of the transformation experimental curves with the h-BN↔c-BN equilibrium line. All of this allows us to conclude that it is impossible to plot an equilibrium phase p,T-diagram for boron nitride from the data on the mutual transformations of its polymorphic modifications, due to the kinetic factors' crucial importance in the BN-phase formation at high pressures and temperatures.

From the above discussion, it is obvious that the correct definition of phase equilibrium lines on the BN diagram is possible only in the framework of the thermodynamic approach.

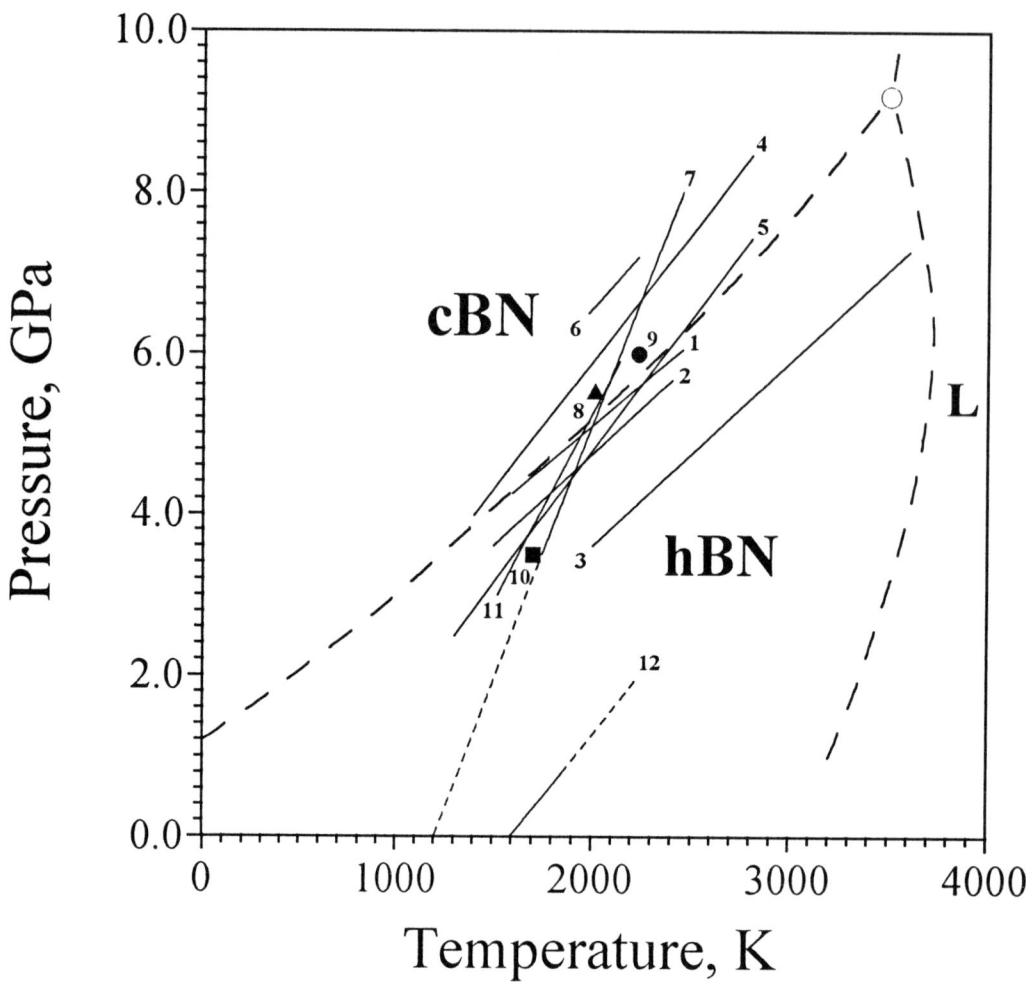

FIGURE 1 Bundy-Wentorf's phase diagram for BN [6] and experimental h-BN↔c-BN boundaries:

1 - h-BN - Li, h-BN - Mg [9];
2 - h-BN - Mg, h-BN - Mg_3N_2 [10];
3 - direct transformation h-BN→c-BN [11,12];
4 - h-BN - Li, h-BN - Li_3N [13];
5 - h-BN - LiH, h-BN - $LiNH_2$ [13];
6 - h-BN - Li_3BN_2 [14];
7 - h-BN - Mg_3BN_3 [15];
8 - h-BN - $LiCaBN_2$ [16];
9 - h-BN - Li_3BN_2 [17];
10 - h-BN - Mg_3N_2 [18];
11 - h-BN - Mg_3BN_3, h-BN - $Me^{II}_3B_2N_4$ [19];
12 - direct transformation c-BN→h-BN [20].

B THERMODYNAMIC ASPECT OF BN POLYMORPHISM

The attempts described in the literature to thermodynamically calculate the h-BN↔c-BN equilibrium line [21-23] can hardly be called successful, as the authors used the values of polymorphic transformation enthalpy that have been found from experimental 'equilibrium curves' mentioned above, and the calculations were practically reduced to extrapolation of the curves to the low-pressure region.

The estimations of transformation enthalpies for mutual transformations of different modifications of BN were made in [24-26], and the corresponding boundary lines were calculated. From the above results, the authors proposed a new phase diagram for boron nitride that differs from the generally accepted one of Bundy-Wentorf [6] with the presence of a p,T-region of w-BN thermodynamic stability. But it follows from the analysis of this phase diagram [27] that the metastable sections of all the three boundary lines are in the region of h-BN thermodynamic stability, contradicting the basic principles of geometric thermodynamics.

Until 1988 there was no experimental data on formation enthalpies for r-BN, c-BN and w-BN and therefore the correct calculations of equilibrium curves for the different modifications of BN were impossible. This explains the flaws of the older analysis as it was based on the incorrect analogy between the p,T-diagrams of carbon and BN [28,29]. In the framework of the concept, the h-BN↔c-BN equilibrium line is parallel to the graphite↔diamond equilibrium line and intersects the pressure axis at 1.3 GPa, suggesting metastability of c-BN at low pressures over the whole temperature range.

All the above dictated the need for a systematic study of thermodynamic properties of all boron nitride polymorphic modifications with the aim of plotting the BN equilibrium phase p,T-diagram.

B1 Low-Temperature Heat Capacity

All the measurements of low-temperature heat capacities of BN polymorphic modifications have been made by adiabatic calorimetry, with the periodic introduction of heat.

h-BN The heat capacity of graphite-like hexagonal boron nitride at temperatures below 300 K was measured in [30-32]. However, it should be noted that in all the papers there were no data on the degree of three-dimensional ordering of the samples being studied, and that led to some uncertainty in the results obtained.

The heat capacity of high-ordered h-BN ($P_3 = 0.98 \pm 0.02$) (the degree of three-dimensional ordering calculated by the method in ref [33]) in the 15-305 K interval was measured in [34]. The temperature dependence of C_p has no flat abnormality between 60 and 140 K, reported in papers [30-32], and lies lower than the h-BN heat capacity curves obtained by other authors.

r-BN The heat capacity of high-ordered graphite-like rhombohedral boron nitride ($P_3 = 0.93 \pm 0.03$) was measured in [35] over the 15-305 K temperature range and C_p abnormalities have not been found.

c-BN The heat capacity of cubic boron nitride single crystals has been measured in [36] over the 4-302 K temperature range. The analysis of the data obtained shows three slightly pronounced smoothed C_p abnormalities, caused by the contribution of impurities. The total contribution of the observed abnormalities to c-BN enthalpy and entropy does not exceed 0.003 and 0.05 % of the corresponding standard values.

A few measurements of c-BN low-temperature heat capacity were performed for polycrystalline samples [37-40]. Data in [37,39,40] are in good agreement with results obtained in [36], while data in [38] lie essentially (10-30%) lower.

w-BN The heat capacity of wurtzitic boron nitride has been measured in the 5-320 K [41] and 6-305 K [32] temperature intervals. Comparison of the results shows that data in [41] are considerably higher (up to 25%) than data in [32], and even higher than the heat capacity of h-BN [30-32,34]. This demonstrates the low reliability of data in [41] and can be associated with the high level of impurities in the sample used and methodological errors in the measurements.

According to [32], in the C_p curve from 9 to 27 K, a λ-abnormality with extremum at 21 K is observed, while for thermostabilized w-BN, a normal shape of heat capacity curve over the 5-300 K temperature range is obtained [20]. The above abnormality contributions to the w-BN enthalpy and entropy are 0.97 J mol^{-1} and 0.042 J (K mol)$^{-1}$ respectively and are evidently caused by the presence of the ordered point defects system in the lattice of the w-BN sample studied in [32].

t-BN The heat capacity of turbostratic graphite-like boron nitride ($P_3 = 0$) has been measured in [34] between 15 and 305 K. The C_p curve obtained lies considerably higher than the h-BN heat capacity curve and has a pronounced flat abnormality in the 60-140 K region. This permits the assumption that the h-BN samples used in the previous works [30-32] were partially disordered.

The smoothed values of heat capacities for all four boron nitride polymorphic modifications and turbostratic BN are presented in TABLE 1.

B2 High-Temperature Heat Capacity

h-BN The estimation of h-BN heat capacity in the 1300-2200 K region was made in [42] using the comparative cooling rate technique. The experimental C_p points can be approximated using Eqn (1):

$$C_p^0 (T) [J (K mol)^{-1}] = 52.4802 - 9.4203 \times 10^{-4} T - 6487762 T^{-2} \qquad (1)$$

c-BN The heat capacity of cubic boron nitride single crystals has been measured between 300 and 1100 K using scanning adiabatic calorimetry at a continuous heating regime [43]. The coefficients of Eqn (2), approximating the experimental C_p points over the above temperature range, have been determined by the least-squares method.

$$C_p^0 (T) [J (K mol)^{-1}] = 48.404(T^2/(T^2 + 9.706T + 60590.141))^2 \qquad (2)$$

w-BN The heat capacity of the thermostabilized wurtzitic boron nitride has been measured in the 420-980 K range using differential scanning calorimetry under step-by-step heating [20]. The experimental C_p points have been approximated using Eqn (3):

$$C_p^0 (T) [J (K mol)^{-1}] = 48.351(T^2/(T^2 - 8.369T + 68306.334))^2 \qquad (3)$$

TABLE 1 Smoothed values of heat capacities for four boron nitride polymorphic modifications and turbostratic BN.

T, K	$C_p(T)$, J (K mol)$^{-1}$				
	h-BN	r-BN	t-BN	c-BN	w-BN
4				0.00005	
6				0.0007	0.0007
10				0.0033	0.0065
15	0.1473	0.1404	0.340	0.0056	0.0278
20	0.2501	0.2518	0.597	0.0078	0.0720
25	0.3914	0.4198	0.877	0.0158	0.0723
30	0.5632	0.6201	1.152	0.0215	0.0750
35	0.7635	0.8516	1.405	0.0306	0.0887
40	0.9906	1.112	1.648	0.0456	0.1061
45	1.242	1.395	1.962	0.0644	0.1303
50	1.517	1.701	2.308	0.0873	0.1632
60	2.129	2.361	3.022	0.1487	0.2594
70	2.813	3.070	3.751	0.2412	0.3873
80	3.556	3.807	4.481	0.3749	0.5525
90	4.221	4.546	5.203	0.5578	0.7595
100	4.900	5.266	5.918	0.8021	1.019
110	5.602	6.000	6.628	1.104	1.340
120	6.323	6.750	7.344	1.474	1.731
130	7.060	7.514	8.080	1.913	2.195
140	7.809	8.288	8.853	2.415	2.729
150	8.567	9.070	9.684	2.977	3.332
160	9.332	9.858	10.49	3.603	4.005
170	10.10	10.65	11.30	4.291	4.726
180	10.88	11.44	12.11	5.034	5.490
190	11.65	12.24	12.93	5.823	6.313
200	12.43	13.04	13.75	6.655	7.169
210	13.20	13.83	14.57	7.524	8.055
220	13.97	14.62	15.38	8.425	8.966
230	14.74	15.44	16.20	9.346	9.896
240	15.50	16.19	17.01	10.29	10.85
250	16.26	16.97	17.81	11.23	11.81
260	17.02	17.74	18.61	12.19	12.78
270	17.77	18.51	19.40	13.14	13.75
280	18.51	19.27	20.19	14.09	14.72
290	19.25	20.03	20.96	15.03	15.68
298.15	19.85	20.63	21.58	15.79	16.45
300	19.99	20.78	21.72	15.97	16.63

FIGURE 2 gives the experimental values and smoothed heat capacity curves for BN dense modifications at high temperatures.

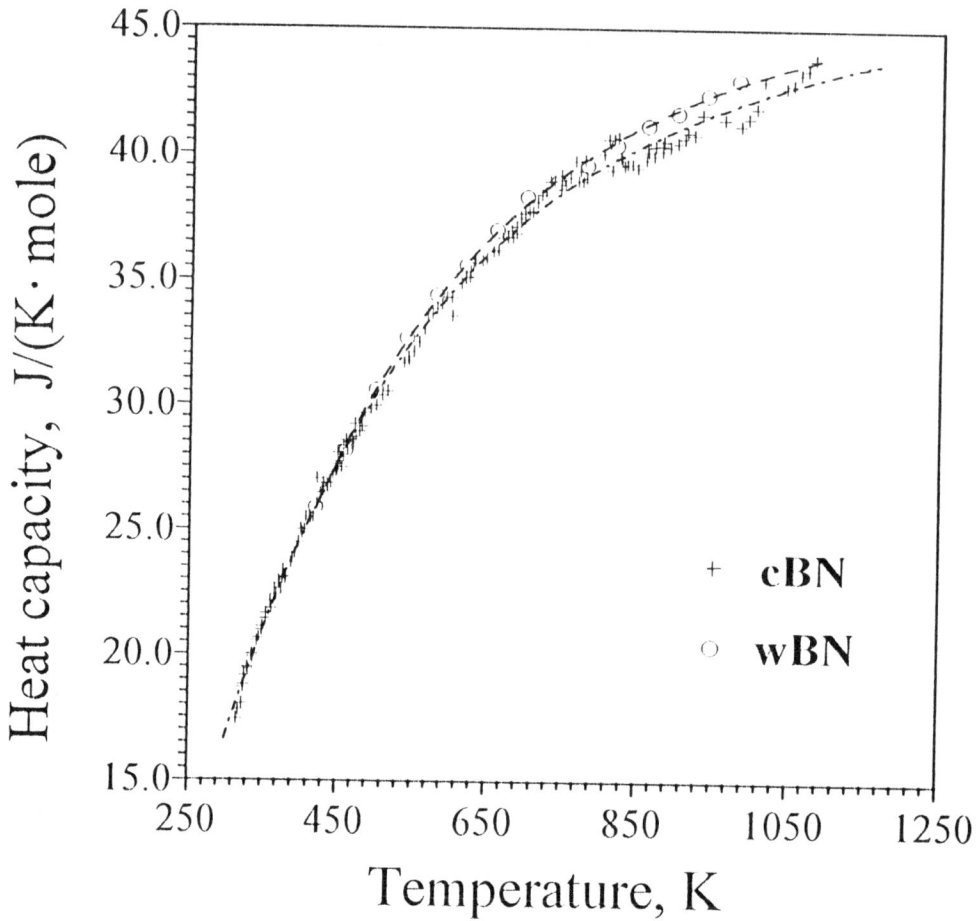

FIGURE 2 The high-temperature heat capacity of BN dense modifications.

B3   High-Temperature Enthalpy

h-BN   The high-temperature enthalpies $H^0(T) - H^0(298.15 K)$ of graphite-like hexagonal boron nitride have been measured by drop-calorimetry in the 298 - 1689 K [44] and 300 - 1200 K [45] intervals.   Enthalpy values of high-ordered h-BN in the 435 - 1652 K interval have been measured in [46] using inverse drop-calorimetry [47].

r-BN   The enthalpies of high-ordered graphite-like rhombohedral boron nitride have been measured in the 435 - 1524 K temperature range [46] using inverse drop-calorimetry.

c-BN   The enthalpies of cubic boron nitride in the high-temperature range have been measured in [45,48-52] for polycrystalline samples and in [52,53] for single crystals.   The measurements [48,49] were performed for polycrystalline sinters, characterized by a high (4 and more mass %) level of impurities of an undetermined nature, as well as by the presence of detectable amounts of h-BN. Hence, the values of enthalpy should be considered as estimations only.

w-BN The enthalpies of wurtzitic boron nitride have been measured in the 300 - 1200 K interval by drop-calorimetry [45] and in the 396 - 1287 K interval by inverse drop-calorimetry [47] (for thermostabilized w-BN).

The experimental enthalpy points for graphite-like modifications, c-BN and w-BN, are shown in FIGURES 3 and 4.

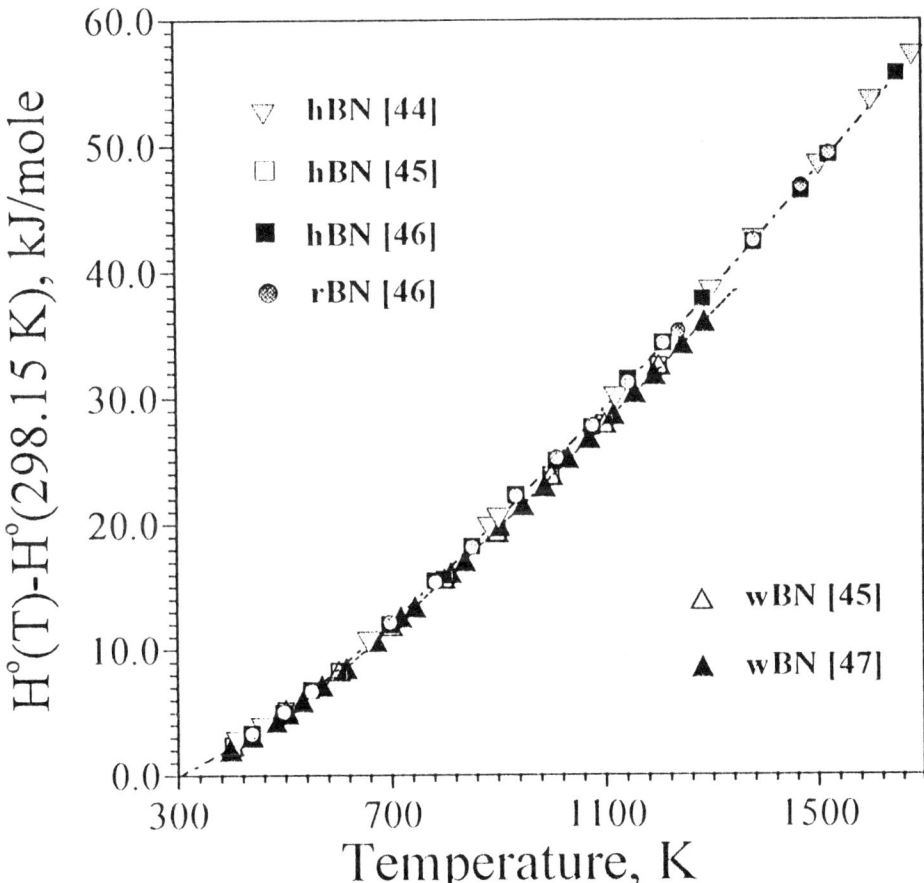

FIGURE 3 The experimental values and smoothed curves of enthalpy for BN graphite-like modifications and w-BN.

The experimental data have been treated mathematically by Shomate's method [54]. When using Shomate's function of the $F(T) = A + BT + CT^2 + DT^3$ type, which allows experimental data to be best approximated, the temperature dependences of the enthalpy are described by Eqn (4) over the temperature ranges under study.

$$H^0(T) - H^0(298.15 \text{ K}) \text{ [J mol}^{-1}] = aT + bT^2 + cT^3 + dT^4 + eT^{-1} + f \qquad (4)$$

The respective coefficients for the four boron nitride modifications are given in TABLE 2.

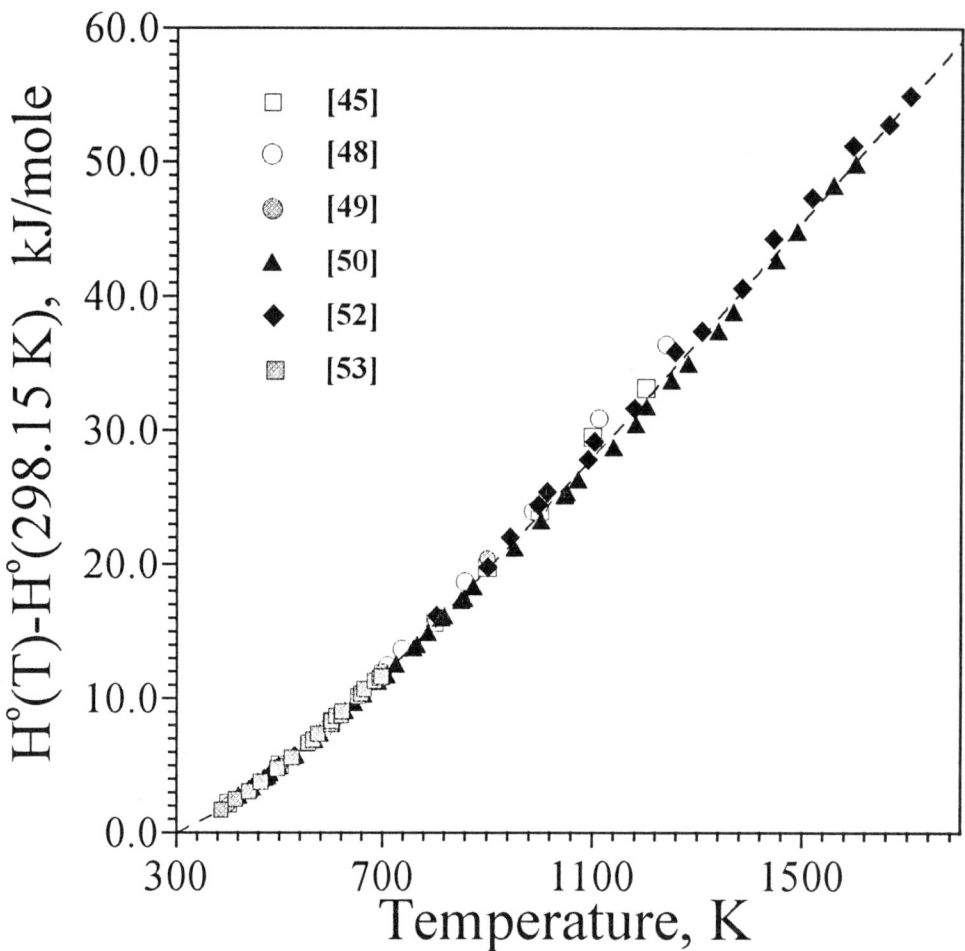

FIGURE 4 The experimental values and smoothed curve of enthalpy for c-BN.

TABLE 2 Coefficients of Eqn (4) approximating temperature dependences of enthalpies for four boron nitride polymorphous modifications.

	a	$b \times 10^2$	$c \times 10^5$	$d \times 10^9$	$e \times 10^{-5}$	f
h-BN [46]	2.0579	4.1285	-1.6977	2.7542	2.2932	-4621.505
r-BN [46]	-3.4955	4.7683	-2.0330	3.4232	0.6754	-2455.129
c-BN [52,53]	6.2507	4.0024	-1.9353	3.7362	8.4850	-7781.667
w-BN [47]	-17.992	7.9652	-4.8917	11.823	1.1095	-882.693

B4   Thermodynamic Functions of BN Modifications

The most reliable experimental data on the thermodynamic properties for each boron nitride modification have been treated [55] in the framework of modified Reshetnikov's approach [56]. TABLE 3 gives the coefficients of Eqn (5) for the four BN modifications, calculated in combination on the base of the heat capacity and enthalpy experimental data.

$$C_p^0 (T) = \delta_0 (T^2/(T^2 + \delta_1 T + \delta_2))^2 \quad [J (K\,mol)^{-1}] \tag{5}$$

TABLE 3 Coefficients of Reshetnikov's equation for four boron nitride modifications.

	δ_0	δ_1	δ_2
h-BN [34,46]	53.63023	68.87958	36927.910
r-BN [35,46]	55.10860	96.31830	27992.881
c-BN [36,43,52,53]	46.83548	-11.66081	66261.937
w-BN [20,32,47]	47.81094	3.45185	61875.367

On the basis of the results obtained, the values of heat capacities, enthalpies, entropies and Gibbs energies for four BN polymorphic modifications over a wide temperature range have been calculated [55].

B5   Enthalpies of Phase Transitions

c-BN→h-BN The enthalpy of the transformation of the cubic BN modification into the graphite-like hexagonal one for high-purity polycrystalline c-BN has been determined using high-temperature scanning heat flux calorimetry under stepped heating conditions [57]. It has been shown that the studied polymorphic transformation is an endothermic one and, at 1800 K, the corresponding enthalpy value can be estimated as

$$\Delta_{tr}H^0(1800\,\text{K}) = +22^{+5}_{-11}\ \text{kJ mol}^{-1}$$

w-BN→h-BN According to the data obtained by differential scanning calorimetry for different w-BN samples, the polymorphic transformation is an endothermic one, and its enthalpy at 1500 K can be evaluated as $+13 \pm 4\,\text{kJ mol}^{-1}$ irrespective of the degree of phase stabilization [58].

The transformation enthalpy over the 1250 - 1400 K temperature range has been measured by inverse drop-calorimetry [58], and at 1380 K a value of $\Delta_{tr}H^0$ has been found to be $+14 \pm 2\,\text{kJ mol}^{-1}$. The transformation enthalpy value obtained is in good agreement with the $\Delta_{tr}H^0(1500\,\text{K}) = +13 \pm 4\,\text{kJ mol}^{-1}$ value obtained by the DSC method.

B6   Formation Enthalpies

h-BN The adopted value of h-BN formation enthalpy was determined in [59,60] using fluorine combustion calorimetry. The obtained value $\Delta_f H^0$ (h-BN, k, 298.15 K) = $-250.6 \pm 2.1\,\text{kJ mol}^{-1}$ is in good agreement with the corresponding values determined by oxygen combustion calorimetry [30,61], mass-spectrometric studies [62], reaction calorimetry [63], Langmuir studies [64], torsion-effusion studies [65], Knudsen effusion studies [61] and fluorine bomb calorimetry [66].

r-BN The standard enthalpy of the rhombohedral boron nitride formation was calculated in [58] to be equal to $\Delta_f H^0$ (r-BN, 298.15 K) = $-247.6 \pm 3.5\,\text{kJ mol}^{-1}$ on the basis of the experimental value of the w-BN→r-BN transformation enthalpy.

c-BN   The thermal effect of the reaction c-BN (k) + 1.5 F_2 (g) = BF_3 (g) + 0.5 N_2 (g) has been measured in [67] using the precision two-chamber bomb calorimeter, and the enthalpy of the reaction under study has been established to be $\Delta_r H^0$ (298.15 K) = - 869.2 $\pm$ 2.0 kJ mol^{-1}. From this result and using the most reliable value of the BF_3(g) formation enthalpy ($\Delta_f H^0$ (298.15 K) = - 1135.95 $\pm$ 0.80 kJ mol^{-1} [68]), the standard enthalpy of cubic boron nitride formation has been calculated as $\Delta_f H^0$ (298.15 K) = - 266.8 $\pm$ 2.2 kJ mol^{-1}. The above value is in agreement with $\Delta_f H^0$ (c-BN, k, 298.15 K) = - 269 $\pm$ 7 kJ mol^{-1} calculated from the experimental value of c-BN$\rightarrow$h-BN transformation enthalpy at high temperatures [57].

w-BN   The standard value of w-BN formation enthalpy was determined using fluorine combustion calorimetry [69,70] to be equal to $\Delta_f H^0$ (r-BN, 298.15 K) = - 263.6 $\pm$ 2.3 kJ mol^{-1}. This value is in good agreement with the - 262 $\pm$ 4 kJ mol^{-1} and - 263 $\pm$ 2 kJ mol^{-1} values calculated from experimental w-BN$\rightarrow$h-BN transformation enthalpies [58] obtained using DSC and inverse drop-calorimetry respectively.

TABLE 4   The formation enthalpies of four BN modifications.

	h-BN [59,60]	r-BN [58]	c-BN [67]	w-BN [69,70]
$\Delta_f H^0$ (298.15 K), kJ mol^{-1}	- 250.6 $\pm$ 2.1	- 247.6 $\pm$ 3.5	- 266.8 $\pm$ 2.2	- 263.2 $\pm$ 2.3

B7   Molar Volumes

As a result of precision X-ray structural analysis of h-BN [71,72], r-BN [72], c-BN [73,74] and w-BN [75] the lattice parameters for the four boron nitride crystalline modifications have been refined, and standard values of molar volumes for these phases have been calculated (TABLE 5).

TABLE 5   Lattice parameters and molar volumes of boron nitride polymorphous modifications under standard conditions.

	h-BN [71,72]	r-BN [72]	c-BN [74]	w-BN [75]
a, nm	0.2504(2)	0.2504(5)	0.36160(3)	0.2550(2)
c, nm	0.6660(8)	1.0000(2)	-	0.4213(1)
V_0, cm^3 mol^{-1}	10.892	10.904	7.1183	7.145

C   MELTING POINT OF BN

On the basis of experimental data [4] on BN melting in the 3 - 8 GPa range, the melting temperature of graphite-like hexagonal boron nitride at normal pressure has been estimated to be 3240 $\pm$ 200 K, and the 81 $\pm$ 25 kJ mol^{-1} value of h-BN melting enthalpy has been calculated [68] (the melting entropy value has been adopted to be 25 J (K mol)$^{-1}$). According to [76], the h-BN melting point under nitrogen pressure of 50 MPa is equal to 3290 $\pm$ 90 K, which is in good agreement with the value of 3273 K obtained earlier [77,78].

D   THERMAL STABILITY OF BN

Thermal decomposition of h-BN to the elements at a total pressure of 0.1 MPa has been observed at 2600 ± 100 K [61]. The decomposition process is accompanied by the formation of amorphous boron and molecular nitrogen [62,64]. To our knowledge, there is no available experimental data on thermal decomposition of BN dense modifications; however, it can be assumed that the dense forms of BN decompose by first forming graphite-like boron nitride.

E   EQUILIBRIUM PHASE P,T-DIAGRAM OF BORON NITRIDE

On the basis of new experimental data on c-BN thermodynamic properties [36,52,67], a $\Delta_{tr}G^0$ (298.15 K) = - 13.7 kJ mol^{-1} value has been calculated for the h-BN→c-BN polymorphic transformation [79]. It follows from the value that a low-temperature region of the BN equilibrium phase diagram has a different shape from the corresponding region in the carbon diagram. In reality, as can be seen from TABLE 6, at atmospheric pressure for graphite→diamond transformation, $\Delta_{tr}G^0$ (298.15 K) > 0, i.e., under normal conditions diamond is a metastable phase, and the corresponding equilibrium line intersects the pressure axis at the point p = 1.35 GPa [5,80], while for the h-BN→c-BN transformation, $\Delta_{tr}G^0$ (298.15 K) < 0. Hence it follows that under normal conditions c-BN is the thermodynamically stable boron nitride modification and not h-BN as was previously thought [6,8,28,29].

TABLE 6 Standard values of thermodynamic functions for polymorphous transformations of boron nitride and carbon different modifications calculated from experimental data [32,34-36,40,46,52,58-60,67,69,70].

	$\Delta_{tr}H^0$ (298.15 K)	$\Delta_{tr}S^0$ (298.15 K)	$\Delta_{tr}G^0$ (298.15 K)
	kJ mol^{-1}	J (K mol)$^{-1}$	kJ mol^{-1}
h-BN→c-BN	- 16.2 ± 3.0	- 8.24 ± 0.11	- 13.7 ± 3.0
h-BN→w-BN	- 13.3 ± 3.1	- 7.26 ± 0.15	- 10.8 ± 3.1
r-BN→c-BN	- 19.2 ± 4.1	- 9.11 ± 0.11	- 16.5 ± 4.1
r-BN→w-BN	- 16.0 ± 4.2	- 8.13 ± 0.15	- 13.6 ± 4.2
r-BN→h-BN	- 3.0 ± 2.8	- 0.87 ± 0.14	- 3 ± 3
w-BN→c-BN	- 3.2 ± 3.1	- 0.98 ± 0.12	- 3 ± 3
graphite→diamond [a]	+ 1.85	- 3.38	+ 2.26

[a] The standard values of thermodynamic functions for graphite and diamond recommended in [68] were used in calculations.

The temperature dependence of $\Delta_{tr}G^0$ (T) for the h-BN→c-BN polymorphic transformation in the 0 - 1700 K interval calculated in [79] from the above data (FIGURE 5) shows that at normal pressure cubic boron nitride remains the thermodynamically stable modification up to temperatures of the order of 1600 K, and due to this the h-BN→c-BN equilibrium line cuts not the pressure axis, but the temperature one at the point T = 1570 K, drastically changing the accepted notions of thermodynamic stability regions for c-BN and h-BN.

Gibbs energy variations at polymorphic transformations of the different BN modifications from 0 to 10 GPa between 0 and 4000 K have been calculated [81-83] using Eqn (6):

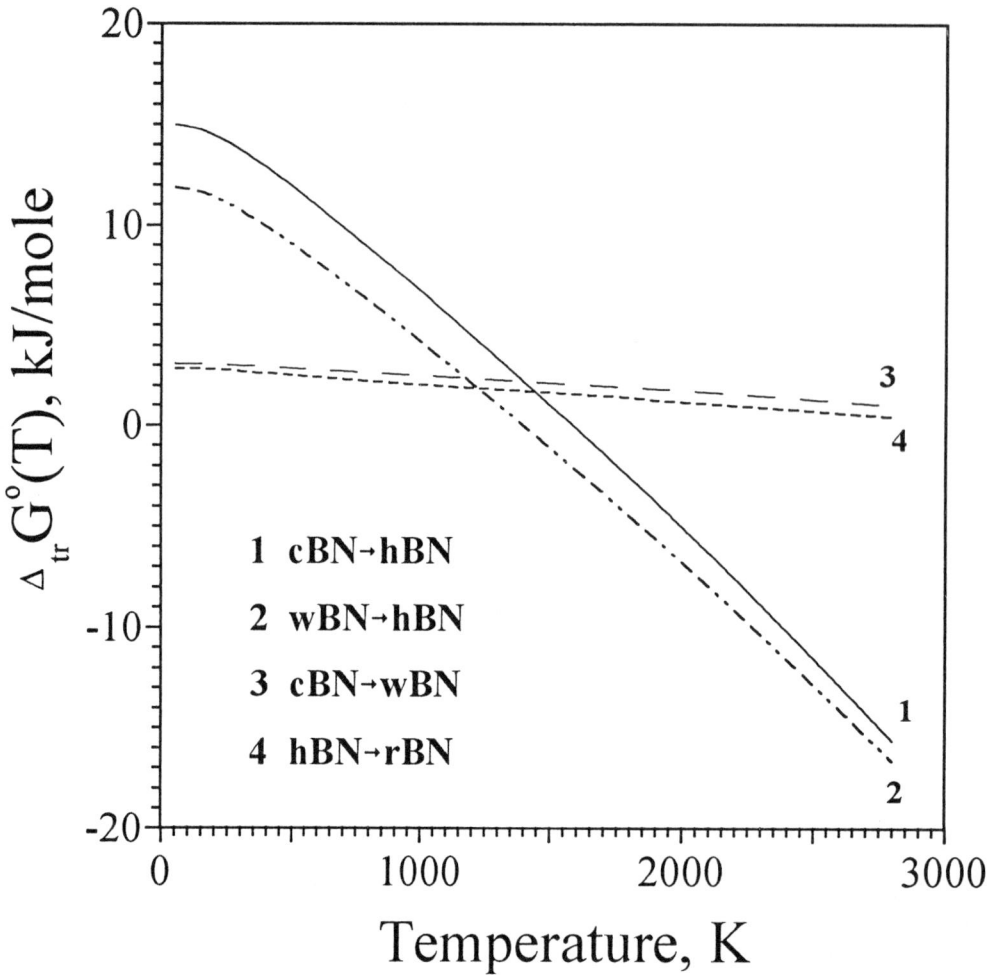

FIGURE 5 Temperature dependences of Gibbs energy for polymorphic transformations of BN modifications.

$$\Delta_{tr}G(T,p) = \Delta_{tr}G^0(298.15\ K) + \int_{298.15}^{T} \Delta_{tr}C_p^0(T)\,dT - T \int_{298.15}^{T} \frac{\Delta_{tr}C_p^0(T)}{T}\,dT$$

$$+ \int_0^p \Delta_{tr}V(T,p)\,dp \qquad (6)$$

The molar volumes of BN modifications at high pressures and temperatures have been calculated in Murnaghan's approximation [84].

From the experimental data on thermodynamic properties [32,34-36,40,46,52,58-60,67,69,70], thermal expansion [71,85-88] and compressibility [89-93] of different boron nitride modifications, the w-BN↔h-BN, w-BN↔c-BN, w-BN↔r-BN, r-BN↔h-BN, c-BN↔r-BN and h-BN↔c-BN phase equilibrium curves have been calculated [52,81-83,94]. The results obtained allow us to conclude that there are no regions of w-BN and r-BN thermodynamic stability on the boron nitride phase diagram, at least in the studied p,T-region. It has been established [81,83,94] that the h-BN↔c-BN equilibrium line cuts the h-BN melting curve at the point with p = $6.9^{+0.5}_{-0.9}$ GPa and T = 3700^{+40}_{-60} K.

FIGURE 6 shows the boron nitride equilibrium phase p,T-diagram [81-83] as well as the generally accepted one [6].

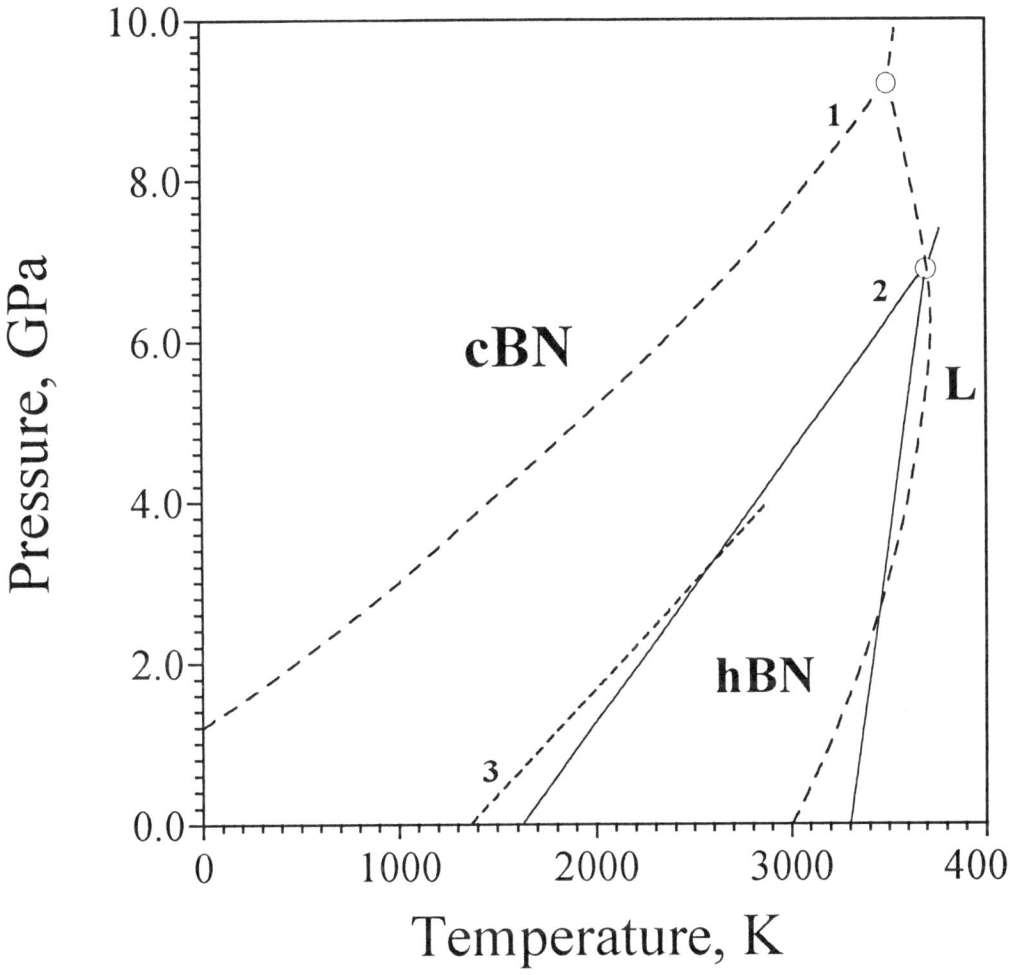

FIGURE 6 Phase p,T-diagrams for boron nitride:
1 - Bundy-Wentorf's diagram [6];
2 - equilibrium diagram [81];
3 - h-BN↔c-BN boundary line according to [15].

The calculations of the BN crystalline modifications' phase equilibria with vapour (BN (ideal gas)) and liquid [20,94] (the necessary thermodynamic data for BN(g) and BN(l) were taken from [68]) have shown that the h-BN-V-L triple point is at p = 120 ± 20 Pa and T = 3250 ± 70 K, so the vapour thermodynamic stability region on the BN phase diagram is extremely small (FIGURE 7).

Recently Maki et al [15] have calculated the h-BN↔c-BN equilibrium curve, using the $\Delta_f H_0$ (c-BN, 298.15 K) = - 266.1 kJ mol^{-1} value, found in [95] from spectroscopic data in the framework of the zone theory of chemical bonding. A minor deviation of their curve from

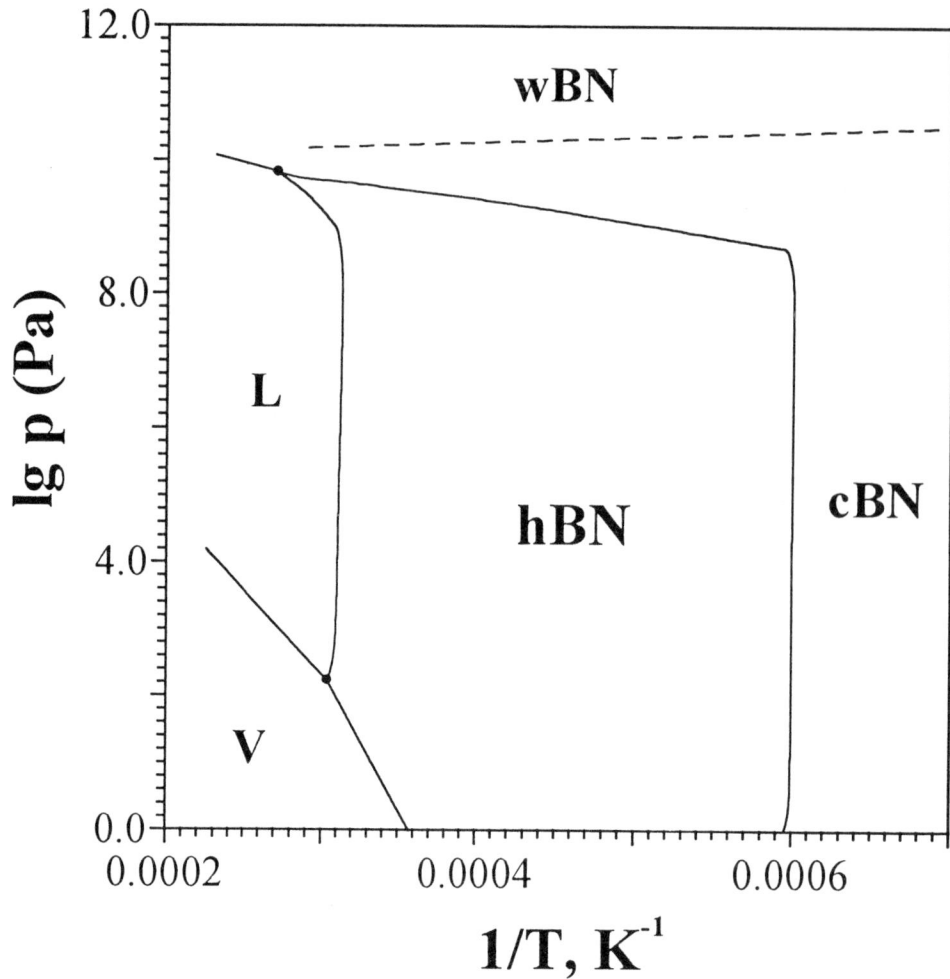

FIGURE 7 Equilibrium phase p,T-diagram of BN in lg p vs. T^{-1} coordinates.

the equilibrium line, calculated in [81-83] (FIGURE 6), is caused by the authors assuming the h-BN formation enthalpy is $-254.0\,kJ\,mol^{-1}$ instead of $-250.5\,kJ\,mol^{-1}$ as recommended in [68].

F   KINETIC ASPECT OF BN POLYMORPHISM

The analysis of various publications dealing with mutual transformations of BN modifications from the point of view of the boron nitride equilibrium phase diagram [81] is indicative of the decisive role of kinetic factors of boron nitride formation, both in direct solid-solid phase transformations and in catalytic systems.

F1   BN Phase Transitions at Atmospheric Pressure

The analysis of literature on the processes of polymorphic transformations of BN dense modifications into the graphite-like ones [57,58,96-105] indicates that experimental data obtained by different authors vary widely. Evidently, this is caused by differences in both the structural state and dispersion of the samples being studied, as well as in the experimental conditions.

c-BN→h-BN   Summarizing the results [96,97,99-103,105] one can conclude that the starting temperature (T_0) for the c-BN→h-BN polymorphic transformation can vary between 1350 and 1900 K. This process occurs over a wide temperature range and completely terminates only at temperatures of the order of 2100 K. For disperse samples, the transition is observed at much lower temperatures (1910 - 2100 K) than for polycrystalline sinters (2120 - 2470 K) (for a scanning rate of 80 K min^{-1} in each case) [57]. This is caused by different conditions of mechanical stress relaxation at the c-BN/h-BN interface due to the large volume effect of the transformation (+51.5%). The heating rate also affects the T_0 observed. Thus, at a scanning rate of 5 K min^{-1} T_0 is 1670 ± 5 K, while at the heating rate of 80 K min^{-1}, the corresponding value of T_0 is 2160 ± 5 K [57].

The presence of even small amounts (of the order of 0.1 vol%) of oxygen in an inert gas atmosphere is accompanied by a distinct (150 K and more) reduction of T_0, and the resulting h-BN contains detectable amounts of boron oxide [20], indicating a possible chemical interaction in the c-BN - B_2O_3 system under the conditions of the experiment.

Using high temperature X-ray diffractometry, it has been shown [20] that for dispersed samples the transformation being studied occurs through the stage of formation of the graphite-like rhombohedral modification, which at the temperature of the experiment transforms into the hexagonal one.

All this makes us view critically the previous results [99-103] for the problem under study as they have not allowed for the above factors affecting the kinetics and, sometimes, the mechanism of this transformation.

Kinetic studies [96,105] of the c-BN→h-BN transformation under isothermal conditions have shown this process is reconstructive diffusion controlled with $E_a = 209 \pm 84$ kJ mol^{-1}. A non-isothermal kinetic study [20] has shown that at scanning rates from 10 to 80 K min^{-1} and transformation degrees from 0.1 to 0.8 the process can be satisfactorily described by the N_2 macrokinetics model (square-law nucleation; nucleation rate dictates the rate of the whole process) with the $E_a = 395 \pm 40$ kJ mol^{-1} and ln A = 20 ± 2 parameters.

w-BN→h-BN   According to [58,96,98,104,106] the starting temperature of the transformation varies in the 500 - 1000 K interval depending on grain size and stacking-fault content in the initial w-BN: the finest and most defective grains are the first to undergo the transformation. The process occurs over a wide temperature range and is fully completed at temperatures of the order of 1600 K (for a textured sample obtained in the r-BN→w-BN transformation under high pressure - high temperature conditions the transformation is completed at 710 K [107]).

During prolonged heating of w-BN in a vacuum with temperature gradually increasing at a rate of 1 K min^{-1} between 500 and 1300 K, w-BN phase stabilization occurs, manifesting itself in a significant increase (up to 1300 K) of the starting temperature of its transformation into h-BN [108]. The structural analysis of w-BN samples of different stabilization degrees has shown [109,110] that the rise of starting temperature for the w-BN→h-BN transformation due to phase stabilization is accompanied by a decrease in point defect concentration and in c/a ratio of wurtzite structure lattice parameters. It allows us to explain the observed effect of w-BN phase stabilization in the framework of Lavaets' concept [111], by the c/a ratio decrease and approximation to the value of 1.633 for perfect wurtzite structure as a result of the annihilation of w-BN lattice point defects during annealing [75].

Isothermal kinetic studies [96,104,106] have shown that the above transformation consists of two stages: an initial rapid stage with $E_a = 20 \pm 12 \, kJ \, mol^{-1}$ and a subsequent slow stage in which generated local stresses essentially decrease the transformation rate. As was shown in [112,113] the studied transformation is a martensitic one with the $(001)_h \parallel (100)_w$ and $[0001]_h \parallel [10\bar{1}0]_w$ orientation relations.

As in the case of c-BN, the presence of even small amounts (<0.1 vol%) of oxygen impurity results in a noticeable intensification of w-BN→h-BN transformation [108] that indicates an increase in the rate of formation and/or growth of new phase nuclei at chemical interaction in the system under study.

r-BN→h-BN   The starting temperatures of the r-BN→h-BN transformation can vary from 2100 K for a highly dispersed powder [114] to 2700 K for whiskers [115] and high-ordered pyrolytic material [114], and the transformation itself occurs in a wide temperature range.

F2   High-Pressure Phase Transitions of BN

h-BN→c-BN   The direct phase transformation of h-BN into c-BN occurs at static pressures above 6.5 GPa in the 1500-2500 K temperature interval [6,8,11,116-121] (FIGURE 8). From the kinetic studies at 6.5 GPa [8] the activation energy for the process has been estimated to lie between about 630 and 1050 $kJ \, mol^{-1}$, and it was supposed that the transformation is a diffusion-controlled one. The pressure threshold for the process and transformation degree vary with structural state, dispersion, purity etc. of the starting h-BN.

r-BN→c-BN   The direct phase transformation of disperse r-BN into c-BN takes place at static pressures above 6.5 GPa and temperatures above 1600 K [122-124]. In situ X-ray studies [125] demonstrated that at room temperature the above transformation takes place starting from 8 GPa. However, similar experiments in the 0-14 GPa range have shown that the cubic phase formation is not observed over the whole pressure range being studied [93]. Formation of the cubic modification from the rhombohedral one has been observed also under explosive shock compression at pressures above 40 GPa [126,127].

h-BN→w-BN The direct phase transformation of h-BN into w-BN has been observed at static pressures above 8.5 GPa in the 1000-2000 K interval [6,8,117,121,128] (FIGURE 8). Above 13 GPa, w-BN formation takes place even at room temperature [8,121,128]. The transformation is a martensitic one with $(001)_w \parallel (001)_h$, $[10\bar{1}0]_w \parallel [10\bar{1}0]_h$ orientation relations [113,128,129]. Such a displacive-diffusionless mechanism requires highly-crystallized h-BN to realize a coherent interface between the parent phase and produced w-BN, and therefore transformation degree depends sharply on crystallinity of the starting material.

The above polymorphic transformation also takes place under shock-wave compression in the 13-55 GPa pressure range [130-134].

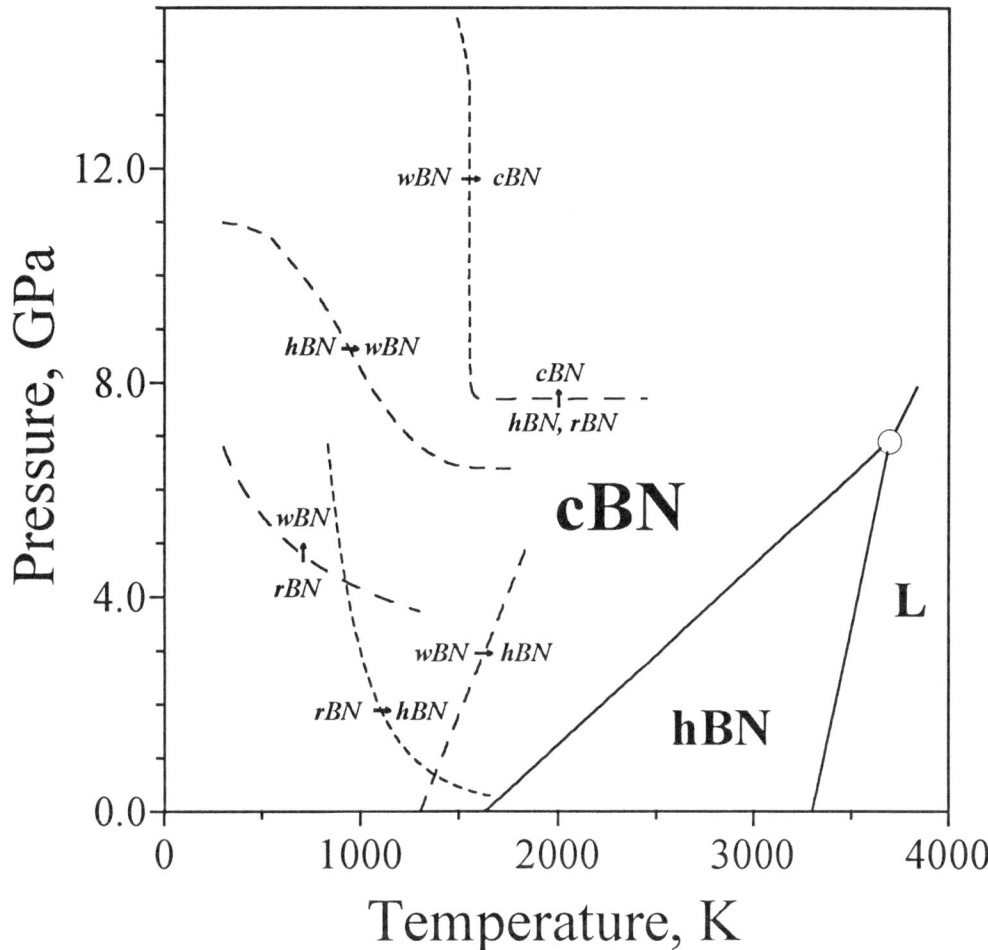

FIGURE 8 Boron nitride equilibrium phase p,T-diagram and kinetic boundaries for direct transformations of BN modifications.

r-BN→w-BN   Recently [107,122,124,125] it was shown that under static pressures above 4 GPa, a dense pyrolytic rhombohedral boron nitride transforms into the wurtzitic phase (FIGURE 8). The transformation degree increases with the pressure increase and reaches its maximum at a temperature of the order of 1300 K. The studied transformation is of a martensitic nature and occurs by shear mechanism through buckling of the hexagon layers of the intermediate ADAD structure [135].

w-BN→c-BN   The direct phase transformation of w-BN into c-BN has been observed at static pressures above 6 GPa and temperatures above 1800 K [118,121,136-140]. A transformation boundary exhibits a broad temperature maximum around 12 GPa [139] (FIGURE 8). From the kinetic studies at 6.5 and 13 GPa [8] the activation energies for the process have been estimated to be 840 and 920 kJ mol^{-1} respectively. In view of the high E_a values, it has been concluded that the transformation is a diffusion-controlled one and occurs with disruption of the w-BN lattice [8]. It has also been shown that diffusionless shear of the wurtzitic structure into a cubic one is possible at the same p,T-parameters [138,140].

r-BN→h-BN   According to [123], at 6 GPa disperse r-BN transforms into h-BN at temperatures above 900 K, while for highly-ordered pyrolytic material the transformation has been observed at 2700 K only [114].

c-BN→h-BN   A surprising result obtained in [15,141] was that polycrystalline c-BN converts to h-BN within the c-BN thermodynamic stability region on the equilibrium phase diagram [81]. This was attributed [142] to oxygen impurities in the studied samples and metastable crystallization of h-BN in the B-N-O system under experimental conditions.

w-BN→h-BN   In the pressure region 5.5 - 6.7 GPa wurtzitic boron nitride undergoes the transformation into the graphite-like hexagonal form at 1800 - 2400 K [136,137,143] (FIGURE 8).

All the transformations described above occur in non-equilibrium conditions. Therefore, the location of boundary lines in FIGURE 8 would be affected by the kinetics of the above processes, which depends sharply on grain size, defect content, purity etc. of the starting materials, and experimental conditions.

F3   Metastable Crystallization of BN

The experimental data on boron nitride crystallization at high pressures and temperatures available in the literature hold true only for a limited number of systems, including the so-called catalytic h-BN→c-BN transformation [1-3,9,10,13-19,144-166].

h-BN metastable crystallization in the c-BN thermodynamic stability region on the equilibrium phase diagram [81] has been observed from ammonium borate flux in the 1500 - 1900 K interval below 4.2 GPa [151,155].

BN-ammonia and BN-hydrazine systems which are characteristic of reversible solution of boron nitride modifications have been studied in a wide p,T-range [167-169]. In these systems at pressures above 4 GPa and high (of the order of 1500 K) temperatures, a recrystallization is observed of graphite-like hexagonal and wurtzitic BN modifications into the cubic one. The recrystallization rate increases greatly in the presence of conventional catalysts of c-BN synthesis. On the other hand, the products of cubic boron nitride interaction with ammonia over the whole pressure range (and, in particular, in the c-BN thermodynamic stability region on the generally accepted BN phase diagram [6]) contain h-BN and r-BN, and with the increase in NH_3:BN ratio the graphite-like modification content rises and can reach 70 - 80 %. The latter is, evidently, caused by the h-BN and r-BN metastable crystallization from boron nitride gaseous solution when rapidly cooled (of the order of 100 K s^{-1}) in the usual finishing of the experiment by switching off the heating current. This explains the presence of trace amounts of BN graphite-like modifications in the products of crystallization up to pressures of 8 GPa [167,169].

A fact worthy of note is that at low temperatures BN crystallization occurs with the formation of mesographitic boron nitride (m-BN), three-dimensional degree of ordering which can vary from 0.2 to 0.7 and depends on pressure, temperature and solution concentration [169].

FIGURE 9 shows the p,T-diagram of crystallization in the BN - ammonia (hydrazine) system at a cooling rate of $100\,K\,s^{-1}$ [169]. The diagram represents the basic regularities of BN crystallization from gaseous solutions. When analyzing the plotted diagram, the existence of the p,T-region of rhombohedral BN crystallization should be noted. In this region, r-BN can be obtained with 98 % basic phase content and this is of certain interest as until recently the r-BN has been met only as an impurity (no more than 30 %) in the graphite-like hexagonal modification [170] or as whiskers [115]. Also, the absence of a w-BN crystallization region in the p,T-diagram is significant. This can be explained by the martensite nature of the w-BN formation process, which cannot be realized under the above experimental conditions.

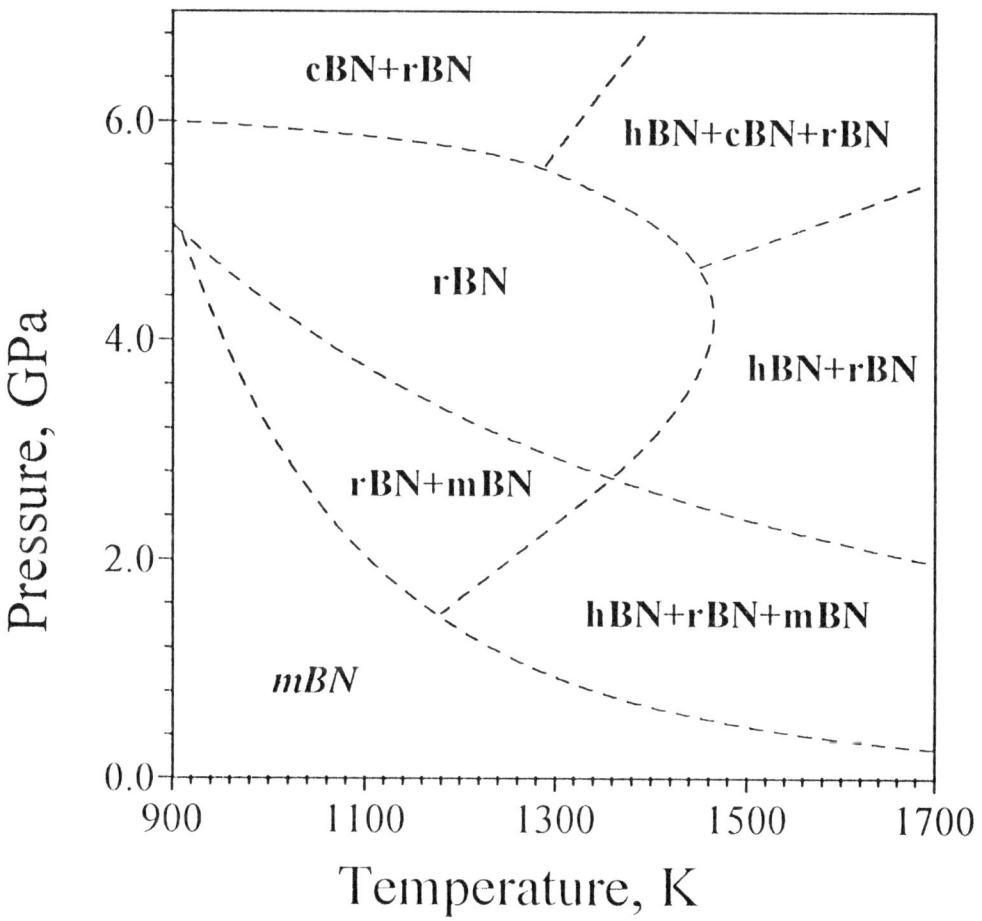

FIGURE 9 Crystallization p,T-diagram of the BN - ammonia system.

F4   BN Phase Formation at Chemical Interaction in a B-N System

Chemical interaction in a B-N system at atmospheric pressure and temperatures above 1200 K is accompanied by the formation of graphite-like hexagonal or mesographitic boron nitride [171].

The peculiarities of boron nitride phase formation in the chemical interaction of boron and its compounds (oxide, phosphide, carbide etc.) with nitrogen, ammonia and hydrazine at pressures 2 - 7 GPa and temperatures 900 - 1500 K have been studied in [167,169].

The results obtained for boron as well as for boron carbide and boron phosphide are the same as for the BN-ammonia (hydrazine) system. In the case of B_2O_3, there is no r-BN formation region in the reaction diagram, due to a rapid transformation of this modification into the hexagonal one in the presence of ammonium borate, forming in the system.

The interaction of boron and its compounds with nitrogen over the whole range of the realized pressures (and in the c-BN thermodynamic stability region of diagram [6], in particular) is accompanied by the formation of mesographitic boron nitride. The crystallization of the cubic modification is observed in the MgB_2 - N_2 system at pressures of 4.2 - 7 GPa and temperatures exceeding 1500 K only, when the eutectic melt of boron nitride and magnesium bornitride acts, obviously, as the solvent.

The arrangement of lines restricting the formation regions of different modifications of BN on crystallization p,T-diagrams and reaction p,T-diagrams depends on the solvent concentration, reagent stoichiometry and cooling rate. The above regularities of boron nitride phase formation are, however, retained, and that allows us to speak about alternative metastable BN behaviour in the studied systems over a wide range of p,T-parameters. That a part of the experimental data obtained can not be explained in the framework of the generally accepted notions of boron nitride polymorphism, and is in good agreement with the BN equilibrium phase diagram [81], is worthy of note.

F5   c-BN Low-Pressure Synthesis

Until very recently c-BN spontaneous crystallization has been observed only at pressures above 4 GPa [1-3,9,10,13-19,144,146-166]. The possibility of c-BN synthesis in the p,T-region restricted by h-BN↔c-BN boundary lines of Bundy-Wentorf's diagram [6] and equilibrium phase diagram [81] (see FIGURE 6) has been studied [172-176].

FIGURE 10 represents experimental data for h-BN - MgB_2, h-BN - Li_3N, h-BN - AlN and h-BN - Mg_3N_2 growth systems, which are commonly used to synthesize cubic boron nitride, in the presence of ammonia.

It follows from the data analysis that in the systems under study when the interaction time is between 300 and 500 s, cubic BN spontaneous crystallization is observed down to pressures of the order of 2.0 GPa. At pressures above 2.5 GPa, the degree of h-BN→c-BN transformation attains 0.96 - 0.98. Both the high transformation degree and the small sizes (20 - 30 μm) of the crystals being formed indicate that the c-BN spontaneous crystallization is going in its thermodynamic stability region away from the equilibrium line.

With the aim of refinement of the threshold pressure value for c-BN spontaneous crystallization in similar systems, the interaction in the h-BN - MgB_2-NH_3 system has been studied using a piston-cylinder type high-pressure apparatus [174]. Analysis of this data shows that in the studied system at temperatures of 1300 - 1500 K and reaction time of 1000 s, c-BN formation is observed down to pressures of 2.1 ± 0.1 GPa, which thus can be considered as the threshold pressure of c-BN spontaneous crystallization in the system.

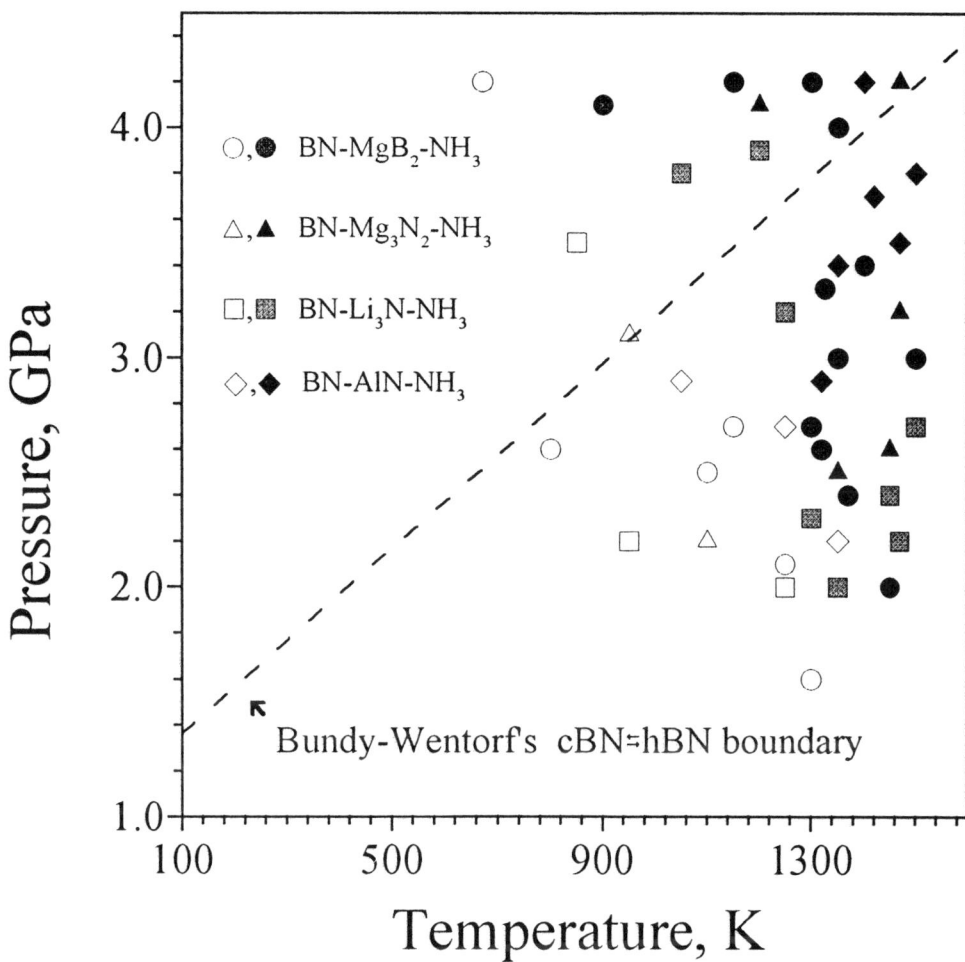

FIGURE 10 c-BN spontaneous crystallization with supercritical ammonia presence (the light signs show the absence of c-BN in the reaction products, the dark ones are indicative of its presence).

Up to now at pressures lower than 2 GPa, c-BN spontaneous crystallization has not been recorded. However, the experiments [175,176] at pressures of 0.5 - 2.0 GPa and temperatures of 1200 - 1300 K using seed diamond and c-BN single crystals have shown the possibility of c-BN crystal growth on the surface of the introduced seeds in the studied systems. A special set of experiments [172,176] has shown that in the h-BN - Li_3N - NH_3 system, the growth of c-BN crystals on the octahedral faces of diamond single crystals at temperatures of the order of 1200 K is observed down to atmospheric pressure.

Thus, according to the BN equilibrium phase diagram [81], synthesis of the cubic modification in the presence of supercritical fluids expands the p,T-region of c-BN formation, opening new fields of developing methods for cubic boron nitride low-pressure synthesis.

The above results have initiated experiments on reducing pressure of cubic boron nitride synthesis in a number of research centres. Thus, in the latest works of Fukunaga et al

[19,177,178] pressure of c-BN crystallization in the Mg_3BN_3 - H_2O system has been reduced to 3.8 GPa.

G   CONCLUSION

1. Analysis of experimental data forming the basis of the generally accepted Bundy-Wentorf's diagram of boron nitride permits us to conclude that it is impossible to plot an equilibrium p,T-diagram of the compound from the data on mutual transformations of its polymorphic modifications due to the critical role of kinetic factors in the process of BN phase formation over the wide ranges of pressures and temperatures.

2. The correct calculation of the phase equilibrium lines on the BN p,T-diagram is possible only in the framework of the thermodynamic approach that demands numerical data on thermodynamic and thermophysical properties of different modifications of BN and, above all, their polymorphic transformation enthalpies.

3. From the recent experimental data on thermodynamic properties of boron nitride, phase equilibrium lines for BN modifications as well as lines of equilibria for crystalline phases with a vapour and liquid up to 10 GPa and 4000 K have been calculated. The BN equilibrium phase diagram has been plotted, which differs from the generally accepted Bundy-Wentorf's one. At atmospheric pressure cubic boron nitride is the thermodynamically stable modification up to temperatures of the order of 1600 K, which drastically changes the established notions of BN polymorphism, based on assumed analogy of phase diagrams for carbon and boron nitride.

4. The analysis of various publications dealing with mutual transformations of different modifications of BN from the point of view of the boron nitride equilibrium phase diagram [81] is indicative of the decisive role of kinetic factors in boron nitride formation, both in direct solid-solid phase transformations and in systems, including so called catalysts of the above transformations.

5. The changes in previously existing notions of c-BN and h-BN thermodynamic stability regions make one expect a noticeable pressure decrease of cubic boron nitride crystallization when using non-traditional growth systems. The studies performed have shown that the threshold pressure of c-BN crystallization can be reduced from 4 down to 2 GPa with the supercritical fluids present, which opens wide avenues for the development of new technologies of low-pressure production of cubic boron nitride.

REFERENCES

[1]   R.H. Wentorf [*J. Chem. Phys. (USA)* vol.26 (1957) p.956]
[2]   R.H. Wentorf [*J. Chem. Phys. (USA)* vol.34 (1961) p.809-12]
[3]   R.H. Wentorf [*Chem. Eng. (USA)* vol.68 (1961) p.177-86]
[4]   R.H. Wentorf [*J. Phys. Chem. (USA)* vol.63 (1959) p.1934-40]
[5]   F.P. Bundy [*J. Chem. Phys. (USA)* vol.38 (1963) p.631-43]
[6]   F.P. Bundy, R.H. Wentorf [*J. Chem. Phys. (USA)* vol.38 (1963) p.1144-9]
[7]   R.E. Lorent [*Rev. Sci. Instrum. (USA)* vol.44 (1973) p.1691-3]
[8]   F.R. Corrigan, F.P. Bundy [*J. Chem. Phys. (USA)* vol.63 (1975) p.3812-20]

[9]   K. Kudaka, H. Konno, T. Matoba [*J. Chem. Soc. Jpn. (Japan)* vol.69 (1966) p.365-9]

10]   L.I. Fel'dgun, V.N. Krylov [*Tr. VNIIASh (USSR)* no.7 (1968) p.13-5]

[11]   A.M. Mazurenko, A.A. Leusenko [*Izv. AN BSSR, Ser. Fiz.-Mat. Nauk (USSR)* no.6 (1978) p.108-10]

[12]   N.N. Sirota, A.M. Mazurenko [*Dokl. Akad. Nauk SSSR (USSR)* vol.241 (1978) p.884-7]

[13]   L.F. Vereshchagin, I.S. Gladkay, G.A. Dubitski, V.N. Slesarev [*Izv. Akad. Nauk SSSR. Neorg. Mater. (USSR)* vol.15 (1979) p.256-9]

[14]   M. Kagamida, H. Kanda, M. Akaishi, A. Nukui, T. Osawa, S. Yamaoka [*J. Cryst. Growth (Netherlands)* vol.94 (1989) p.261-9]

[15]   J. Maki, H. Iwata, O. Fukunaga [*Proc. 2nd Int. Conf. on New Diamond Science and Technology*, Washington, USA, 23-27 Sept 1990, Eds R. Messier, J.T. Glass, J.E. Batler, R. Roy (MRS, Pittsburg, USA, 1991) p.1051-5]

[16]   R.C. DeVries, J.F. Fleischer [*J. Cryst. Growth (Netherlands)* vol.13/14 (1972) p.88-92]

[17]   O. Mishita, S. Yamaoka, O. Fukunaga [*J. Appl. Phys. (USA)* vol.61 (1987) p.2822-5]

[18]   S. Yamaoka et al [*Physica B (Netherlands)* vol.139/140 (1986) p.668-70]

[19]   S. Nakano, O. Fukunaga [*Diam. Relat. Mater. (Switzerland)* vol.2 (1993) p.1409-13]

[20]   V.L.Solozhenko [unpublished results]

[21]   L.I. Fel'dgun, V.M. Davidenko [*Abrazivy (USSR)* no.10 (1975) p.1-5]

[22]   N.N. Sirota, N.A. Kofman [*Sov. Phys.-Dokl. (USA)* vol.24 (1979) p.1001-2]

[23]   S.V. Pyaternev [*Zh. Fiz. Khim. (Russia)* vol.65 (1991) p.1373-4]

[24]   V.D. Andreev, V.R. Malik, A.V. Bochko [*Sov. J. Superhard Mater. (USA)* vol.11 (1989) p.8-15]

[25]   V.D. Andreev [*High Press. Res. (UK)* vol.5 (1990) p.950-1]

[26]   V.D. Andreev, V.R. Malik [*Sov. J. Superhard Mater. (USA)* vol.12 (1990) p.10-8]

[27]   V.L. Solozhenko [*Sov. J. Superhard Mater. (USA)* vol.13 (1991) p.65]

[28]   F.P. Bundy, J.S. Kasper [*J. Chem. Phys. (USA)* vol.46 (1967) p.3437-46]

[29]   A.V. Kurdyumov, A.N. Pilyunkevich [*Phase transformations in carbon and boron nitride* (Naukova Dumka, Kiev, 1979)]

[30]   A.S. Dvorkin, D.J. Sasmor, E.R. Van Artsdalen [*J. Chem. Phys. (USA)* vol.22 (1954) p.837-42]

[31]   N.N. Sirota, N.A. Kofman, Zh.K. Petrova [*Izv. AN BSSR, Ser. Fiz.-Mat. Nauk (USSR)* no.6 (1975) p.75-8]

[32]   V.E. Gorbunov, K.S. Gavrichev, G.A. Totrova, A.V. Bochko, V.B. Lazarev [*Russ. J. Phys. Chem. (UK)* vol.62 (1988) p.9-12]

[33]   A.V. Kurdyumov [*Sov. Phys.-Crystallogr. (USA)* vol.17 (1972) p.534-8]

[34]   K.S. Gavrichev, V.L. Solozhenko, V.E. Gorbunov, L.N. Golushina, G.A. Totrova, V.B. Lazarev [*Thermochim. Acta (Netherlands)* vol.217 (1993) p.77-89]

[35]   K.S. Gavrichev, V.E. Gorbunov, V.L. Solozhenko, G.A. Totrova, L.N. Golushina [*Zh. Fiz. Khim. (Russia)* vol.66 (1992) p.2824-8]

[36]   V.L. Solozhenko, V.E. Yachmenev, V.A. Vil'kovsky, A.N. Sokolov, A.A. Shul'zhenko [*Russ. J. Phys. Chem. (UK)* vol.61 (1987) p.1480-2]

[37]   T. Atake, A. Honda, Y. Saito, K. Saito [*Jpn. J. Appl. Phys. (Japan)* vol.29 (1990) p.1869-70]

[38]   N.N. Sirota, N.A. Kofman [*Sov. Phys.-Dokl. (USA)* vol.20 (1975) p.861-2]

[39]   V.E. Gorbunov, K.S. Gavrichev, G.A. Totrova, A.V. Bochko, V.B. Lazarev [*Russ. J. Phys. Chem. (UK)* vol.61 (1987) p.3357-60]

[40]   V.L. Solozhenko, V.E. Yachmenev, V.A. Vil'kovsky, I.A. Petrusha [*Inorg. Mater. (USA)* vol.25 (1989) p.134-6]

[41]   N.N. Sirota, N.A. Kofman [*Sov. Phys.-Dokl. (USA)* vol.21 (1976) p.516-7]

[42]   H. Prophet, D.R. Stull [*J. Chem. Eng. Data (USA)* vol.8 (1963) p.78-81]

[43]   V.E. Lusternik, V.L. Solozhenko [*Zh. Fiz. Khim. (Russia)* vol.66 (1992) p.1186-91]

[44]   R.A. McDonalds, D.R. Stull [*J. Phys. Chem. (USA)* vol.65 (1961) p.1918]

[45]   V.M. Agoshkov, S.V. Bogdanova [*Sov. J. Superhard Mater. (USA)* vol.12 (1990) p.26-30]

[46]   V.L. Solozhenko [*Zh. Fiz. Khim. (Russia)* vol.67 (1993) p.1580-2]

[47]   V.L. Solozhenko [*Thermochim. Acta (Netherlands)* vol.218 (1993) p.395-400]

[48]   R. Mezaki, E.W. Tilleux, D.W. Barnes, J.L. Margrave [in *Thermodynamics of nuclear materials* (IAEA, Vienna, 1962) p.494-5]

[49]   I.A. Kiseleva, L.V. Mel'chakova, N.D. Topor [*Izv. Akad. Nauk SSSR. Neorg. Mater. (USSR)* vol.9 (1973) p.256-9]

[50]   V.L. Solozhenko, I.Ya. Chaikovskaya, I.A. Petrusha [*Inorg. Mater. (USA)* vol.25 (1989) p.1414-7]

[51]   A.F. Mayorova, S.N. Mudretsova, V.L. Solozhenko, L.A. Ryabova [*Proc. Int. Ceramic Congress*, Istanbul, Turkey, 19-23 Oct 1992 (TCS, Istanbul, 1992) p.659-69]

[52]   V.L. Solozhenko [*Thermochim. Acta (Netherlands)* vol.218 (1993) p.221-7]

[53]   V.L. Solozhenko, I.Ya. Chaikovskaya, A.N. Sokolov, A.A. Shul'zhenko [*Russ. J. Phys. Chem. (UK)* vol.61 (1987) p.412-4]

[54]   C.H. Shomate [*J. Am. Chem. Soc. (USA)* vol.66 (1944) p.928-9]

[55]   V.L. Solozhenko, V.Z. Turkevich [unpublished results, 1993]

[56]   M.A. Reshetnikov [*Zh. Neorg. Khim. (USSR)* vol.11 (1966) p.1489-96]

[57]   V.L. Solozhenko, V.Z. Turkevich [*J. Therm. Anal. (UK)* vol.38 (1992) p.1181-8]

[58]   V.L. Solozhenko [*J. Therm. Anal. (UK)*, in press]

[59]   V.Ya. Leonidov, I.V. Timofeev, V.L. Solozhenko [in *Thermodynamics of Chemical Compounds* (GGU, Gorky, 1988) p.16-8]

[60]   V.Ya. Leonidov, I.V. Timofeev [*Zh. Neorg. Khim. (USSR)* vol.34 (1989) p.2701-4]

[61]   D.R. Stull, H. Prophet (Eds) [*JANAF Thermochemical Tables*, 2nd edition (NBS, Washington, 1971)]

[62]   P. Schissel, W. Williams [*Bull. Am. Phys. Soc. (USA)* vol.4 (1959) p.139]

[63]   G.L. Galchenko, A.N. Kornilov, S.M. Skuratov [*Zh. Neorg. Khim. (USSR)* vol.5 (1960) p.2651-4]

[64]   L.H. Dreger, V.V. Dadape, J.L. Margrave [*J. Phys. Chem. (USA)* vol.66 (1962) p.1556-9]

[65]   D.L. Hildenbrand, W.F. Hall [*J. Phys. Chem. (USA)* vol.67 (1963) p.888-93]

[66]   S.S. Wise, J.L. Margrave, H.M. Feder, W.N. Hubbard [*J. Phys. Chem. (USA)* vol.70 (1966) p.7-10]

[67]   V.Ya. Leonidov, I.V. Timofeev, V.L. Solozhenko, I.V. Rodionov [*Russ. J. Phys. Chem. (UK)* vol.61 (1987) p.1503-4]

[68]   V.P. Glushko, L.V. Gurvich, G.A. Bergman et al (Eds) [*Thermodynamic properties of individual substances* (Nauka, Moscow, 1981)]

[69]   V.Ya. Leonidov, I.V. Timofeev, V.B. Lazarev, A.V. Bochko [*Zh. Neorg. Khim. (USSR)* vol.33 (1988) p.1597-600]

[70]   V.Ya. Leonidov, P.A.G. O'Hare [*Pure Appl. Chem. (UK)* vol.64 (1992) p.103-10]

[71]   R.S. Pease [*Acta Crystallogr. (Denmark)* vol.5 (1952) p.356-61]
[72]   A.V. Kurdyumov, V.L. Solozhenko [unpublished results]
[73]   G. Will, A. Kirfel, B. Josten [*J. Less-Common Met. (Switzerland)* vol.117 (1986) p.61-71]
[74]   V.L. Solozhenko, V.V. Chernyshev, G.V. Fetisov, V.B. Rybakov, I.A. Petrusha [*J. Phys. Chem. Solids (UK)* vol.51 (1990) p.1011-2]
[75]   A.V. Kurdyumov, V.L. Solozhenko, W.B. Zelyavsky, I.A. Petrusha [*J. Phys. Chem. Solids (UK)* vol.54 (1993) p.1051-3]
[76]   V.L. Vinogradov, A.V. Kostanovski [*Teplofiz. Vys. Temp. (Russia)* vol.29 (1991) p.1112-20]
[77]   E. Friderich, L. Sittig [*Z. Anorg. Allg. Chem. (Germany)* vol.143 (1925) p.293-320]
[78]   I.E. Campbell, C.F. Powell, D.H. Nowicki, B.W. Gonser [*J. Electrochem. Soc. (USA)* vol.96 (1949) p.318-33]
[79]   V.L. Solozhenko, V.Ya. Leonidov [*Russ. J. Phys. Chem. (UK)* vol.62 (1988) p.1646-7]
[80]   F.P. Bundy [*J. Chem. Phys. (USA)* vol.38 (1963) p.618-30]
[81]   V.L. Solozhenko [*Doklady Phys. Chem. (USA)* vol.301 (1988) p.592-4]
[82]   V.L. Solozhenko [in *Application of mathematical methods for describing and studying physico-chemical equilibria* (Novosibirsk, 1989) part 2, p.69-70]
[83]   V.L. Solozhenko [*High Press. Res. (UK)* vol.7-8 (1991) p.201-3]
[84]   F.D. Murnaghan [*Proc. Natl. Acad. Sci. USA (USA)* vol.30 (1944) p.244-7]
[85]   B. Yates, M.J. Overy, O. Pirgon [*Philos. Mag. (UK)* vol.32 (1975) p.847-57]
[86]   G.A. Slack, S.F. Bartram [*J. Appl. Phys. (USA)* vol.46 (1975) p.89-98]
[87]   Z.I. Kolupayeva, M.Ya. Fuks, L.I. Gladkikh, A.V. Arinkin, S.V. Malikhin [*J. Less-Common Met. (Switzerland)* vol.117 (1986) p.259-63]
[88]   V.L. Solozhenko [*J. Superhard Mater. (USA)*, in press]
[89]   R.W. Lynch, H.G. Drickamer [*J. Chem. Phys. (USA)* vol.44 (1966) p.181-4]
[90]   E.V. Yakovenko, I.V. Aleksandrov, A.F. Goncharov, S.M. Stishov [*Sov. Phys.-JETP (USA)* vol.68 (1989) p.1213-5]
[91]   E. Knittle, R.M. Wentzcivitch, R. Jeanloz, M.L. Cohen [*Nature (UK)* vol.337 (1989) p.349-52]
[92]   P.E. Van Camp, V.E. Van Doren, J.T. Devreese [*High Press. Res. (UK)* vol.5 (1990) p.944-6]
[93]   V.L. Solozhenko, G. Will, H. Hupen, F. Elf [*Solid State Commun. (USA)* vol.90 (1994) p.65-7]
[94]   V.L. Solozhenko [*Abstr. 31st EHPRG Annual Meeting*, Belfast, Northern Ireland, 30 Aug - 3 Sept 1993 (Belfast, 1993) p.G-5]
[95]   J.C. Phillips, J.A. Van Vechten [*Phys. Rev. B (USA)* vol.2 (1970) p.2147-60]
[96]   A.V. Kurdyumov, A.N. Pilyankevich [*Phase Transformations in Carbon and Boron Nitride* (Naukova Dumka, Kiev, 1979)]
[97]   E. Rapoport [*Ann. Chim. (France)* vol.10 (1985) p.607-38]
[98]   R.R. Wills [*Int. J. High Technol. Ceram. (UK)* vol.1 (1985) p.139-53]
[99]   H.J. Milledge, E. Nave, F.H. Weller [*Nature (UK)* no.4687 (1959) p.715]
[100]   V.P. Bondarenko, A.P. Khalepa, E.S. Cherepenina [*Sint. Almazy (USSR)* no.4 (1970) p.22-5]
[101]   M.S. Druy, M.I. Sokhor, L.I. Fel'dgun, V.A. Ponomarenko, M.V. Kharitonova [*Trudy VNIIASh (USSR)* no.13 (1971) p.6-18]

[102] M.S. Druy, M.I. Sokhor, B.F. Mgeladze, L.I. Fel'dgun, A.A. Lavrinovich, L.V. Burtseva [*Abrazivy (USSR)* no.5 (1972) p.1-3]

[103] E.G. Gatilova, V.G. Malogolovets, G.A. Kolesnichenko, B.D. Kostyuk [in *Adhesion of Melts* (Naukova Dumka, Kiev, 1974) p.135-8]

[104] A.V. Kurdyumov, N.F. Ostrovskaya [in *Hexanit and Hexanit-R* (IPM AN UkrSSR, Kiev, 1975) p.29-33]

[105] A.V. Kurdyumov, N.F. Ostrovskaya, A.N. Pilyankevich, T.R. Balan, A.V. Bochko [*Dokl. AN Ukr. SSR (USSR)* no.10 (1976) p.937-40]

[106] A.V. Kurdyumov, N.F. Ostrovskaya, A.N. Pilyankevich, T.R. Balan, A.V. Bochko [*Poroshk. Metall. (USSR)* no.1 (1976) p.64-9]

[107] V.L. Solozhenko, I.A. Petrusha, A.A. Svirid [*Abstr. 31ˢᵗ EHPRG Annual Meeting*, Belfast, Northern Ireland, 30 Aug - 3 Sept 1993 (Belfast, 1993) p.E-5]

[108] V.L. Solozhenko [*React. Solids (Netherlands)* vol.7 (1989) p.371-4]

[109] W.B. Zelyavsky, A.V. Kurdyumov, V.L. Solozhenko [*Sov. Phys.-Dokl. (USA)* vol.35 (1990) p.911-2]

[110] V.L. Solozhenko, A.V. Kurdyumov, I.A. Petrusha, W.B. Zelyavsky [*J. Hard Mater. (UK)* vol.4 (1993) p.107-11]

[111] G. Lavaets [*Phys. Rev. B (USA)* vol.5 (1972) p.4039-45]

[112] T.R. Balan, A.V. Kurdyumov, N.F. Ostrovskaya, A.N. Pilyankevich [*Dokl. AN Ukr. SSR (USSR)* no.12 (1976) p.1109-11]

[113] A.V. Kurdyumov, N.F. Ostrovskaya [*Fiz. Tekh. Vysokikh Davlenii (Ukraine)* no.3 (1992) p.5-18]

[114] V.L. Solozhenko, I.A. Petrusha, A.A. Svirid [*J. Mater. Sci. (UK)*, in press]

[115] T. Ishii, T. Sato, Y. Sekikawa, M. Iwata [*J. Cryst. Growth (Netherlands)* vol.52 (1981) p.285-9]

[116] K. Ichinose, M. Wakatsuki, T. Aoki, Y. Maeda [*Proc. 4ᵗʰ Int. Conf. on High Pressure*, Kyoto, Japan, 25-29 Nov 1974 (Kawakita Printing Co., Kyoto, Japan, 1975) p.436-40]

[117] M. Wakatsuki, K. Ichinose [*Proc. 4ᵗʰ Int. Conf. on High Pressure*, Kyoto, Japan, 25-29 Nov 1974 (Kawakita Printing Co., Kyoto, Japan, 1975) p.441-4]

[118] A. Sawaoka, S. Saito, M. Araki [*Proc. 6ᵗʰ AIRAPT Int. High Pressure Conf.*, Boulder, USA, 25-29 July 1977 (Plenum Press, New York, 1979) vol.1 p.986-93]

[119] F.R. Corrigan [*Proc. 6ᵗʰ AIRAPT Int. High Pressure Conf.*, Boulder, USA, 25-29 July 1977 (Plenum Press, New York, 1979) vol.1 p.994-9]

[120] M. Wakatsuki, K. Ichinose, T. Aoki [*Mater. Res. Bull. (USA)* vol.7 (1972) p.999-1004]

[121] O. Shimomura et al [*Proc. 32ⁿᵈ High Pressure Conf. of Japan* (Japan Society for High Pressure Science and Technology, Yokkaichi, 1991) p.136]

[122] I.A. Petrusha, A.A. Svirid, A.N. Lutsenko, B.V. Sharupin, S.M. Potekhin [in *High Pressure Influence on Materials Properties* (IPM AN UkrSSR, Kiev, 1990) p.22-7]

[123] A. Onodera, K. Inoue, H. Yoshihara, H. Nakae, T. Matsuda, T. Hirai [*J. Mater. Sci. (UK)* vol.25 (1990) p.4279-84]

[124] I.A. Petrusha, A.A. Svirid [*High Press. Res. (UK)* vol.9 (1992) p.136-9]

[125] M. Ueno et al [*Phys. Rev. B (USA)* vol.45 (1992) p.10226-30]

[126] T. Sato, T. Ishii, N. Setaka [*J. Am. Ceram. Soc. (USA)* vol.65 (1982) p.C-162]

[127] T. Sekine, T. Sato [*J. Appl. Phys. (USA)* vol.74 (1993) p.2440-4]

[128] A.V. Kurdyumov, N.F. Ostrovskaya, A.N. Pilyankevich, G.A. Dubitsky, V.N. Slesarev [*Dokl. Akad. Nauk SSSR (USSR)* vol.229 (1976) p.338-40]

[129] A.V. Kurdyumov, N.F. Ostrovskaya, A.N. Pilyankevich, I.N. Frantsevich [*Sov. Phys.-Dokl. (USA)* vol.19 (1974) p.232-3]

[130] G.A. Adadurov et al [*Sov. Phys.-Dokl. (USA)* vol.12 (1967) p.173-5]

[131] T. Soma, A. Sawaoka, S. Saito [*Mater. Res. Bull. (USA)* vol.9 (1974) p.755-62]

[132] A. Sawaoka, T. Soma, S. Saito [*Jpn. J. Appl. Phys. (Japan)* vol.13 (1974) p.891-2]

[133] T. Soma, A. Sawaoka, S. Saito [*Proc. 4th Int. Conf. on High Pressure*, Kyoto, Japan, 25-29 Nov 1974 (Kawakita Printing Co., Kyoto, Japan, 1975) p.446-53]

[134] A. Sawaoka [*Am. Ceram. Soc. Bull. (USA)* vol.62 (1983) p.1379-83]

[135] V.F. Britun, A.V. Kurdyumov, I.A. Petrusha, A.A. Svirid [*J. Superhard Mater. (USA)* vol.14 (1992) p.3-7]

[136] E. Tani, T. Soma, A. Sawaoka, S. Saito [*Jpn. J. Appl. Phys. (Japan)* vol.14 (1975) p.1605-6]

[137] T. Akashi, A. Sawaoka, S. Saito [*J. Am. Ceram. Soc. (USA)* vol.61 (1978) p.245-6]

[138] V.A. Pesin, M.I. Sokhor, L.I. Fel'dgun [*Russ. J. Phys. Chem. (UK)* vol.53 (1979) p.1602-3]

[139] A. Onodera, H. Miyazaki, N. Fujimoto [*J. Chem. Phys. (USA)* vol.74 (1981) p.5814-6]

[140] A.V. Kurdyumov, I.N. Frantsevich, S.S. Dzhamarov, A.V. Bochko [*Fiz. Tekh. Vysokikh Davlenii (USSR)* no.4 (1981) p.35-46]

[141] V.D. Andreev, V.R. Malik, A.V. Bochko, V.S. Silvestrov, T.R. Balan [*Fiz. Tekh. Vysokikh Davlenii (Ukraine)* no.3 (1992) p.126-31]

[142] V.L. Solozhenko, V.Ya. Leonidov [*J. Superhard Mater. (USA)* vol.15 (1993) p.62-4]

[143] H. Hiraoka, O. Fukunaga, M. Iwata [*Yogyo-Kyokai-Shi (Japan)* vol.84 (1976) p.163-70]

[144] R.C. DeVries, J.F. Fleischer [*Mater. Res. Bull. (USA)* vol.4 (1969) p.433-42]

[145] O. Fukunaga, T. Sato, M. Iwata, H. Hiraoka [*Proc. 4th Int. Conf. on High Pressure*, Kyoto, Japan, 25-29 Nov 1974 (Kawakita Printing Co., Kyoto, Japan, 1975) p.454-9]

[146] K. Susa, T. Kobayashi, S. Taniguchi [*Proc. 4th Int. Conf. on High Pressure*, Kyoto, Japan, 25-29 Nov 1974 (Kawakita Printing Co., Kyoto, Japan, 1975) p.429-35]

[147] T. Kobayashi, K. Susa, S. Taniguchi [*Mater. Res. Bull. (USA)* vol.10 (1975) p.1231-6]

[148] T. Kobayashi, K. Susa, S. Taniguchi [*Mater. Res. Bull. (USA)* vol.12 (1977) p.847-52]

[149] A.R. Badzian, T. Kieniewicz-Badzian [*Proc. 7th Int. AIRAPT Conf.*, Le Creusot, France, 30 July - 3 Aug 1979 (Pergamon Press, Oxford, 1980) vol.2 p.1087-91]

[150] T. Kobayashi [*Mater. Res. Bull. (USA)* vol.14 (1979) p.1541-51]

[151] T. Kobayashi [*J. Chem. Phys. (USA)* vol.12 (1979) p.898-905]

[152] T. Endo, O. Fukunaga, M. Iwata [*J. Mater. Sci. (UK)* vol.14 (1979) p.1375-80]

[153] T. Endo, O. Fukunaga, M. Iwata [*J. Mater. Sci. (UK)* vol.16 (1981) p.2227-32]

[154] T. Sato, T. Endo, S. Kashima, O. Fukunaga, M. Iwata [*J. Mater. Sci. (UK)* vol.18 (1983) p.3054-62]

[155] H. Sumiya, T. Iseki, A. Onodera [*Mater. Res. Bull. (USA)* vol.18 (1983) p.1203-7]

[156] S. Hirano, T. Yamaguchi, S. Naka [*J. Am. Ceram. Soc. (USA)* vol.64 (1981) p.734-6]

[157] G. Demazeau, G. Biardeau, L. Vel [*Mater. Lett. (Netherlands)* vol.10 (1990) p.139-44]

[158] L. Vel, G. Demazeau [*Solid State Commun. (USA)* vol.79 (1991) p.1-4]

[159] G. Demazeau, G. Biardeau, L. Vel [*High Press. Res. (UK)* vol.7 (1991) p.210-2]

[160] T. Yogo, S. Naka, H. Iwahara [*J. Mater. Sci. (UK)* vol.26 (1991) p.3758-62]

[161] M.M. Bindal, S.K. Singhal, B.P. Singh, R.K. Nayar, R. Chopra, A. Dhar [*J. Cryst. Growth (Netherlands)* vol.112 (1991) p.368-401]

[162] R.C. DeVries [in *Diamond and Diamond-Like Films and Coatings*, Eds R.E. Clausing, L.L. Norton, J.C. Angus, P. Koidl (Plenum Press, New York, 1991) p.151-72]

[163] V. Gonnet, L. Vel, G. Demazeau [*Mater. Lett. (Netherlands)* vol.14 (1992) p.43-9]

[164] G. Bocquillon, C. Loriers-Susse, J. Loriers [*J. Mater. Sci. (UK)* vol.28 (1993) p.3547-56]

[165] S. Nakano, O. Fukunaga [*Diam. Relat. Mater. (Netherlands)* vol.2 (1993) p.1168-74]

[166] S. Nakano, H. Ikawa, O. Fukunaga [*Diam. Relat. Mater. (Netherlands)* vol.3 (1994) p.75-82]

[167] V.L. Solozhenko, V.A. Mukhanov, N.V. Novikov [*Dokl. Chem. (USA)* vol.312 (1990) p.125-7]

[168] V.L. Solozhenko [*Abstr. 13th AIRAPT Int. Conf. on High Pressure Science and Technology*, Bangalore, India, 1991 (Elsevier) p.K-4]

[169] V.L. Solozhenko [*High Press. Res. (UK)*, in press]

[170] A. Herold, B. Marzluf, P. Perio [*C.R. Acad. Sci. (France)* vol.246 (1958) p.1866-8]

[171] G.V. Samsonov, L.Ya. Markovski, A.F. Zhigach, M.G. Valyashko [*Boron, its compounds amd alloys* (AN UkrSSR, Kiev, 1960)]

[172] V.L. Solozhenko, V.A. Mukhanov, N.V. Novikov [*Dokl. Phys. Chem. (USA)* vol.308 (1989) p.728-30]

[173] V.L. Solozhenko [*High Press. Res. (UK)* vol.9 (1992) p.140-3]

[174] V.L. Solozhenko, A.B. Slutsky, Yu.A. Ignatiev [*J. Superhard Mater. (USA)* vol.14 (1992) p.61]

[175] V.L. Solozhenko [*Abstr. 3rd Int. Conf. on Diamond Science and Technology, Heidelberg*, Germany, 31 Aug - 4 Sept 1992, p.8.134; reported in *Diam. Relat. Mater. (Netherlands)* vol.2 no.2-4]

[176] V.L. Solozhenko [*Diam. Relat. Mater. (Netherlands)*, in press]

[177] S. Nakano, H. Ikawa, O. Fukunaga [*J. Am. Ceram. Soc. (USA)* vol.75 (1992) p.240-3]

[178] O. Fukunaga, H. Fujioka, S. Nakano [*Abstr. 3rd Int. Conf. on Diamond Science and Technology*, Heidelberg, Germany, 31 Aug - 4 Sept 1992, p.7.2; reported in *Diam. Relat. Mater. (Netherlands)* vol.2 no.2-4]

2.2 Phase diagram of AlN

S. Porowski and I. Grzegory

June 1994

A INTRODUCTION

We discuss here some thermodynamical properties of AlN which are critical for crystal growth of this important electronic material. Its very high melting temperature is the main obstacle in obtaining large single crystals of AlN. Additionally, N_2 pressure of the order of 100 bar is necessary to suppress decomposition of AlN at temperatures approaching the melting point.

B THERMAL STABILITY OF AlN

FIGURE 1 shows the temperature dependence of N_2 pressure in equilibrium with the system consisting of solid AlN and liquid Al reported by Slack and McNelly [1]. The results are in good agreement with earlier calculations of Pastrnak and Roskovcova [2].

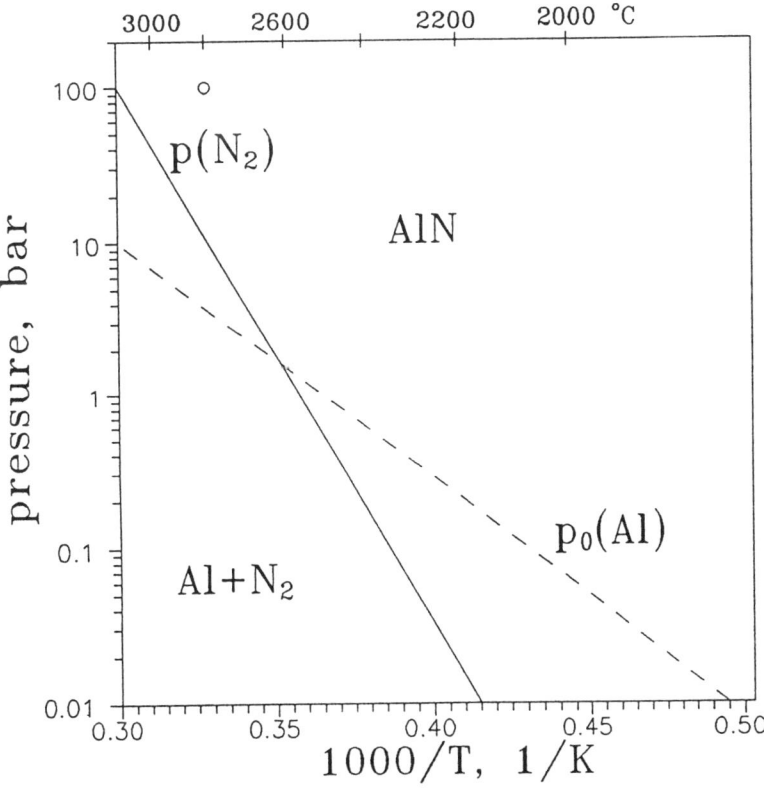

FIGURE 1 Equilibrium N_2 pressure over AlN(s) + Al(l) according to [1]; dashed line - Al vapour pressure over pure liquid Al, o - experimental result of Class [4].

Slack and McNelly [1] have calculated the N_2 pressures over AlN in equilibrium with liquid Al, from the JANAF data [3]. The calculated values are 1, 10 and 100 atm at 2563, 2815 and 3117 °C, respectively. The pressure of Al vapour over the Al liquid was assumed to be the same as that over pure liquid Al. This is valid if the solubility of AlN in the liquid is small, i.e. at temperatures far from the melting point (see Sections C and D). The partial pressure of Al following from the above approximation is marked in FIGURE 1 by the dashed line. The diagram indicates that Al pressure exceeds N_2 pressure, in the system considered, over a very large temperature range. According to the data presented, the N_2 pressure dominates only at temperatures higher than 2600 °C.

In FIGURE 1 we have also shown the experimental point corresponding to 2800 °C and 100 bar N_2 pressure. Under these conditions the appearance of the liquid phase has been observed by Class [4] during his attempts to melt the nitride with a graphite filament (see Section C).

C   MELTING POINT OF AlN

The most extensive experimental study of AlN melting was performed by Class [4]. Two approaches, melting with resistively heated filaments (W, Re and graphite) and arc melting, were used to evaluate the melting behaviour and to determine the melting temperature of AlN. The melting of AlN has been observed only during heating with graphite filaments, at 2750 - 2850 °C, at nitrogen pressures of 100 and 200 bar. However, the melted products were heavily contaminated with graphite. Graphite X-ray diffraction peaks were stronger than the AlN peaks.

The melting temperature for AlN (and also for other group III nitrides) was evaluated by Van Vechten [5] with the use of his semiempirical theory of electronegativity. Both the enthalpy and entropy of melting have been calculated with the assumption that melted AlN is a metallic liquid. The calculated values are:

$$\Delta H^M = 57.919 \, \text{kcal mol}^{-1}$$
$$\Delta S^M = 16.61 \, \text{cal mol}^{-1} \text{K}^{-1}$$
$$T^M = 3487 \, \text{K}$$

D   LIQUIDUS FOR THE Al-AlN SYSTEM

The solubility of nitrogen in molten aluminium, at 1 bar N_2 pressure, was evaluated by Chernega et al [6] from the measurements of diffusion coefficient in the temperature range 1073 - 1373 K. The resulting data have been described by the following dependence:

$$\lg[N] = (-1036/T) - 2.25 \tag{1}$$

where [N] is nitrogen solubility in at %. The solubility of nitrogen in liquid Al, evaluated in this work, varied from 6×10^{-4} to 10^{-3} at % in the investigated temperature range. Very low solubility was also suggested in earlier work by Long and Foster [7] and Evans and Pehlke [8].

Experiments with crystal growth of AlN [9] from liquid Al at high N_2 pressure, at 1600-1800 °C, show that crystallization of AlN from the solution is much less effective than in the case of GaN where solubilities of the order of 1 % over the same temperature range were found.

In FIGURE 2(a) we have shown the liquidus curve for the Al - AlN system calculated from the ideal solution approximation assuming the melting temperature and melting entropy of AlN equal to those obtained by Van Vechten and specified in Section C. A comparison of the results of Chernega et al (FIGURE 2(b)) with the calculated data for low temperatures indicates a wide discrepancy between the calculated and experimental data. It may be due to the positive deviation of the solution from ideality as well as due to uncertainty in determination of AlN melting conditions.

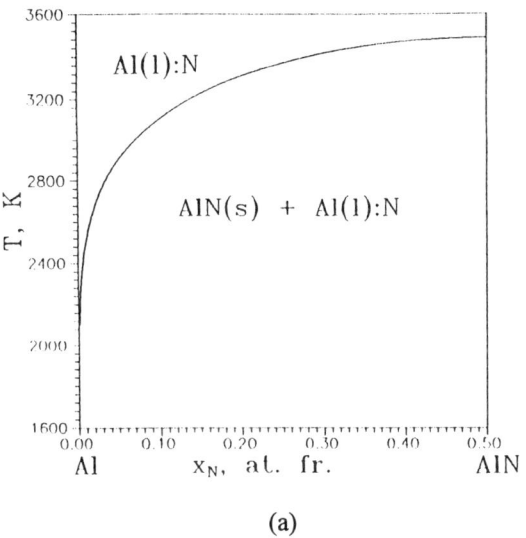

(a)

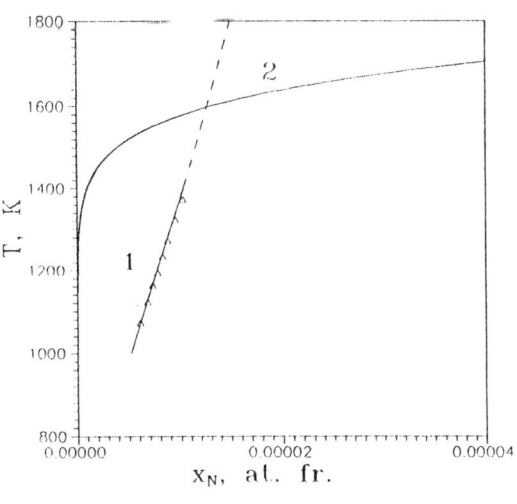

(b)

FIGURE 2 Liquidus of the Al - AlN system: (a) calculated in ideal solution approximation, (b) Δ - experimental results of [6], solid line - calculated in ideal solution approximation.

E   HIGH PRESSURE PHASE TRANSITION

The transformation of AlN into a high pressure phase has been observed by Perlin et al [10] by the registration of Raman spectra from AlN crystals in a diamond anvil cell. The disappearance of the Raman peaks and the change of colour indicate a phase transition at 165 kbar at room temperature. A transition pressure of 229 kbar has been reported by Ueno et al [11]. It was shown by X-ray diffraction measurements that AlN transforms into the cubic NaCl structure. This transition has also been observed by Vollstaedt et al [12] for AlN powder at high temperatures. The reverse transition has not been detected, indicating metastability of AlN in its high pressure phase.

The first principle total energy calculations performed for several possible high pressure structures are in good agreement with experiment, giving a wurtzite-rocksalt transition pressure of 166 kbar (Gorczyca and Christensen [13]) or 129 kbar (Van Camp et al [14]).

F   CONCLUSION

Nitrogen pressures in equilibrium with the system consisting of Al(l) and AlN(s) are orders of magnitude less than N_2 pressures necessary for stability of GaN (see Datareview 2.3) and especially InN (see Datareview 2.4), at high temperatures. This is due to the very high bonding energy in the AlN crystal lattice which greatly lowers the chemical potential of N in the crystal. The energy of the Al-N bond in a tetrahedrally coordinated crystal is 2.88 eV/bond [15] whereas for GaN and InN it is 2.2 and 1.93 eV/bond, respectively.

The melting temperature is very high, probably in excess of 3000 K. The experimentally determined value of T^M = 3073 K [4] seems to be lowered by the contamination of the liquid phase by carbon. Nitrogen pressure of a few hundred bar (or excited nitrogen atmosphere) is necessary for AlN melting.

At a pressure of 165 kbar, AlN transforms from the hexagonal wurtzite to the cubic NaCl phase which is metastable at normal pressure.

REFERENCES

[1]   G.A. Slack, T.F. McNelly [*J. Cryst. Growth (Netherlands)* vol.34 (1976) p.263-79]
[2]   J. Pastrnak, L. Roskovcova [*Phys. Status Solidi (Germany)* vol.7 (1964) p.331-8]
[3]   R.D. Stull, H. Prophet [*JANAF Thermochemical Tables*, 2nd edition (NSRDS-NBS37, USA, 1971)]
[4]   W. Class [NASA CR-1171, USA (1968)]
[5]   J.A. Van Vechten [*Phys. Rev. B (USA)* vol.7 (1973) p.1479-505]
[6]   D.F. Chernega, W.G. Mogilatenko, A.P. Dyatlov [*Izv. Vyssh. Uchebn. Zaved. Chern. Metall. (USSR)* no.6 (1985) p.31-4]
[7]   G. Long, L.M. Foster [*J. Am. Ceram. Soc. (USA)* vol.42 (1959) p.53-61]
[8]   D.B. Evans, R.D. Pehlke [*Trans. Metall. Soc. AIME (USA)* vol.230 (1964) p.1651-6]
[9]   I. Grzegory, J. Jun, St. Krukowski, M. Bockowski, S. Porowski [*Physica B (Netherlands)* vol.185 (1993) p.99-102]

[10]   P. Perlin, I. Gorczyca, S. Porowski, T. Suski, N. E. Christensen, A. Polian [*Jpn. J. Appl. Phys. (Japan)* vol.32 (1993) p.334-9]

[11]   M. Ueno, A. Onodera, O. Shimomura, K. Takemura [*Phys. Rev. B (USA)* vol.45 (1992) p.10123-6]

[12]   H. Vollstaedt, E. Ito, M. Akaishi, S. Akimoto, O. Fukunaga [*Proc. Jpn. Acad. B (Japan)* vol.7 (1990) p.7-9]

[13]   I. Gorczyca, N.E. Christensen, P. Perlin, I. Grzegory, J. Jun, M. Bockowski [*Solid State Commun. (USA)* vol.79 (1991) p.1033-7]

[14]   P.E. Van Camp, V.E. Van Doren, J.T. Devreese [*Phys. Rev. B (USA)* vol.44 (1991) p.9056]

[15]   W.A. Harrison [in *Electronic Structure and Properties of Solids* Ed. Freeman (San Francisco, 1980)]

2.3   Phase diagram of GaN

S. Porowski and I. Grzegory

June 1994

A   INTRODUCTION

GaN crystals are usually grown in the form of thin epitaxial layers on foreign substrates, by low temperature methods like MBE or MOCVD. The lack of large, substrate quality crystals is the main obstacle in the development of device technology based on GaN. This is due to the very high N_2 pressure required for GaN stability at high temperatures. Especially, it concerns the melting point of this nitride which makes impossible the application of the standard crystallization techniques like Czochralski or Bridgman growth from the stoichiometric melt.

B   THERMAL STABILITY OF GaN

The most extensive study on GaN stability at high temperatures has been performed by Karpinski et al [1] by the use of the gas pressure technique (up to 20 kbar) and a tungsten carbide anvil cell (up to 60 kbar). Earlier, the results of GaN stability investigations, up to 8 kbar, were reported by Madar et al [2]. The results of [1] were consistent with the lower pressure data of [2].

The p_{N_2}-T curve separating GaN and Ga + N_2 phases proposed by Karpinski et al is shown in FIGURE 1. In the high pressure range, the curve strongly deviates from the linear dependence proposed by Thurmond and Logan [3] who had calculated the equilibrium N_2 pressure from the measurements of the equilibrium pressure of NH_3 over GaN, assuming nitrogen to be an ideal gas. A linear dependence for high pressure data is obtained transforming nitrogen pressure into its activity, a_{N_2}, via the equation of state of real nitrogen [4]. This dependence is shown in FIGURE 2 (for the pressure range where activity data are available). The curve is described by the expression:

$$\ln a_{N_2}(p_{eq}, T_{eq}) = 32.45 - 37604/T \ (\pm 0.44) \tag{1}$$

Standard thermodynamical functions for GaN synthesis have been evaluated:

$$\Delta G^\circ = 32.43T - 37\,700\,\text{cal mol}^{-1} \ (\pm 700\,\text{cal mol}^{-1}) \tag{2a}$$

$$\Delta H^\circ = -37.7\,\text{kcal mol}^{-1} \tag{2b}$$

$$\Delta S^\circ = -32.43\,\text{cal mol}^{-1}\,\text{K}^{-1} \tag{2c}$$

The value of the enthalpy ΔH° is in excellent agreement with the one estimated by Madar et al [2] and ΔG° agrees very well with the results of Thurmond and Logan [3].

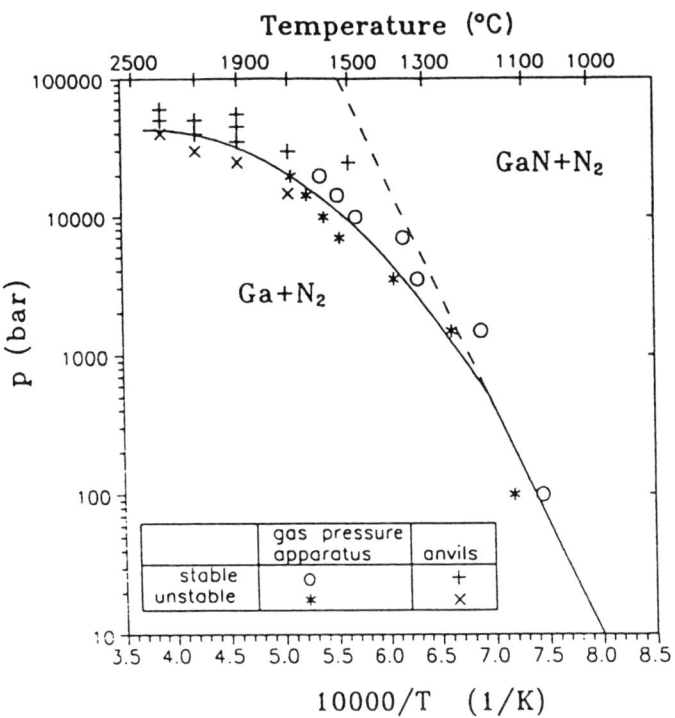

FIGURE 1 Equilibrium N_2 pressure over GaN(s) + Ga(l) according to [1].

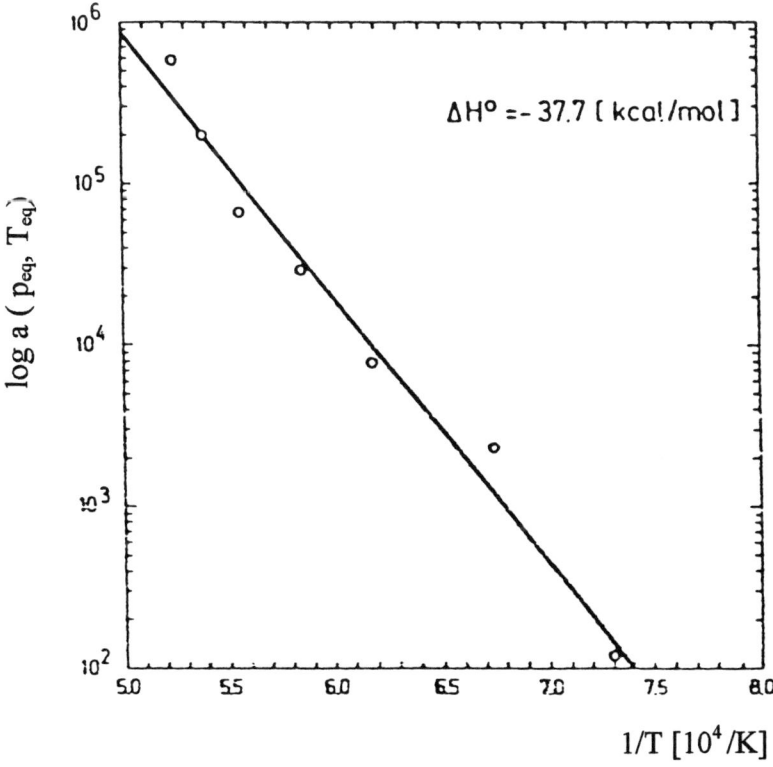

FIGURE 2 Equilibrium activity of N_2 over GaN(s) + Ga(l), [1], $a_{N_2}(p) = f_{N_2}(p)/f_{N_2}$ (1 bar).

C   MELTING POINT OF GaN

The melting point of GaN is not known due to experimental difficulties related to very high temperature and N_2 pressure necessary for melting. The experiments in the high pressure tungsten carbide anvil cell performed by Karpinski et al [1] have shown that GaN does not melt at a temperature as high as 2573 K, at 60 kbar. This is consistent with the theoretical estimation of T^M = 2791 K (normal pressure) based on Van Vechten's semiempirical theory of electronegativity [5]. The melting entropy ΔS^M = 16.01 cal mol^{-1} K^{-1} and melting enthalpy ΔH^M = 44.68 kcal mol^{-1} resulted from the calculations. At high pressure, the value of T^M should be lowered due to decrease of the volume of tetrahedrally coordinated semiconductors upon melting. The calculated slope of the melting curve of GaN is 2.5 K kbar^{-1} [5].

D   LIQUIDUS FOR THE GaN-Ga SYSTEM

The liquidus curve has been studied by Grzegory et al [6]. In this work the nitrogen content in Ga was measured by the use of LECO nitrogen determinator TN-14 based on a thermal conductivity cell sensitive to the difference in thermal conductivity of gases. Ga samples were quenched from three phase equilibrium conditions. These conditions have been determined on the basis of the experimental data of Karpinski et al [1].

The experimental results concerning nitrogen solubility in the liquid Ga are shown in FIGURE 3. The solid line in the figure was calculated assuming ideal behaviour of the solution and a melting temperature for GaN of 2791 K.

The experimental curve should be regarded as the projection onto the T-atomic nitrogen concentration, x, plane since each T-x point corresponds to the minimum N_2 pressure at which GaN is stable. Taking into account that tetrahedrally coordinated III-V compounds decrease their volume upon melting by approximately 15 % [5], the influence of hydrostatic pressure on the liquidus curve can be roughly estimated. FIGURE 4 shows the pressure dependence of nitrogen content in the liquid Ga calculated from an ideal solution approximation for 1600 K [6].

Solubility of nitrogen in liquid gallium has been determined experimentally for 1150 °C [3], 1200 °C [2] and 1500 °C [1] by weighing GaN crystallized during cooling of the system Ga + N_2 (or Ga + NH_3 in [3]) at N_2 (or NH_3) pressure corresponding to three phase equilibrium for the given temperatures. The results of these measurements were 3×10^{-5}, 5×10^{-4} and 10^{-2} atomic fraction respectively, and are in reasonable agreement with the data of FIGURE 3.

The liquidus curve for the Ga-N system has been calculated by Jones and Rose [7] by Quasi Chemical Equilibrium theory. Resulting N concentrations highly exceeded the experimental values. The most important reason for this large discrepancy is the melting temperature of 1700 °C assumed for the calculations which is inconsistent with the experimental results (see Sections B and C).

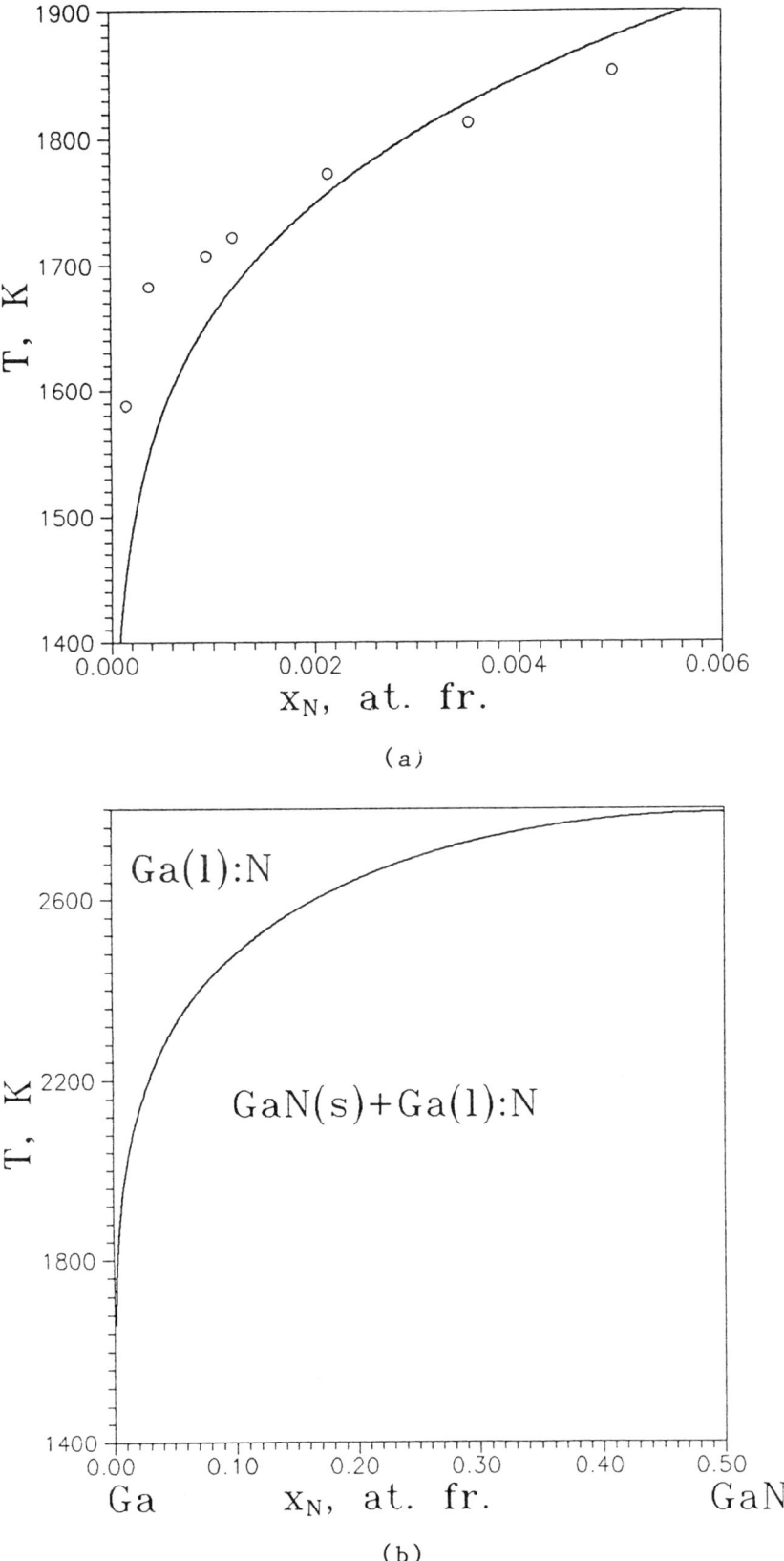

(a)

(b)

FIGURE 3 Liquidus of the Ga - GaN system: (a) o - experimental results of [6], solid line on (a) and
(b) - calculated in ideal solution approximation.

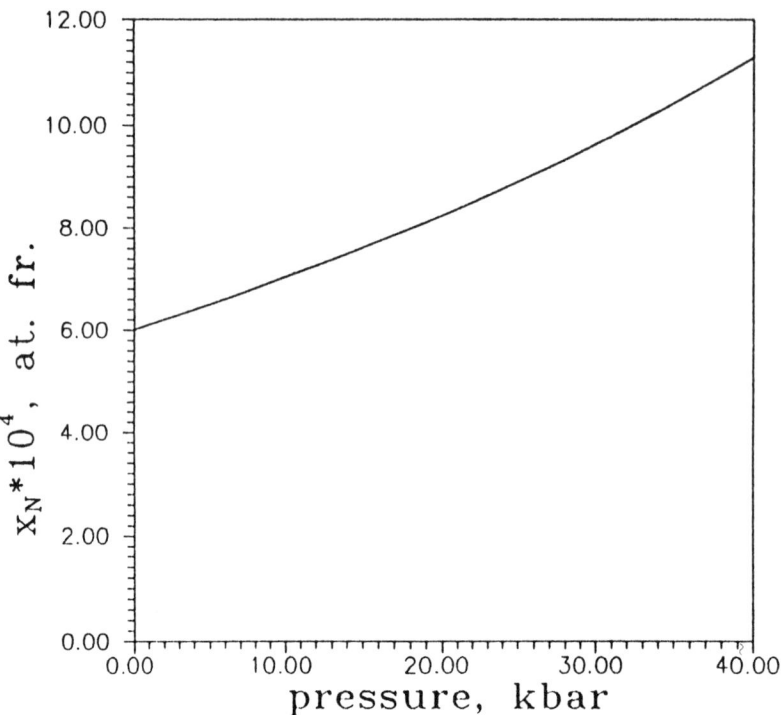

FIGURE 4 Pressure dependence of N solubility in the liquid Ga at 1600 K, calculated as described in the text.

E HIGH PRESSURE PHASE TRANSITION

The phase transition in GaN has been detected by Perlin et al [8] by X-ray absorption spectroscopy, Raman spectroscopy and direct observation (change in colour) of GaN single crystals in a diamond anvil cell. GaN transforms into its high pressure phase (NaCl) at 470-500 kbar. The reverse transition occurs at 300-200 kbar, reflecting large metastability of the high pressure structure. First principle total energy calculations predict the transformation into the NaCl semiconducting phase at 550 kbar [9] or 650 kbar [10].

F CONCLUSION

Temperature in excess of 2500 K and N_2 pressure on the order of tens of kbars are necessary for the melting of GaN. The solubility of nitrogen in liquid Ga on the order of 1 % can be reached by the application of N_2 pressure of 10-20 kbar at temperatures 1500-1600 °C. This allows the growth of high quality GaN single crystals from the solution in temperature gradients [11]. Crystals grown at 10 kbar in a 20 hr process are shown in FIGURE 5.

At pressures of 470-500 kbar, the hexagonal wurtzite phase of GaN transforms into the cubic NaCl modification. This transition is reversible.

|—————————————| 1 mm

FIGURE 5 GaN crystals grown at 10 kbar from the solution in the liquid Ga.

REFERENCES

[1]   J. Karpinski, J. Jun, S. Porowski [*J. Cryst. Growth (Netherlands)* vol.66 (1984) p.1-10]

[2]   R. Madar, G. Jacob, J. Hallais, R. Fruchart [*J. Cryst. Growth (Netherlands)* vol.31 (1975) p.197-203]

[3]   C.D. Thurmond, R.A. Logan [*J. Electrochem. Soc. (USA)* vol.119 (1972) p.622]

[4]   J. Karpinski, S. Porowski [*J. Cryst. Growth (Netherlands)* vol.66 (1986) p.11-20]

[5]   J.A. Van Vechten [*Phys. Rev. B (USA)* vol.7 (1973) p.1479-505]

[6]   I. Grzegory, M. Bockowski, S. Krukowski, J. Jun, S. Porowski [to be published]

[7]   R.D. Jones, K. Rose [*CALPHAD (UK)* vol.8 (1984) p.343-54]

[8]   P. Perlin, C. Jauberthie-Carillon, J.P. Itie, A. San Miguel, I. Grzegory, A. Polian [*Phys. Rev. B (USA)* vol.45 (1992) p.83]

[9]   P.E. Van Camp, V.E. Van Doren, J.T. Devreese [*Phys. Rev. B (USA)* vol.44 (1991) p.9056]

[10] P. Perlin, I. Gorczyca, S. Porowski, T. Suski, N.E. Christensen, A. Polian [*Jpn. J. Appl. Phys. (Japan)* vol.32 (1993) p.334-9]

[11] I. Grzegory, J. Jun, St. Krukowski, M. Bockowski, S. Porowski [*Physica B (Netherlands)* vol.185 (1993) p.99-102]

2.4 Phase diagram of InN

S. Porowski and I. Grzegory

June 1994

A INTRODUCTION

The range of uncertainty in the available experimental data concerning N_2 pressure necessary for InN stability reaches several orders of magnitude. This large discrepancy is connected with the very high equilibrium pressures and extremely slow kinetics of InN decomposition under experimentally available conditions.

B THERMAL STABILITY OF InN

FIGURE 1 shows the reported N_2 pressures over InN. Especially large disagreement is noted between the results of high pressure (up to 1 kbar) experiments [1] and the low pressure data obtained by mass spectroscopic evaluation [2,3] of decomposition pressures.

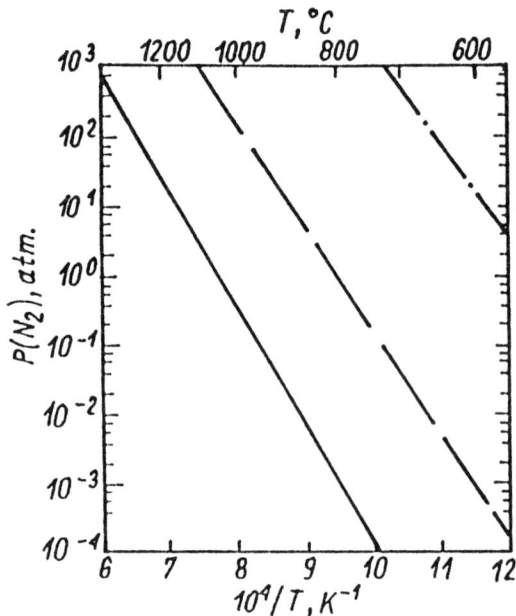

FIGURE 1 Equilibrium N_2 pressure over the InN(s) + In(l) system: ——— [2], — — — [3], · — · — · — [1].

Recently [4], InN thermal stability was investigated at N_2 pressures extended up to 18.5 kbar. It was shown by Differential Thermal Analysis that, over the whole investigated pressure range of 0.1 - 18.5 kbar, rapid decomposition occurs at $710 \pm 10\,°C$. Therefore it was concluded that the observed decomposition was controlled by kinetics and that the equilibrium temperatures are lower than $710\,°C$ for the investigated pressures. The long time (24 hr) annealing of InN at 18.5 kbar showed that InN decomposes at temperature as low as $600\,°C$.

Additionally, the InN powder has been annealed at pressures of 35 and 50 kbar at 700 °C for 30 min in a belt type high pressure cell. It follows, from these experiments, that pressures higher than 35 kbar are necessary to suppress decomposition of InN at 700 °C.

FIGURE 2 summarizes the results of the experiments described above. The solid curve was calculated with the assumption that at 10 kbar N_2 pressure, InN is in equilibrium with its constituents at 570 °C, and that the enthalpy of nitride formation is equal to that of GaN, i.e. -37.7 kcal mol^{-1} [5,6]. The equation of state for N_2 [7] has been used to transform N_2 fugacity into pressure. The results of MacChesney et al [1], obtained from InN decomposition experiments at N_2 pressures up to 1 kbar, are also included in FIGURE 2. The dashed curve is the extrapolation of these data to higher pressure.

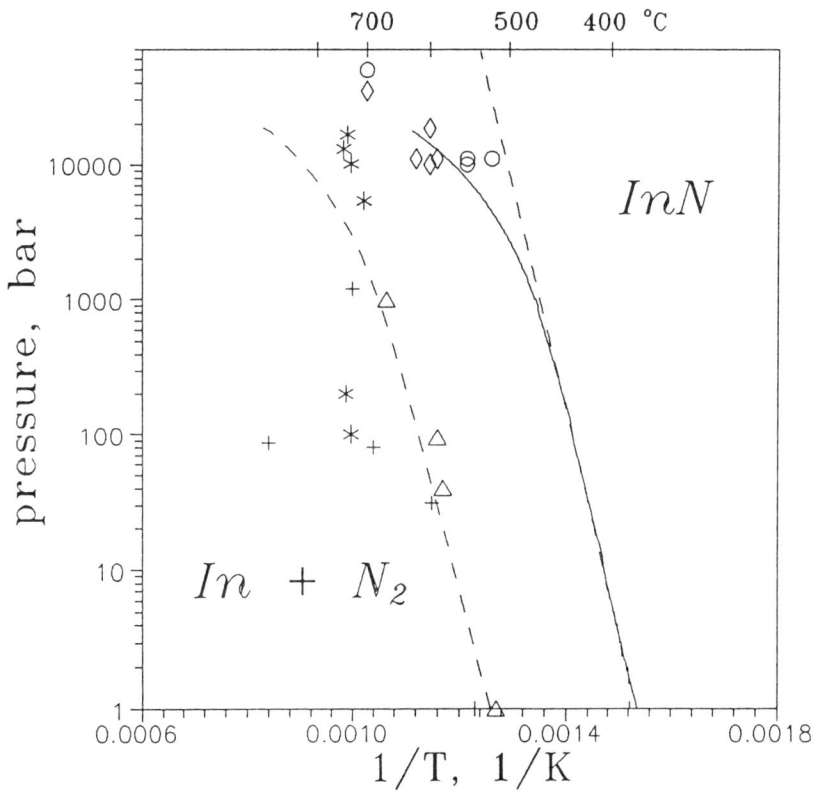

FIGURE 2 Equilibrium N_2 pressure over the InN(s) + In(l) system from high pressure experiments. Ref [4]: * - DTA, o - annealing: InN undecomposed, ◊ - annealing: InN decomposed. Ref [1]: + - InN instability points, Δ - InN stability points. Solid curve: calculated as described in the text; dashed line: ideal gas approximation of the solid curve; dashed curve: extrapolation of data from [1].

Trainor and Rose [8] have observed the dissociation of thin InN films at 500 °C, at 1 atm of N_2. The position of this point is consistent with the solid curve (FIGURE 2) since it lies in the InN instability field (to the left of the curve).

We suggest that the equilibrium N_2 pressures over InN can be even higher than predicted by the solid curve in FIGURE 2. This can be verified at a higher pressure range where the kinetic effects would not distort the experimental results.

C   MELTING POINT

As discussed in Section B, the N_2 pressure required for InN stability at high temperatures is so high that the melting point of the compound has not been determined to date.

The estimation of T^M [4] based on the theory of electronegativity of Van Vechten [9] gives 2146 K for atmospheric pressure. The melting entropy $\Delta S^M = 14.1 \, cal \, mol^{-1} \, K^{-1}$ estimated by the method suggested by Marina [10] has been used for the calculation.

At high pressure, the melting temperature should be lowered since tetrahedrally coordinated semiconductors decrease their volume upon melting (the calculated slope of the liquidus curve for GaN is $2.5 \, K \, kbar^{-1}$ [9]).

D   LIQUIDUS OF THE InN-In SYSTEM

To our knowledge, there have been no experimental results on nitrogen solubility in liquid In.

Assuming a melting temperature of InN close to that estimated by Van Vechten's theory, very low solubilities of N in the liquid indium in the In - InN system should be expected at N_2 pressures up to 20 kbar. This was confirmed by the results of crystal growth experiments [4] from the solution in liquid In at high N_2 pressure where only very small crystals (5 - 50 μm) have been obtained in a 24 hr process in a temperature gradient of 100 °C cm^{-1}.

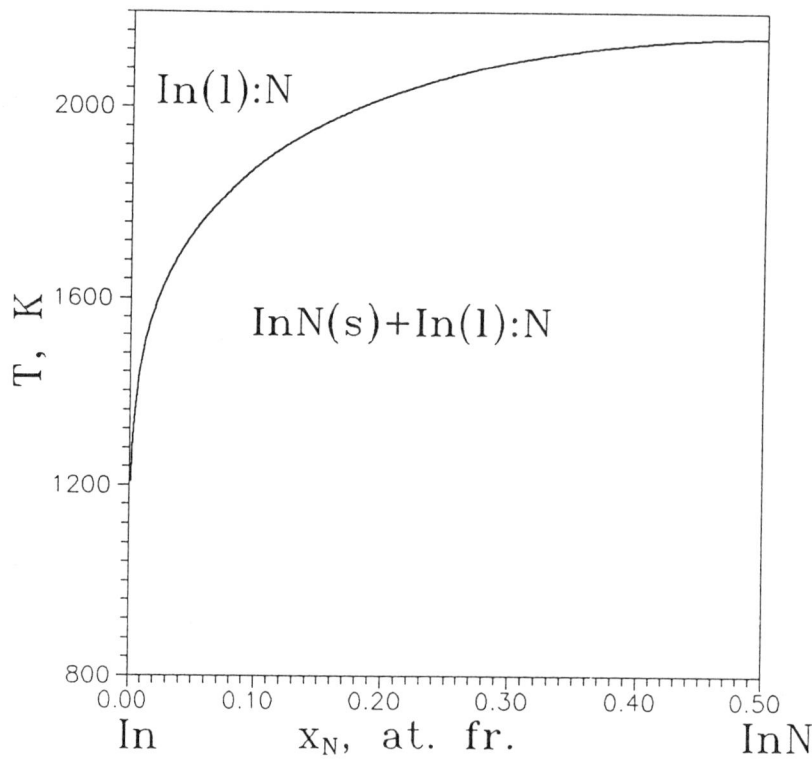

FIGURE 3 Liquidus curve of the In - InN system calculated in ideal solution approximation.

In FIGURE 3, we have shown the liquidus curve for the In-InN system calculated from an ideal solution approximation assuming $T^M = 2146\,K$ [4] and $\Delta S^M = 14.1\,cal\,mol^{-1}\,K^{-1}$ [10].

E HIGH PRESSURE PHASE TRANSITION

Total energy calculations performed by the LMTO method for InN [11] indicate that InN should transform from the hexagonal wurtzite into the NaCl phase at a pressure of 254 kbar. Perlin et al [11] have observed the change in the colour of InN crystals at 230 kbar.

F CONCLUSION

For temperatures lower than 700 °C the kinetics of InN decomposition is very slow. Therefore the verification of the existing stability data at higher temperatures (and corresponding pressures) is necessary. The same is needed for the effective crystallization of InN from solution in liquid In.

REFERENCES

[1]   J.B. MacChesney, P.M. Bridenbaugh, P.B. O'Connor [*Mater. Res. Bull. (USA)* vol.5 (1970) p 783-92]

[2]   R.D. Jones, K. Rose [*J. Phys. Chem. Solids (UK)* vol.48 (1987) p.587-90]

[3]   A.M. Vorob'ev, G.V. Evseeva, L.V. Zenkevich [*Russ. J. Phys. Chem. (UK)* vol.47 (1973) p.1616]

[4]   I. Grzegory, S. Krukowski, J. Jun, M. Bockowski, M. Wroblewski, S. Porowski [to be published in *High Pressure Res. (UK)*]

[5]   J. Karpinski, S. Porowski [*J. Cryst. Growth (Netherlands)* vol.66 (1986) p.11-20]

[6]   R. Madar, G. Jacob, J. Hallais, R. Fruchart [*J. Cryst. Growth (Netherlands)* vol.31 (1975) p.197-203]

[7]   R.T. Jacobsen, R.B. Stewart [*J. Phys. Chem. Ref. Data (USA)* vol.2 (1973) p.757-922]

[8]   J.W. Trainor, K. Rose [*J. Electron. Mater. (USA)* vol.3 (1974) p.821]

[9]   J.A. Van Vechten [*Phys. Rev. B (USA)* vol.7 (1973) p.1479-505]

[10]   L.I. Marina, A.Ya. Nashel'skij [*Russ. J. Phys. Chem. (UK)* vol.43 (1969) p.963]

[11]   P. Perlin, I. Gorczyca, S. Porowski, T. Suski, N.E. Christensen, A. Polian [*Jpn. J. Appl. Phys. (Japan)* vol.32 (1993) p.334-9]

CHAPTER 3

ELECTRICAL TRANSPORT PROPERTIES

3.1 Electrical transport properties of BN

V.V. Lopatin

July 1994

A INTRODUCTION

All modifications of boron nitride (hexagonal (h-BN), rhombohedral (r-BN), wurtzite (w-BN) and cubic (c-BN)) are narrow band insulators (wide band semiconductors) according to their electronic band structure. The conductivity of boron nitride can range from 10^{-6} to $10^{-16}\,(\Omega\,\text{cm})^{-1}$ depending on the modification, the impurity content, and how it is prepared and processed (vapour phase deposition - pyrolytic boron nitride (PBN) [1], plasma deposition, ion implantation, powder synthesis with subsequent sintering, or structural ordering with compression and heating). At present, all processing techniques produce only polycrystalline BN with crystallite sizes not exceeding several microns, except c-BN which has been produced in single crystal form up to 3 mm in diameter [2]. Therefore, the optical, electrical and dielectric properties are not only affected by atomic point defects and their complexes, but also by the structural hierarchy, crystallite sizes, textures, stacking faults, dislocations (e.g. up to $10^{10}\,\text{cm}^{-2}$ in PBN [3]), pores, and other macrodefects.

B INFLUENCE OF THE STRUCTURAL HIERARCHY ON THE ELECTRICAL PROPERTIES OF h-BN AND r-BN

The structural hierarchy, the geometric relationship between individual crystal grains, is particularly significant for h-BN and r-BN which have pronounced anisotropic properties in the 'a' and 'c' directions due to anisotropic electronic state densities. When h-BN and r-BN have axial orientation and three dimensional ordering, as is observed in PBN and highly oriented PBN (HOPBN), the anisotropic conductivity is readily apparent. In materials produced from BN powder, the texture is weaker as shown in FIGURE 1 [4,5] and therefore the anisotropy of the properties, e.g. electrical conductivity, is weaker too. The spread of the texture increases as the density of h-BN and r-BN decreases from $2.2\,\text{g cm}^{-3}$ to $1.7\,\text{g cm}^{-3}$. For density below $2.1\,\text{g cm}^{-3}$, a texture-free component ($I_{\text{F.T.}}$, FIGURE 1) appears. Its fraction attains about 25 % of the bulk of the material with a density of $1.8\,\text{g cm}^{-3}$. The presence of this component is believed to be associated with the amorphous-like crystallite interfaces. At the PBN crystallite boundaries there are regions with high stacking faults, i.e. the material is turbostratic BN. Expressions relating the conductivity (σ) of h-BN polycrystals in the 'a' and 'c' directions (σ_a and σ_c) to the sum of the conductivities of a single crystallite and its interfaces (σ_{ka} and σ_{kc}) were obtained in [5]:

$$\sigma_c = \sigma_{ka}\,(1-b) + \sigma_{kc}\,b \tag{1a}$$

$$\sigma_a = \sigma_{ka}\left[\frac{1+b}{2}\right] + \sigma_{kc}\left[\frac{1-b}{2}\right] \tag{1b}$$

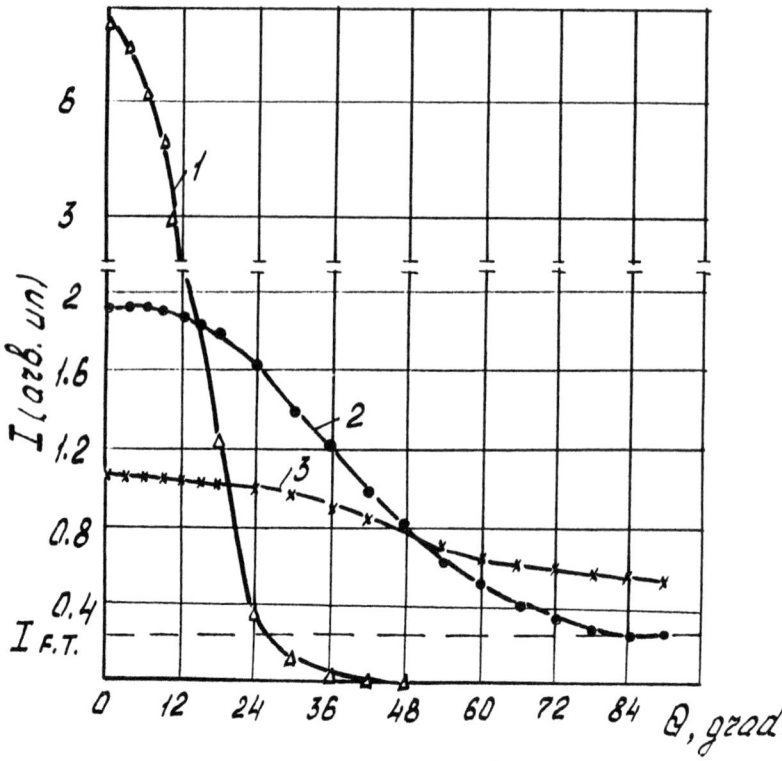

FIGURE 1   Angular distribution of boron nitride orientations.   Maximum on the curves coincides with 'c'-direction of the most crystallites (axial texture). 1 - PBN density is $2.15\,\mathrm{g\,cm^{-3}}$; 2 - PBN density is $1.9\,\mathrm{g\,cm^{-3}}$; 3 - sintered BN density is $1.9\,\mathrm{g\,cm^{-3}}$; $I_{F.T.}$ - texture-free fraction.

Measured values of the texture (parameter b), σ_a and σ_c for h-BN and sintered BN powders are listed in TABLE 1.

TABLE 1   Variation of the anisotropic electrical conductivity of polycrystalline h-BN with texture (b).

Material	b	Density g cm^{-3}	Conductivity x 10^{17} $(\Omega\,cm)^{-1}$					
			polycrystal			crystallite		
			σ_a	σ_c	σ_a/σ_c	σ_{ka}	σ_{kc}	σ_{ka}/σ_{kc}
PBN	0.97	2.1	82	5.5	15	83	3.3	25
PBN	0.72	1.8	17	7	2	19	5.2	3.6
BN	0.58	1.9	50	28	1.8	62	3.4	18

The ratios of the conductivities in specific directions, σ_a/σ_c (macroscopic) and σ_{ka}/σ_{kc} (in crystallites) for PBN ($2.1\,\mathrm{g\,cm^{-3}}$) and BN correlate with b. The anomalous decrease in σ_a and σ_{ka} for PBN ($1.8\,\mathrm{g\,cm^{-3}}$) was explained by the increased concentration of twins and dislocations (small angular interfaces) in the basal crystal plane [3,7]. In PBN, all the grain boundaries are weakly conductive; therefore, the conductivity depends not only on the texture, but also on the concentration of grain boundaries. For low density h-BN, the effect of the concentration of boundaries (twins) on the anisotropy of conductivity may prevail over the texture.

In rhombohedral PBN (r-PBN) the influence of the texture is stronger than the boundary anisotropy because of the larger crystallite sizes and greater three-dimensional ordering as compared with hexagonal PBN (h-PBN).

C   ELECTRICAL PROPERTIES OF h-BN

The electrical conductivity of h-BN has two distinct temperature regimes. The plot of the DC conductivity (σ) vs 1/T in the 'c' - direction has a breakpoint when T ~ 800 K (FIGURE 2). In the first approximation, over the temperature range of 80 to 2000 K, the conductivity is well fitted by the exponential function:

$$\sigma = \sigma_1 \exp\left[-\frac{E_1}{kT}\right] + \sigma_2 \exp\left[-\frac{E_2}{kT}\right] \qquad (2)$$

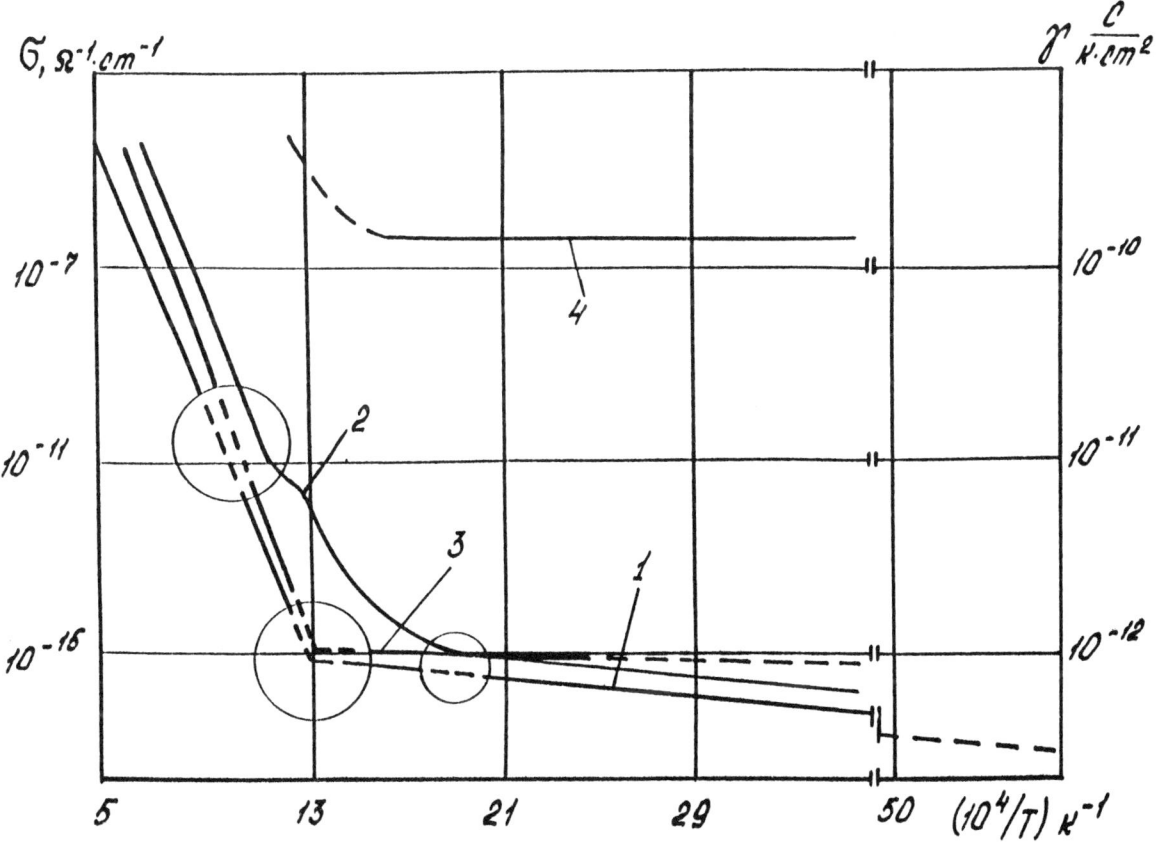

FIGURE 2 Temperature dependences of conductivity: 1,2 - h-PBN; 3 - r-PBN; 1,3 - c-direction; 2 - a-direction. Fine line denotes the regions of $\sigma(T)$ where deviations from linearity are usually observed; 4 - temperature dependence of pyroelectrical static coefficient γ of r-PBN. Dotted lines denote the regions of ambiguous determination of (σ) and (γ) values.

For temperatures less than 700 K (the region of extrinsic conductivity), the parameters E_1 and σ_1 are sensitive to the structure, type and impurity content of the h-BN. Linear regions of log σ vs. (1/T) only occur over short temperature ranges in this region, although the total conductivity trend is the same. For nominally pure PBN (the impurity content is less than 10 ppm, except for carbon which is difficult to assess), $E_1 \simeq 0.2\,eV$ [8], and for sintered BN, $E_1 = 0.2 - 1.0\,eV$ [9]. At room temperature, the values for σ_1 and σ_2 are $\sim 10^{-16}\,(\Omega\,cm)^{-1}$ and 10^{-13} to $10^{-10}\,(\Omega\,cm)^{-1}$, respectively. Intrinsic conductivity occurs for temperatures greater than 800 K, and $E_2 = 2.5 \pm 0.1\,eV$ for PBN and 2 to 6 eV for sintered BN. Since twice the value of E_2 is close to the bandgap calculated and estimated by optical absorption [10], this suggests that the conductivity at this temperature is intrinsic: due to the creation of electrons and holes from band to band transitions. The same conclusion was reached in [11] albeit with higher values of σ_2 and E_2 (3.55 eV). For temperatures less than 800 K, electron transport occurs through electron-hole transitions such as band-defect-induced level-band or by intraband hopping.

Impurity induced energy states in the bandgap were detected by cathodoluminescence in PBN powders [12] and h-BN powder compacts [13], and by photoluminescence of h-BN powders [14,15]. Peaks were also observed in the spectra of PBN by diffusive reflection absorption. However, insufficient control of the structural hierarchy, stoichiometry and impurities makes it difficult to determine the exact cause of the observed spectra.

Much more reliable identification of defects and the activation energies they produce has been achieved by alternative techniques [18-20]. From measurements by photoluminescence (PL), thermally stimulated luminescence (TSL), thermally stimulated conductivity (TSC), and electron paramagnetic resonance (EPR), the assignment of signals to two defects associated with nitrogen vacancies (V_N) and vacancy complexes such as V_N - B and V_N - 3B were ascertained. A scheme of transitions was proposed assuming the energy levels of those defects were donors. Experiments [20] confirmed the conclusions of [16] on the significant influence of carbon impurities in stabilizing anion vacancies and three-boron centres. The calculated position of the dopant carbon atom was not quite at a substitutional position, but located perpendicular to the BN hexagonal plane at a distance of 0.076 nm under V_N.

The results of synchronic TSL and TSC measurements on different species of PBN with fractional annealing [22-24] indicated the activation character of electron-hole transport including the defect-induced localized states in the range of extrinsic (impurity dominated) conduction (T < 800 K). At low temperature (T < 280 K), the effect of hopping conductivity, which occurs by transitions between localized states near the Fermi level and state transitions near the band edges, could not be ruled out. A continuous energy distribution of traps occurs in the section where the activation energies vary only slightly with temperature. FIGURES 3(a), 3(b), 3(c) and 3(d) show their temperature correspondence to TSL and TSC peaks; they are local levels. Depending on the PBN sample, the TSL and TSC characteristically have three general groups of levels: (i) 0.5 to 0.7 eV (300 to 400 K); (ii) 0.8 to 1.0 eV (400 to 500 K); and (iii) 1.0 to 1.33 eV (T > 500 K) with a small degree of population n/N = 10^{-7} to 10^{-2} (n is the concentration of occupied traps and N is the total concentration of traps) [23,24]. By comparing the TSL intensities, TSC and thermal polarization currents, frequency factors and n/N, the donor levels E = 0.5 to 1 eV contributed mainly to the luminescence of PBN, with higher concentration of anion vacancies V_N and σ due to the deeper levels. For PBN with exact stoichiometry, both luminescence and σ are mainly determined by donor and acceptor states with E = 1 - 1.3 eV. By using the results of the definition of the current carrier sign (from the thermodiffusive current, thermo-electric DC, and the thermo-depolarization current, FIGURE 3(a)) and suggesting the existence of the direct transitions through the 5 eV bandgap the positions of the energy levels were evaluated and an approximate scheme of the states and transitions was determined as shown in FIGURE 4.

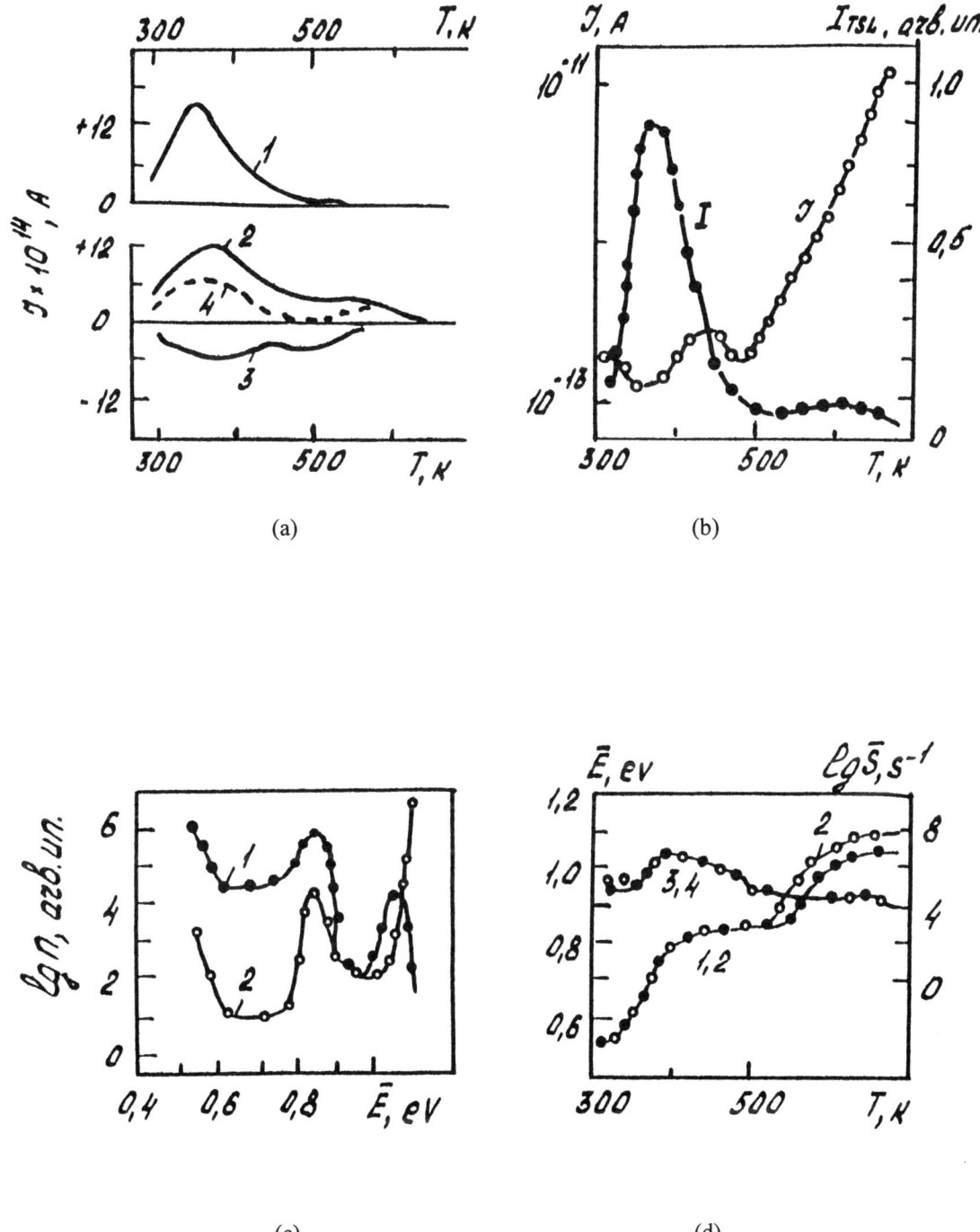

FIGURE 3 Data of thermally activated spectroscopy of PBN: (a) - thermal diffusion (1) and polarization (2,3) currents, total hole current (4); (b) - TSL intensity (I) and TSC current (J) as a function of T; (c) - function of the centre distribution over TSL(n_l) - (1) and TSC(n_l) - (2); (d) - activation energies of trapping centres in TSL - (1) and TSC - (2), and frequency factors in TSL - (3) and TSC - (4) [23,24].

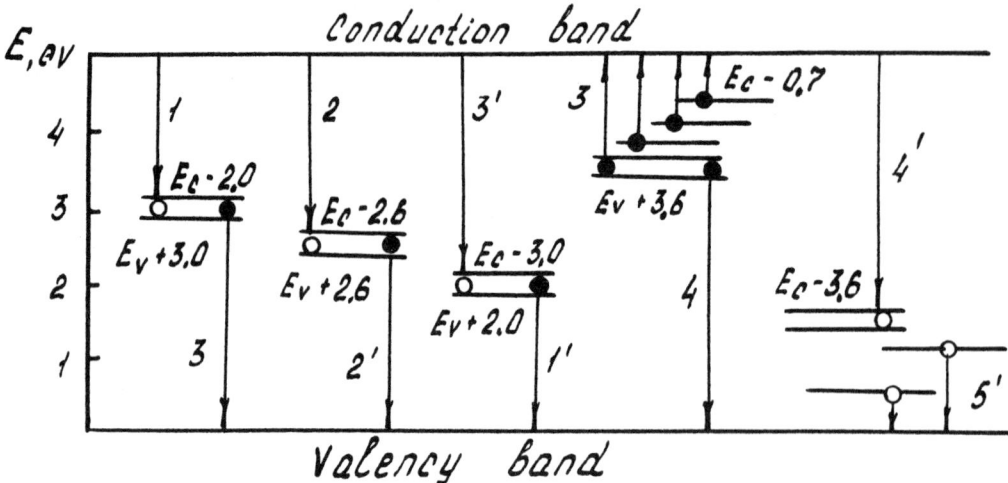

FIGURE 4 The approximate scheme of levels and transitions in n (1-5) and p (1'-5') type PBN [23,24]. Arrows denote the directions of electron transitions at radiation recombination (1-4), (1'-4') and thermally activated (5,5') transitions of electrons to CB and holes to VB. Transitions (1,2,3',4') correspond to the recombination of free electrons with the centres containing localized holes (○-open circles), transitions (1',2',3,4) correspond to the recombination of the holes with the centres containing localized electrons (● - closed circles).

Analysis of semiempirical calculations of the primary vacancy defects [25,26] and of a set of results [17-23] made it possible to classify the observed defects as vacancy complexes V_N - B (nitrogen vacancy bonded with a single boron atom) and V_N - 3B (nitrogen vacancy bonded with three boron atoms); impurity complexes $(V_B$ - 3B) - C; intercalated carbon atoms C_i or $(- C - C)_i$ which are similar to Li intercalates in BN [27] and V_B - 2N (boron vacancy -two nitrogen interstitials) described in [28]; possibly V_B; $(V_B$ - N) and $(V_B$ - 3N); substitutional defects C_N and C_B (carbon on nitrogen or boron sites respectively); V_N - 2B, C_B - V_N and V_N - V_N - C complexes. Matching of the observed energy levels to specific defects has been debated in the literature, but most remain hypothetical and will require further experimental evidence for definitive assignments. There is a tendency that the more perfectly ordered the h-BN structure and the higher the content of r-BN in the boron nitride the more significant the effect of acceptor levels and the greater the contribution of the hole component to the conductivity. In contrast, the greater the V_N concentration, the more donor levels and the higher the electron concentration.

Data from [8] indicate that the dielectric constant, ε, in the direction coinciding with axial texture ('c' lattice direction) ranges from $\varepsilon = 3.8$ to 4.2 (FIGURE 5) depending on the specific material and 'b' parameter in Eqn (1), but it is practically independent of the frequency up to 3×10^{10} Hz [29]. In the azimuthal direction (perpendicular to the 'c' lattice direction) $\varepsilon_a = 4.2 - 4.8$. The dielectric losses from all types of h-BN species have small $\tan \delta \leq 10^{-3}$. The frequency dependence of relaxation loss is also weak, but there is a maximum relaxation loss at 10^6 - 10^9 Hz. The materials based on h-BN are unique high temperature insulators with energy loss factor $\varepsilon \tan \delta$ in fact independent of the frequency and temperature to 800 K (FIGURE 5). Lag in the growth of $\varepsilon(T)$ with respect to $\tan \delta(T)$ indicates that conductivity losses dominate relaxation ones.

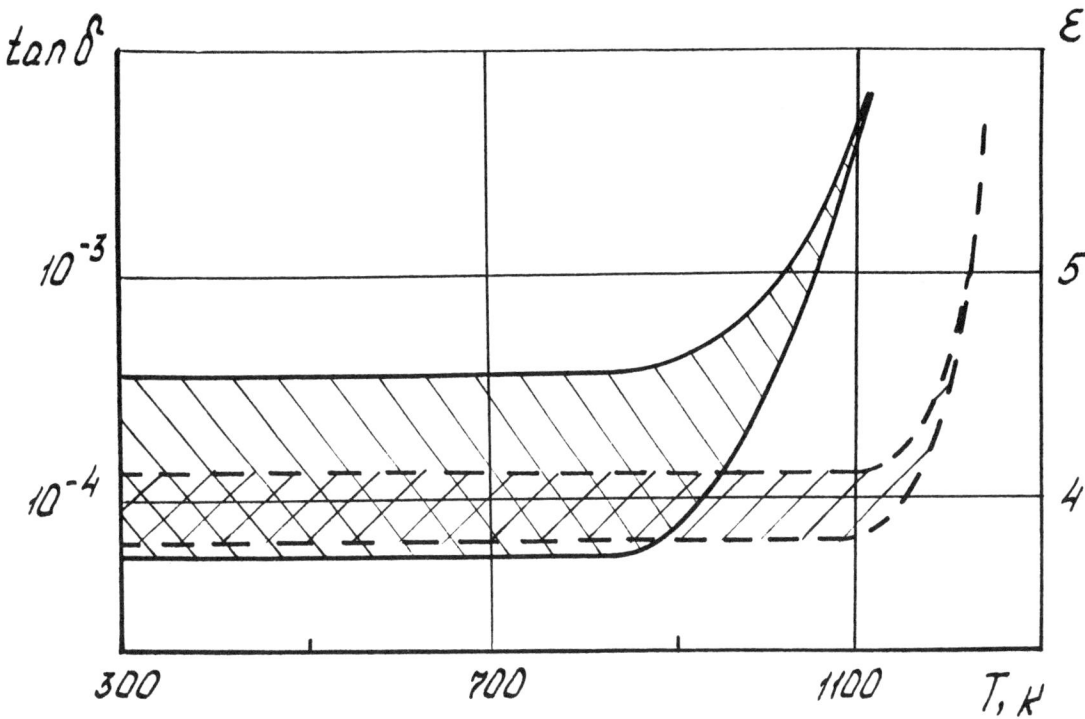

FIGURE 5 Temperature dependences of the most probable values of dielectric loss (tanδ) and permittivity
(ε) (shaded-broken lines) of PBN_h measured at frequencies of 1 to 50 kHz.

The electrical strength of h-PBN is significantly higher than that of other materials. At room
temperature its DC value is 220 kV mm^{-1}. Its AC (50 Hz) value is 100 kV mm^{-1}, and is also
independent of the temperature to T = 800 K (FIGURE 6).

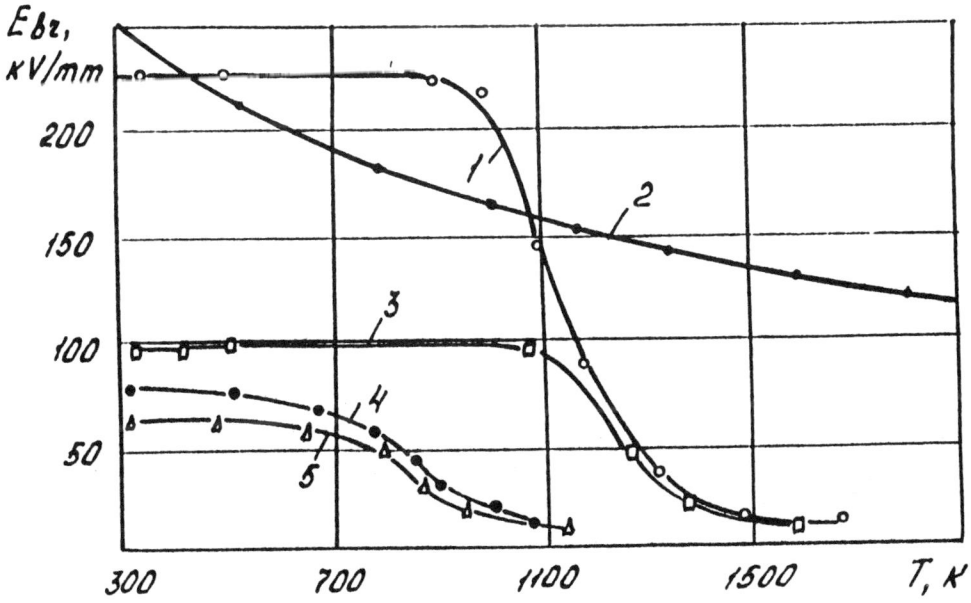

FIGURE 6 Temperature dependence of electrical strength: 1,2,3 - PBN; 4 - Si_3N_4; 5 - AlN; 1,4,5 - direct
current, 3 - alternating current at 50 Hz, 2 - damping pulse with a period of oscillations of 500 ns,
damping factor is 2.

D   ELECTRICAL PROPERTIES OF r-BN

Analysis of the properties of rhombohedral boron nitride (r-BN) suggests its promising application as a highly thermally conductive insulator, a high temperature luminophor, a radiation detector, and a pyroelement. Sublimation growth produced r-BN fibres [30], whereas vapour-phase deposition yields compacts of r-BN and h-BN compositions [1,31] where r-BN content can be controlled to 90%.

In spite of significant differences in their crystal structures, the electronic structures of r-BN and h-BN are very similar. For r-BN, anisotropy in the charge distribution is somewhat higher and the bandgap is slightly larger [32]. The bandgap of r-BN was calculated to be 4.8 eV in comparison to 4.65 eV for h-BN as calculated by the same technique [32]. The noncentrosymmetrical crystalline lattice causes a spontaneous polarization, i.e. r-BN has pyroelectrical properties up to at least 650 K (at higher temperatures, the TSC currents in r-PBN prevented measurement of the pyroelectric static coefficient γ [33] (FIGURE 2)). Pyroelectric currents prevented a correct measurement of σ at T < 500 K; however, in this range its value is not smaller than $10^{-15} (\Omega \, cm)^{-1}$. In the range of intrinsic conductivity there is an exponential increase in σ with activation energy of 2.7 eV and at T = 1700 K, $\sigma = 10^{-5} (\Omega \, cm)^{-1}$.

FIGURE 7 shows that the anisotropy of σ, tan δ, and especially ϵ is even more pronounced in r-PBN than in h-PBN. The electric strength of r-PBN and its temperature dependence are approximately the same as for h-PBN. Therefore, r-BN is also a high temperature insulator.

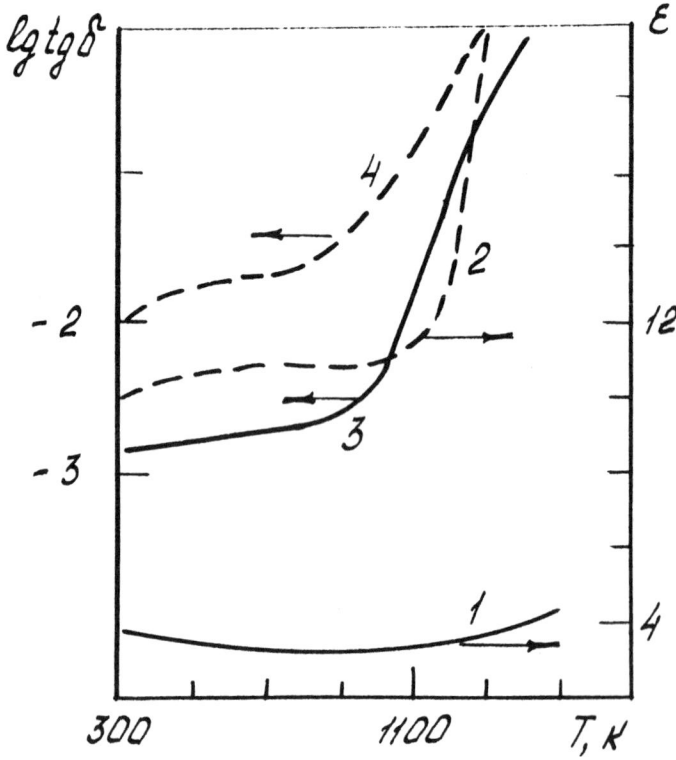

FIGURE 7 Temperature dependence of permittivity ϵ (1,2) and dielectric loss tanδ (3,4) for PBN$_r$ in c- (1,3) and a-lattice directions (2,4).

By means of fractional annealing of r-PBN, from TSL, the energy levels induced by defects in the bandgap were determined [33]. There are two groups of traps with activation energies ranging from 0.2 to 0.5 eV. The second group of traps resulted from the acceptor levels which give rise to more intense r-PBN luminescence than in h-PBN especially at T > 500 K.

E RADIATION EFFECTS ON THE ELECTRICAL PROPERTIES OF GRAPHITIC BN

Irradiation of boron nitride by electrons, γ-quanta, ions and neutrons generates radiation defects which colour the material, and increase its electrical conductivity and dielectrical loss. However, the accumulation of radiation defects does not lead to a dramatic deterioration of its properties. This is due to a high concentration of sinks and to annihilation of the induced defects such as crystallite surfaces, dislocations, and stacking faults. Otherwise, a high dispersion of boron nitride (PBN may be considered as even a nanocrystalline material [34]) provides high resistance to radiation. Heating anneals the induced defects and restores the absorption spectra and dielectrical properties at moderate absorbed doses.

The most intense defect formation takes place with irradiation by ions and neutrons which at doses greater than 10^{20} cm^{-3} can cause irreversible changes of the properties. The results of $\varepsilon(T)$, $\tan\delta(T)$ and $\sigma(T)$ measured on PBN as a function of neutron fluences of various spectra were reported in [35]. Increase in electrical conductivity and dielectrical loss was already observed at 10^{16} n cm^{-2}. The induced defects at fluences up to 10^{19} n cm^{-2} as well as postradiation defects generated by electrons and γ-quanta are fully annealed at $T \leq 1700$ K. Dielectrical properties are herein restored. An irreversible deterioration of dielectric properties starts only at fluences higher than 10^{20} n cm^{-2} and with domination of thermal neutrons in the spectra. This results from a high concentration of V_B induced because of the large cross-section of the threshold free reaction $(^{10}\text{B} + \text{n}) \rightarrow {}^{7}\text{Li} + {}^{4}_{2}\text{He}$. Using EPR and electron-positron annihilation the neutron-generated defect was identified as a boron vacancy which traps unpaired electrons delocalized at two equivalent boron atoms $(V_B\text{-}2N)$ [28]. The annealing temperature of these defects is again 1400 K.

Dielectrical and optical properties of the materials based on h-BN and r-BN and their composites may be modified by ion implantation and post-implantation treatment. Thus ion-thermal treatment (implantation of Li, B, C ions) allows us to create thermostable (to 1500 K) resistive surface coatings with controlling conductivity in the range 10^{-15} to 10^{-3} (Ω cm)$^{-1}$ [36,37]. The modified coatings have electronic conductivity [38]. Such a treatment gives rise to inverse conductivity in the modified part of the material. Thus there is hope of producing p-n structures in high thermo-conductive substrates of h-BN and r-BN directly.

F ELECTRICAL PROPERTIES OF c-BN

Shortly after first synthesizing cubic boron nitride (c-BN) Wentorf [39] demonstrated that its conductivity could be altered by the addition of impurities. The addition of 0.01 to 1.0 weight % beryllium produced blue, p-type c-BN with typical conductivities of 10^{-3} (Ω cm)$^{-1}$ and a maximum of 5×10^{-3} (Ω cm)$^{-1}$ at room temperature. Adding sulphur produced pale yellow, n-type c-BN with conductivities in the 10^{-5} to 10^{-7} (Ω cm)$^{-1}$ range. Similar results were reported by Mishima et al [40], who were able to produce a p-n junction diode by growing a Si-doped crystal on a Be-doped seed crystal.

The only electrical transport properties which have been reported for single crystal c-BN are those of Bam et al [41]. Hall effect measurements were taken on undoped c-BN over the temperature range of 500 to 900 K, and an increasing electron mobility of 0.2 to 4 cm^2 V^{-1} s^{-1}

was measured. Carrier activation energies varied over the range 0.09 to 0.17 eV for single crystals, and 0.23 to 0.7 eV for polycrystalline c-BN.

Taniguchi et al [42] studied Be-doped polycrystalline c-BN and concluded that the activation energy was a function of the amount of Be incorporated. The activation energy decreased with increasing Be concentration from 1.0 eV at 800 ppm to 0.25 - 0.35 eV at 1700 - 4500 ppm. The absolute resistivity of polycrystalline c-BN was ten times larger than that of single crystal c-BN [42].

Shishonok and Shipilo [43] observed that the electrical conductivity of undoped polycrystalline c-BN was a sensitive function of post-synthesis annealing. Initially after synthesis, the conductivity was $10^{-10} (\Omega\,cm)^{-1}$, but by annealing at 870 K, the conductivity was increased by almost four orders of magnitude. This change was attributed to an increase in vacancies and acceptor levels.

A summary of the results found on the electrical properties of c-BN by these various groups is given in TABLE 2.

TABLE 2 Electrical properties of c-BN.

Conductivity $(\Omega\,cm)^{-1}$	Dopant	Conductivity type	Activation energy (eV)	Structure	Ref
$1 - 5 \times 10^{-3}$	Be	p	0.19 - 0.23	single crystal	[39]
$1 - 10 \times 10^{-4}$	S	n	0.05	single crystal	[39]
$10^{-7} - 10^{-5}$	C	n	0.28 - 0.41	single crystal	[39]
$10^{-2} - 1$	Be	p	0.23	single crystal	[40]
$10^{-3} - 10^{-1}$	Si	n	0.24	single crystal	[40]
$10^{-10}*$	none	p	0.36	polycrystalline	[43]
$10^{-8}\wedge$	Be	p	see text	polycrystalline	[42]

* Initial value. Optimal annealing increased the conductivity by several orders of magnitude.
^ With approximately 500 ppm Be.

REFERENCES

[1]   B.N. Sharupin, A.E. Kravchik, M.M. Efremenko, R. Mametev, E.V. Tupitsina, A.S. Osmanov [*J. Appl. Chem. USSR (USA)* vol.63 (1990) p.1569-72]

[2]   M. Kagamida, H. Kanda, M. Akaishi, A. Nukui, T. Osawa, S. Yamaoka [*J. Cryst. Growth (Netherlands)* vol.94 (1989) p.261-9]

[3]   V.V. Lopatin, V.S. Dedkov, Yu.F. Ivanov, A.V. Kabyshev [in *Boron Nitride: Production, Properties, Applications* (Obninsk, Russia 1994) in press]

[4] K.P. Arefiev, V.V. Lopatin, Yu.P. Surov [*Phys. Status Solidi A (Germany)* vol.98 (1986) p.K27-32]

[5] V.V. Lopatin [*Sov. Phys.-Solid State (USA)* vol.33 (1991) p.1097-9]

[6] V.V. Lopatin, A.V. Kabyshev [*Inorg. Mater. (USA)* vol.25 (1989) p.972-5]

[7] V.S. Dedkov, Yu.F. Ivanov, V.V. Lopatin, B.N. Sharupin [*Crystallogr. Rep. (USA)* vol.38 (1993) p.248-51]

[8] A.V. Butenko, V.V. Lopatin, V.P. Chernenko [*Inorg. Mater. (USA)* vol.20 (1984) p.1428-32]

[9] D.N. Poluboyarinov, N.V. Shishkov, I.G. Kuznetsova [*Inorg. Mater. (USA)* vol.3 (1967) p.1594-9]

[10] D.M. Hoffman, G.L. Doll, P.C. Eklund [*Phys. Rev. B (USA)* vol.30 (1984) p.6051-6]

[11] L.G. Carpenter, P.J. Kirby [*J. Phys. D (UK)* vol.15 (1982) p.1143-51]

[12] A.I. Lukomskij, V.B. Shipilo, L.M. Gameza [*J. Appl. Spectrosc. (USA)* vol.34 (1993) p.607]

[13] S. Larach, R.E. Shrader [*Phys. Rev. B (USA)* vol.104 (1956) p.68-73]

[14] V.A. Krasnoperov, N.V. Vekshina, M.B. Husidman, V.S. Neshpor [*Zh. Prikl. Spektrosk. (USSR)* vol.11 (1969) p. 299-302]

[15] U.D. Djuzeev, P.E. Pamazanov [*Sov. Phys. J. (USA)* vol.12 (1969) p.890-3]

[16] A.W. Moore, L.S. Singer [*J. Phys. Chem. Solids (UK)* vol.23 (1972) p.343-56]

[17] M.B. Husidman [*Fiz. Tverd. Tela (USSR)* no.11 (1972) p. 3287-9]

[18] A. Katzir, J.T. Suss, A. Halperin [*Phys. Lett. A (Netherlands)* vol.41 (1972) p.117-8]

[19] A. Katzir, J.T. Suss, A. Zunger, A. Halperin [*Phys. Rev. B (USA)* vol.11 (1975) p.2370-7]

[20] E.Y. Andrei, A. Katzir, J.T. Suss [*Phys. Rev. B (USA)* vol.13 (1976) p.2831-4]

[21] M.B. Khusidman, V.S. Neshpor, L.I. Feldgun [*Inorg. Mater. (USA)* vol.22 (1986) p.611-3]

[22] Yu.I. Galanov, F.V. Konusov, V.V. Lopatin [*Cryst. Res. Technol. (Germany)* vol.25 (1990) p.1343-6]; Yu.I. Galanov, F.V. Konusov, V.V. Lopatin [*Sov. Phys. J. (USA)* vol.32 (1989) p.926-9]

[23] V.V. Lopatin, F.V. Konusov [*J. Phys. Chem. Solids (UK)* vol.53 (1992) p.847-54]

[24] S.N. Grinyaev, F.V. Konusov, V.V. Lopatin, Yu.P. Surov [in *Boron Nitride: Production, Properties, Applications* (Obninsk, Russia 1994) in press]

[25] A.M. Dobrotvorskij, R.A. Evarestov [*Phys. Status Solidi B (Germany)* vol.66 (1974) p.83-91]

[26] R.A. Evarestov, E.A. Kotomin, A.N. Ermoshkin [*Models of Point Defects in Solids* (Zinatne, Riga, USSR, 1983)]

[27] B.L. Faifel, L.A. Gribov, A.O. Dmitrienko A.F. Bol'shakov [*Sov. Phys.-Crystallogr. (USA)* vol.31 (1986) p.497-500]

[28] A.V. Kabyshev, V.M. Kezkalo, V.V. Lopatin, L.V. Kerikov, Yu.P. Surov, L.N. Shiyan [*Phys. Status Solidi A (Germany)* vol.126 (1991) p.K19-K23]

[29] M.D. Bershadskaya, V.G. Avetikov, B.N. Sharupin [*Elektronnaya Tehknika. Ser. Materiali (USSR)* no.6 (1978) p. 60-6]

[30] T. Ishii, T. Sato, Y. Sekikawa, M. Iwata [*J. Cryst. Growth (Netherlands)* vol.52 (1981) p.285-9]

[31] B.N. Sharupin [in *Chemical CVD Refractory Inorganic Materials* Ed. V.S. Shpak (Leningrad, USSR, 1976) p.66-101]

[32] S.N. Grinyaev, V.V. Lopatin [*Sov. Phys. J. (USA)* vol.35 (1992) p.122-6]

[33]   O.I. Buzhinskij, V.V. Lopatin, B.N. Sharupin [*J. Nucl. Mater. (Netherlands)* vol.196-198 (1992) p.1118-20]

[34]   Yu.F. Ivanov, V.V. Lopatin, V.S. Dedkov [*Izv.VUZov, Fizika (Russia)* no.1 (1994) p.107-13]

[35]   A.V. Kabyshev, V.V. Lopatin, Yu.P .Surov [*Izv. Akad. Nauk SSSR Neorg. Mater. (Russia)* vol.27 (1992) p.1974-9]

[36]   V.V. Lopatin, A.V. Kabyshev, L.S. Bushnev [*Phys. Status Solidi A (Germany)* vol.116 (1989) p.K69-K72]

[37]   O.I. Buzhinskij, I.V. Opimakh, A.V. Kabyshev, V.V. Lopatin [*J. Nucl. Mater. (Netherlands)* vol.173 (1990) p.179-84]

[38]   A.V. Kabyshev, V.V. Lopatin [*Poverkhn. Fiz. Khim. Mekh. (Russia)* no.7 (1994) in press]

[39]   R.H. Wentorf [*J. Chem. Phys. (USA)* vol.36 (1962) p.1990-1]

[40]   O. Mishima, K. Era, J. Tanaka, S. Yamaoka [*Appl. Phys. Lett. (USA)* vol.53 (1988) p.962-4]

[41]   I.S. Bam, V.M. Davidenko, V.G. Sidorov, L.I. Fel'gdun, M.D. Shangalov, Y.K. Shalabutov [*Sov. Phys.-Semicond. (USA)* vol.10 (1976) p.331-2]

[42]   T. Taniguchi, J. Tanaka, O. Mishima, T. Ohsawa, S. Yamaoka [*Appl. Phys. Lett. (USA)* vol.62 (1993) p.576-8]

[43]   N.A. Shishonok, V.B. Shipilo [*Sov. Powder Metall. Met. Ceram. (USA)* vol.38 (1992) p.650-3]

3.2 Electrical transport properties of AlN, GaN and AlGaN

D.K. Gaskill, L.B. Rowland and K. Doverspike

May 1994

A INTRODUCTION

In the last few years, significant progress has been achieved in growing high quality GaN and AlGaN layers. Not only have films with smooth morphology and good crystallinity been grown on sapphire, but intentionally n- and p-type doped layers and structures have been synthesized. This progress has resulted in the successful fabrication of several device types, most notably the light emitting diode [1]. Yet, given these accomplishments, much is still not known about this semiconductor system, such as the effect of compensation on transport properties and under what conditions a native defect may exist and dominate the transport properties. Hence, it is timely to review the transport properties of these semiconductors.

This Datareview is organized into the following sections: a brief summary of Hall effect measurements and analysis; transport studies of wurtzite GaN prior to 1985; transport studies of n- and p-type wurtzite GaN after 1985; theoretical models for the transport properties of wurtzite GaN; experimental transport properties of AlN and AlGaN alloys; magnetotransport results; transport studies of zinc blende GaN; conclusion.

B HALL EFFECT MEASUREMENTS AND ANALYSIS

In laboratory practice, the Hall effect is the simplest and most widely utilized measurement of transport properties and is often used to assess the quality of epitaxial layers. The Hall coefficient, R_H, and the resistivity, ρ, are experimentally determined and then related to the electrical parameters of the semiconductor through

$$R_H = r_H/en \qquad (1a)$$

$$\mu_H = R_H/\rho \qquad (1b)$$

where n is the free carrier concentration, e is the unit of electrical charge, μ_H is the Hall mobility, and r_H is the Hall scattering factor [2]. The drift mobility, μ, is the average velocity per unit electric field in the limit of zero electric field and is related to the Hall mobility through the Hall scattering factor by

$$\mu_H = r_H\mu \qquad (2)$$

The Hall scattering factor is dependent on the energy-weighted averages of the free carrier scattering time using Matthiessen's rule and is, in general, a function of magnetic field, B, and temperature. One of the simplest methods for evaluating r_H is to perform the Hall measurement at a large magnetic field where $R_H = 1/en$ [3]. Typically, the Hall scattering factor is of order unity in III-V semiconductors, e.g. for GaAs the factor is about 1.15 at

300 K for a 0.4 T field [3]. The work quoted in this Datareview assumes the Hall scattering factor to be unity unless otherwise noted.

Many excellent reviews exist on analyzing Hall effect data [4] and Anderson and Apsley [2] and Leitch [5] have written short reviews on using the Hall effect to assess III-V semiconductors. Both Anderson and Apsley and Rode also review the various scattering mechanisms that affect transport measurements [2,4]. But, as the reader will presently see, much of the low temperature GaN transport data available to date is dominated by impurity band conduction, a general discussion of which is reviewed in detail by Mott and Twose [6]. Thus, only limited information will be presented on transport scattering mechanisms in this Datareview.

The van der Pauw clover leaf geometry is frequently used for a Hall measurement because the result is then relatively insensitive to the placement of contacts [7]. Errors in measurement that can occur using other geometries are discussed by Wieder [8]. The effects of sample inhomogeneity can also result in error and have recently been investigated by Koon and Knickerbocker [9]. Most of the results quoted in this Datareview are based on van der Pauw techniques, although geometry was often unspecified. Indium or In-based alloys were mainly used to form ohmic contacts to GaN and AlGaN alloys for the results reported herein. Other metallizations have been reported, as described in Datareview 10.2, yet the use of such metallizations for Hall measurements is not widespread at this time.

Most workers have obtained Hall data with magnetic fields, when specified, ranging from about 0.2 to 0.6 T. Magnetoresistance effects should not be important in this range since $\mu B \ll 1$ for reported values of mobility [8,10].

The effect of space-charge depletion near the surface or the substrate interface of the sample is not known quantitatively at this time. A recent report by Binari et al [11] indicated that the surface Fermi level of GaN is not pinned. This may mean that depletion effects, if present, would depend on sample surface preparation.

C   TRANSPORT STUDIES OF WURTZITE GaN PRIOR TO 1985

Until recently, the best reported transport data on GaN were summarized in the classic paper by Ilegems and Montgomery [12]. Yet, most of the conclusions drawn from analyzing that unintentionally doped n-type material are still true today! Ilegems and Montgomery grew GaN layers on (0001) sapphire substrates by halide vapour phase epitaxy (VPE). Note that the films grown on sapphire substrates are lattice mismatched by about 13 % (as measured along a in the (0001) plane) [13]. VPE layers that were very thick, $>100\,\mu m$, had improved transport properties with electron carrier concentrations in the $10^{17}\,cm^{-3}$ range as compared to thinner layers which had concentrations in the $10^{19}\,cm^{-3}$ range. FIGURE 1 shows the 300 K mobility vs. free electron concentration for several samples. The best mobility (electron concentration) at 298 and 77 K was $440\,cm^2\,V^{-1}\,s^{-1}$ ($1.4 \times 10^{17}\,cm^{-3}$) and about $410\,cm^2\,V^{-1}\,s^{-1}$ (about $8 \times 10^{16}\,cm^{-3}$). Superimposed on the figure are calculations by Rode that include the effects of compensation on mobility [14], which will be discussed in Section E.

Ilegems and Montgomery also performed variable temperature Hall measurements on the best sample and found that the mobility peaked near 150 K with a value of about 700 cm² V⁻¹ s⁻¹. The mobility varied as T⁻² for temperatures, T, in the 300 - 900 K range. The onset of impurity conduction occurred near 150 K and samples with higher carrier concentration displayed higher onset temperatures. Samples with electron concentrations above 4 - 8 x 10¹⁸ cm⁻³ showed metallic impurity conduction behaviour. This behaviour is consistent with the criteria of Mott and Twose [6] where the impurity concentration for metallic conduction occurs when the ratio of the mean distance between donors, R_D, and the hydrogenic radius for a donor

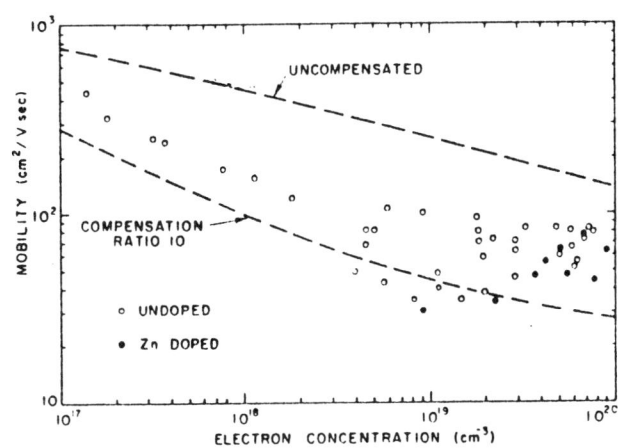

FIGURE 1 Dependence of Hall mobility on 300 K electron concentration for VPE grown GaN. Dashed lines are calculated by [14] and the compensation ratio is defined as $(N_D + N_A)/n$ in this case.

electron, R, is about 1.4 to 1.7, where $R_D = (3/4\pi N_d)^{1/3}$ and N_d is the donor density, and $R = \varepsilon_0 a_B / m^*_e$ where ε_0 is the low-frequency dielectric constant, a_B is the Bohr radius, and $m^*_e = 0.218 \, m_e$ is the effective mass of the electrons. Fits to the carrier concentration above about 150 K yielded a donor level activation energy of 36 meV assuming negligible acceptor compensation or about 14 meV assuming large acceptor compensation.

Native defects were thought to dominate the transport properties for the samples with free electron concentrations greater than 10¹⁹ cm⁻³ [12]. This conclusion was based on two reasons. First, differences in carrier concentration over two orders of magnitude were found from several films grown in a single run, with thinner layers exhibiting the larger carrier concentrations. This suggested that chemical impurities were not responsible for the carrier concentration behaviour and that higher growth rates resulted in improved transport properties. Second, secondary ion mass spectroscopy (SIMS) analysis of layers from the same run, again with carrier concentrations differing by two orders of magnitude, did not show any trends when analyzed for elemental impurities such as C, O and Mg.

Attempts made by various groups to Zn-dope VPE grown GaN p-type were unsuccessful and only the resistivity was affected. The effect of growth temperature on Zn doping was found to be important; at low growth temperatures the layers were highly resistive or insulating [15-17] and resistivities in excess of 10⁹ Ω cm have been reported [16], but at higher growth temperatures no significant reduction of electron carrier concentration was observed [12]. Recently, highly resistive Zn-doped layers have also been synthesized by organometallic vapour phase epitaxy (OMVPE) [18]. The connection between Zn incorporation into a film and the resulting high resistivity is not fully understood at this time.

D   TRANSPORT STUDIES OF WURTZITE GaN AFTER 1985

The transport results of the VPE grown material summarized the state-of-the-art until recently, when the advent of modern epitaxial techniques such as molecular beam epitaxy (MBE) [19] and OMVPE [20], coupled with new methods of minimizing the effects of large lattice mismatch using buffer layers, resulted in significant improvement in transport properties. For the case of OMVPE, GaN has been grown on AlN [21] and GaN [22] buffer layers formed at reduced temperatures and with thicknesses ranging from 200 to 600 Å. The improvements found using either growth method also resulted in the benefits of reducing the growth rates and thicknesses by about one and two orders of magnitude, respectively, as compared to the VPE studies.

D1   n-Type GaN

FIGURE 2 shows the best Hall mobility vs. free electron concentration at 300 K for n-type GaN as reported by various groups [22-31]. The samples were grown on sapphire substrates by OMVPE, using buffer layers grown at 450 to 600 °C, and MBE. The data in FIGURE 2 are very similar to those in FIGURE 1, but close examination reveals two differences. With respect to the data in FIGURE 1, the current growth technology has, in general, resulted in a slight increase in the mobility throughout the carrier concentration range and the range of accessible carrier concentration has broadened. The best mobility and carrier concentration at 300 K were 900 cm^{2}V^{-1}s^{-1} and about 3×10^{16} cm^{-3}, respectively [30]. Note that the lowest carrier concentration and highest mobility data are, in general, from films 3 to 5 μm thick; thicker epilayers often resulted in deleterious cracks.

Several observations can be made from the data shown in FIGURE 2. First, the data from many groups all fall generally in the same pattern, where the spread in values is probably related to insufficient growth optimization and slight differences in Hall measurement technique. For that reason, it is more fruitful to concentrate on the overall trends shown by the data. The most clearly visible trend is that the mobility shows no signs of levelling, down to the lowest carrier concentration reported. (A recent report gives a mobility of about 1700 cm^{2}V^{-1}s^{-1} at 1×10^{15} cm^{-3}, measured at an unspecified temperature [32], which may suggest that only slight levelling of the mobility is occurring at low carrier concentrations.) Second, the dependence of the MBE and OMVPE mobility data on carrier concentration is, for the most part, similar. This implies that the data are reflecting transport properties inherent to GaN and not to the method of synthesis. Third, the data for unintentionally doped samples (open symbols) are very similar to the data for the intentionally doped samples (closed symbols) in the region where the results overlap, approximately 10^{17} to 10^{19} cm^{-3}. This implies that the magnitude of compensation for doped and unintentionally doped samples is approximately the same.

FIGURE 3 shows the electron concentration and mobility as a function of temperature for a high purity sample [30]. The mobility has a peak value of about 3000 cm^{2}V^{-1}s^{-1} with a carrier concentration of about 1.2×10^{16} cm^{-3} at 70 K. Using these data, the conductivity of the sample was evaluated as a function of 1000/T and the onset of impurity band conduction was found to begin near 40-50 K. The temperature dependence of mobility between 70 and 50 K was approximately T$^{3\pm1}$. Thus, the low temperature mobility may be dominated by impurity scattering, which should behave approximately as $T^{1.5}$ [2]. Between 70 and 300 K the mobility varied roughly as T^{-1}, which was attributed to polar phonon scattering. (Polar phonon scattering behaves approximately as $T^{1/2}[\exp(\theta/T) - 1]$ where θ is the LO phonon temperature [2].)

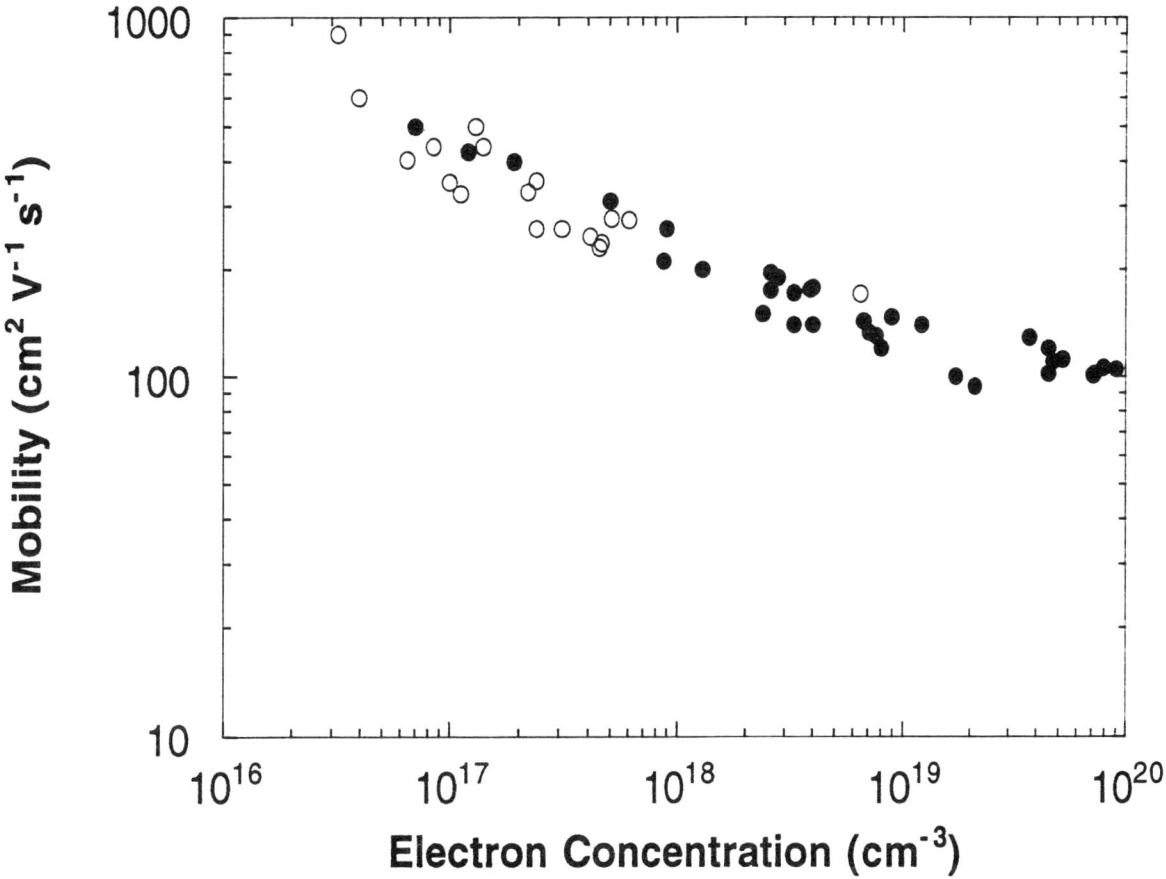

FIGURE 2 The 300 K Hall mobility vs. free electron concentration for GaN from various groups using both OMVPE and MBE [22-31]. The open symbols are from unintentionally doped samples and the closed symbols are from samples doped with either Si or Ge.

FIGURE 4 shows the resistivity of five samples vs. 1000/T [33]. TABLE 1 lists the 300 K free electron concentration, N_0, of the samples. For these samples, which have higher electron concentrations, onsets of impurity band conduction occur at higher temperatures than for the data of FIGURE 3, consistent with the report of Ilegems and Montgomery [12] for the behaviour of samples with higher carrier concentration. Samples with carrier concentrations above 4×10^{18} cm^{-3} showed metallic impurity conduction behaviour, again consistent with Ilegems and Montgomery.

The thermal activation energy of free carriers can be extracted from the data in FIGURES 3 and 4. Fitting the data in FIGURE 3 to the activation energy, E_d, in $N\exp(-E_d/2kT)$, where k is Boltzmann's constant, yields 34 meV between 42 and 100 K and 5 meV between 100 and 300 K [30]. The high temperature activation energy is smaller than the 14 to 36 meV values of Ilegems and Montgomery [12]. Activation energies were also extracted from the data in FIGURE 4 and are listed in TABLE 1. In this case, a two-band model was used to discriminate between the transport of the conduction band and the impurity band. The results of this analysis show that E_d lies in the range 19 to 32 meV [33]. FIGURES 5 and 6 display the temperature dependence of electron concentration and mobility derived from a two-band analysis of Hall data for sample 119. The subscripts meas, c and d refer to the measured (Hall) values, the calculated conduction band values, and the defect (impurity) band values, respectively. Summarizing the recent and previous results, the activation energy of n-type GaN free carriers lies in the range 14 to 36 meV.

Lastly, it should be noted that n-type GaN has been grown on other substrates such as 6H SiC [24] or GaAs [34]. In general, mobilities at a given electron concentration have been less than the best reported for growth on sapphire. This is probably not a fundamental limitation since Morkoc [24] reports a 300 K mobility of 580 $cm^2\,V^{-1}\,s^{-1}$ for a sample with electron concentration $1.8 \times 10^{17}\,cm^{-3}$ grown on 6H SiC.

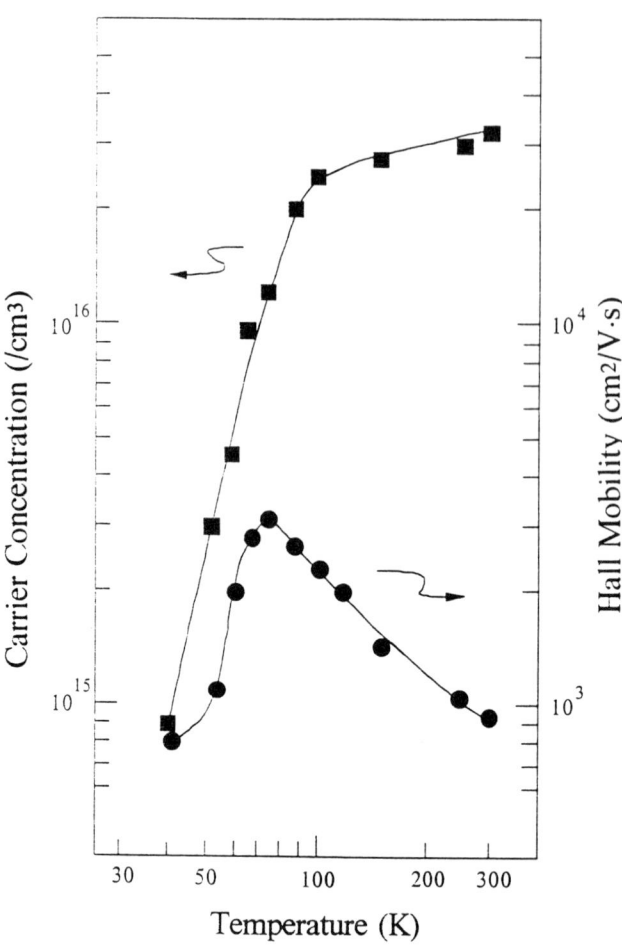

FIGURE 3 The Hall mobility and electron carrier concentration of the best reported GaN film grown with a GaN buffer layer by MOVPE as a function of temperature [30].

D2   p-Type GaN

FIGURE 7 shows the 300 K p-type GaN Hall mobility as a function of hole concentration. The circles are data from Mg-doped layers [35-37]. The Mg dopant required activation by N_2 heat treatment or low-energy electron-beam irradiation. Otherwise the films were resistive. The squares are data from C-doped MBE grown layers, which did not require any activation process [34]. Two observations can be made about the data in FIGURE 7. One, the mobility monotonically increases as the carrier

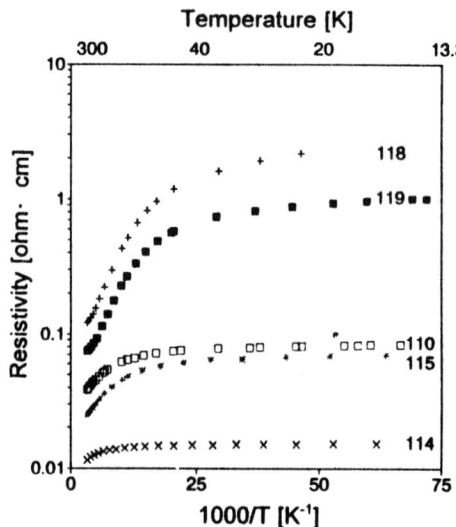

FIGURE 4 The resistivity vs. 1000/T for the samples listed in TABLE 1.

concentration decreases, perhaps levelling at concentrations near mid 10^{16} cm^{-3}. This is consistent with a recent report on an unintentionally doped sample with a 300 K mobility and hole concentration of 150 cm^2 V^{-1} s^{-1} and 3.5×10^{12} cm^{-3}, respectively [32]. Second, the best mobilities have been obtained for C-doped samples where the highest mobility reported was 175 cm^2 V^{-1} s^{-1} for a hole concentration of 3.65×10^{16} cm^{-3} [34]. But, trends in the data for Mg-doped samples suggest that similar values could be achieved for lower hole concentrations.

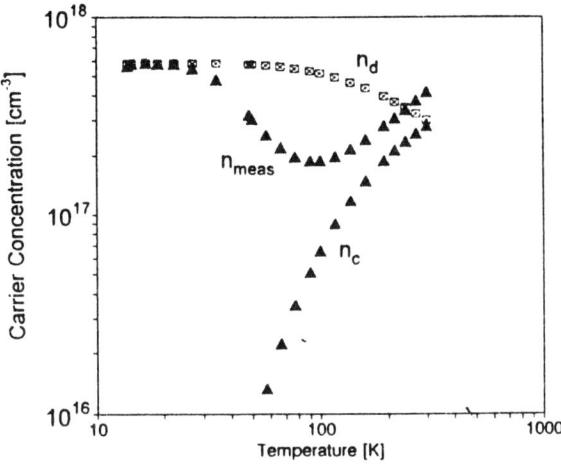

FIGURE 5 Temperature dependence of n_{meas}, n_c and n_d for sample 119 in TABLE 1.

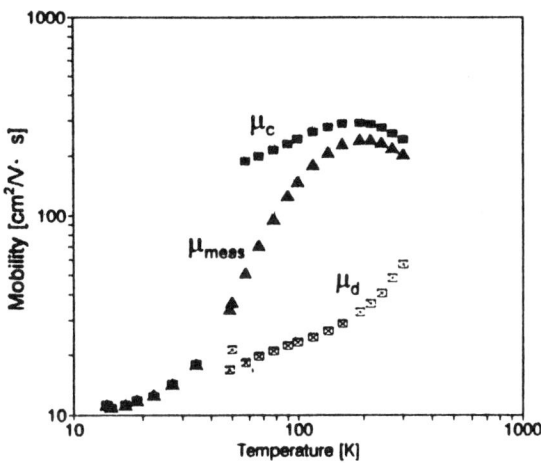

FIGURE 6 Temperature dependence of μ_{meas}, μ_c and μ_d for sample 119 in TABLE 1.

TABLE 1 Values of the 300 K free electron concentration, N_0, and activation energy of donors, E_d, for samples studied in [33] and shown in FIGURES 4, 5, and 6.

Sample	N_0 (cm^{-3})	E_d (meV)
118	4.75×10^{17}	19.09
119	5.80×10^{17}	18.82
110	4.00×10^{18}	32.16
115	5.56×10^{18}	31.31
114	2.19×10^{19}	-

E   THEORETICAL MODELS FOR THE TRANSPORT PROPERTIES OF WURTZITE GaN

The dashed curves of 300 K drift mobility as a function of free electron concentration for n-type GaN in FIGURE 1 are results from theoretical calculations determined from iterative solutions of the Boltzmann equation by Rode [38] assuming $m^*_e = 0.3\,m_e$. FIGURE 8 displays details of this type of calculation for 300 K Hall mobility, where $m^*_e = 0.218\,m_e$, as a function of compensation, defined as the ratio of the acceptor to donor density, N_a/N_d, ranging from 0 to 0.8 in steps of 0.2 [39]. As can be seen by comparing the calculation in FIGURE 8 with the data in FIGURE 2, the curves most consistent with the experimental data have the most

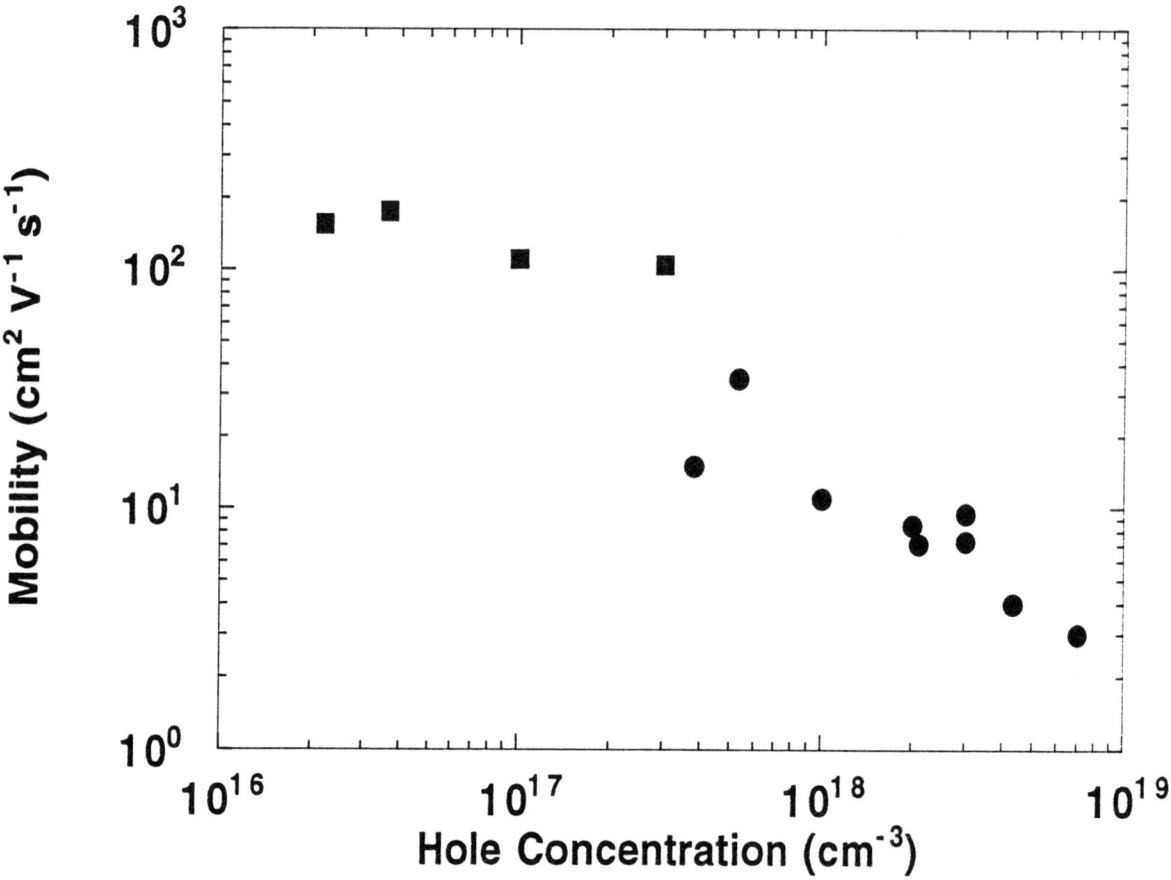

FIGURE 7 The 300 K Hall mobility vs. hole concentration for OMVPE and MBE grown GaN [34-37]. The squares (circles) correspond to C-doped (Mg-doped) samples.

compensation. However, the calculations for drift or Hall mobility are dependent on the input material parameters [40], the accuracy of which may be questioned. For example, the calculations depend on the square of the deformation potential and this quantity is poorly known [41]. Thus, an accurate estimate of compensation of GaN films cannot be made at this time.

Electric-field dependent Monte Carlo simulations of mobility as a function of impurity concentration predicted a peak drift velocity of $2 \times 10^7 \, cm \, s^{-1}$ for a field near $2 \times 10^5 \, V \, cm^{-1}$ [42]. Recently, Monte Carlo based calculations of low and high electric field transport properties of n-type GaN have been performed by Gelmont and co-workers [43,44]. FIGURE 9 shows the behaviour of the low-field mobility as a function of temperature for a free electron concentration of $10^{17} \, cm^{-3}$. The calculations predict a 300 K mobility of about $900 \, cm^2 \, V^{-1} \, s^{-1}$ for uncompensated films. This mobility is similar to that predicted by calculations for lightly compensated material shown in FIGURE 8. FIGURE 10 shows calculations for electron drift velocity as a function of electric field for three temperatures, 400, 500 and 750 K, again for a sample with an electron concentration of $10^{17} \, cm^{-3}$. The peak drift velocity decreases from about $2.5 \times 10^7 \, cm \, s^{-1}$ at 400 K to $2.0 \times 10^7 \, cm \, s^{-1}$ at 750 K for electric fields of about $2 \times 10^5 \, V \, cm^{-1}$. Further, the high-field calculations show the importance of intervalley transitions on electron transport and predict a large negative differential conductance. The Monte Carlo technique has also been used to calculate the temperature dependence of the electron mobility for GaN quantum wells and the predicted mobility was enhanced compared to bulk-like material [45]. Note that all Monte Carlo calculations are dependent upon GaN material parameters that are not well known, and, hence, cannot be used to draw definitive conclusions about compensation.

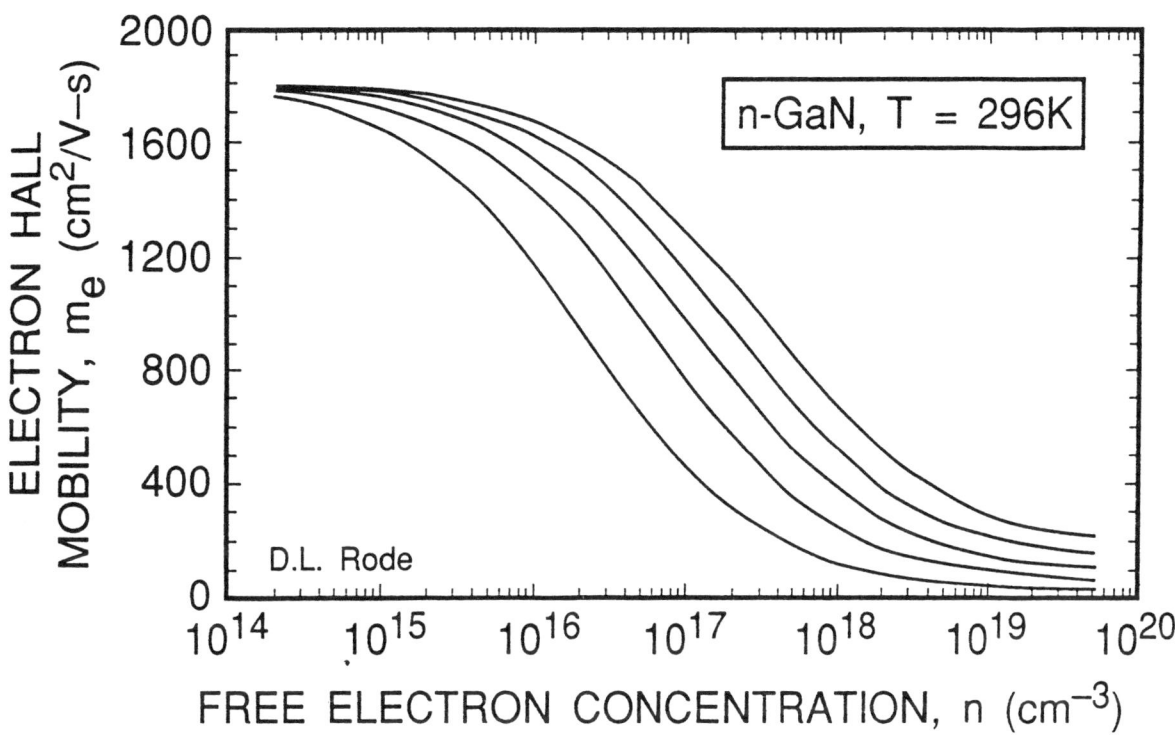

FIGURE 8 The electron concentration dependence of the Hall mobility calculated by iterative solutions of the Boltzmann equation with Fermi statistics. The topmost curve corresponds to a compensation ratio N_a/N_d of 0 with successive curves corresponding to 0.2, 0.4, 0.6, and 0.8, respectively [39].

F EXPERIMENTAL TRANSPORT PROPERTIES OF AlN AND AlGaN ALLOYS

F1 AlN Properties

The electrical properties of AlN are not well known probably because undoped material is highly resistive and intentional doping appears to be difficult. Resistivities in the range $10^7 - 10^{13}\,\Omega\,cm$ have been reported for unintentionally doped single crystals [46-48]. The conductivity exhibited an Arrhenius behaviour for all crystals and the activation energy was reported to be 1.4 eV between 330 and 400 K [47]. Some crystals showed an activation energy of 0.5 eV for temperatures between 300 and 330 K [48]. Unintentionally doped AlN samples grown by a modified physical transport technique by Rutz were n-type and had resistivities as low as $400\,\Omega\,cm$ [49].

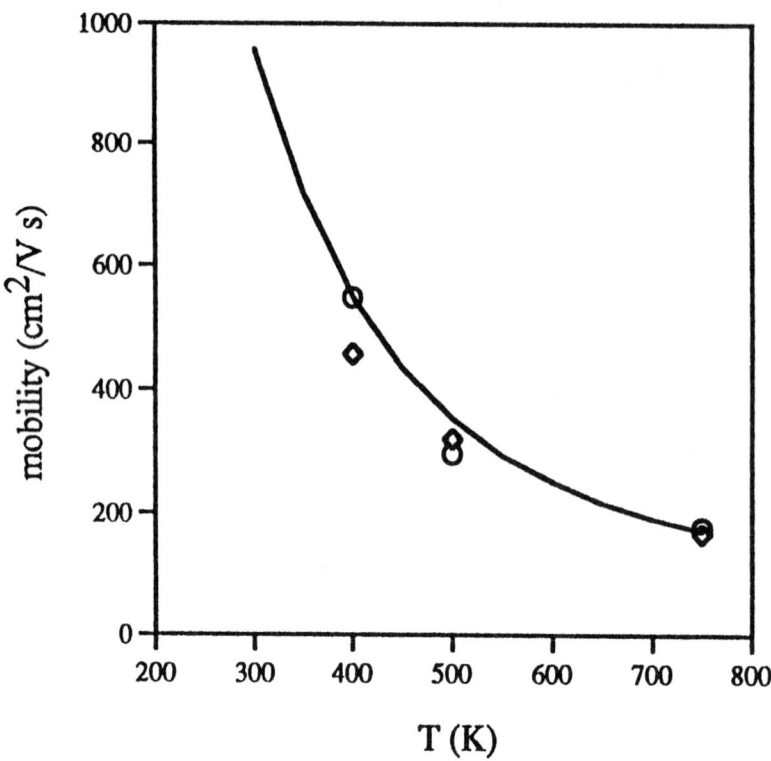

FIGURE 9 The temperature dependence of GaN mobility with an electron concentration of $10^{17}\,cm^{-3}$. Solid line is from the analytical calculation, diamonds are from the slope of the velocity vs. electric field curve extrapolated to zero field (Monte Carlo simulation), and squares are from the diffusion coefficient and the Einstein relation (Monte Carlo simulation).

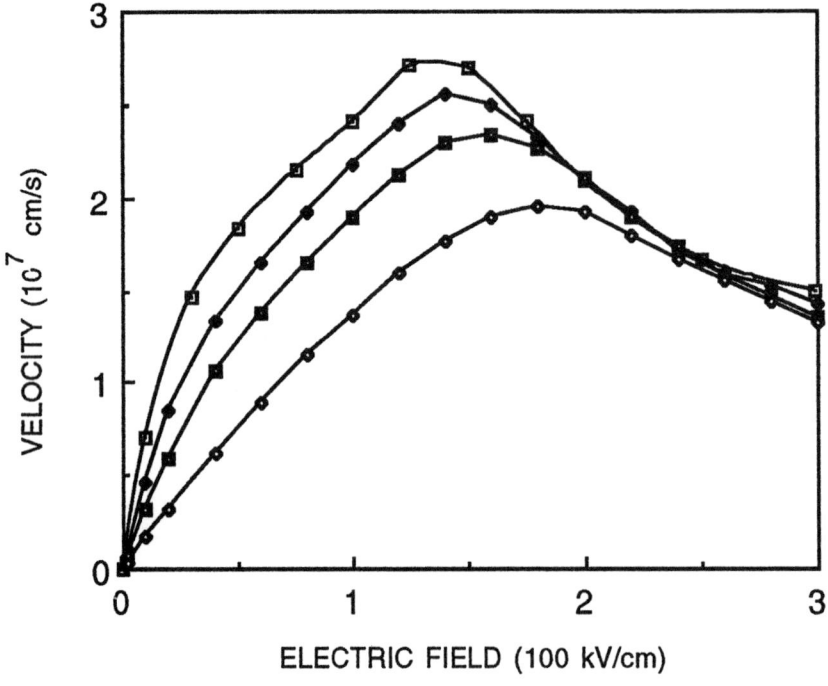

FIGURE 10 The Monte Carlo simulation of the electron drift velocity as a function of electric field in $10^{17}\,cm^{-3}$ n-GaN. The topmost curve is for 300 K and the following successive curves are for 400, 500 and 750 K [45].

Intentional doping of AlN has resulted in both n- [50,51] and p-type [46,50] material. Epitaxial AlN has been doped with Se and Hg resulting in n- and p-type conductivity, respectively [50]. Gorbatov and Kamyshov [51] obtained n-type conductivity of polycrystalline AlN with the incorporation of percent levels of Si. However, the addition of dopants in these studies did not change the resistivities of the material. The only Hall measurements reported for AlN were from material thought to be doped with traces of Al_2OC, giving it a blue colour [46]. These samples were p-type and had resistivities in the range 10^3-$10^5 \, \Omega \, cm$. A hole mobility of $14 \, cm^2 \, V^{-1} \, s^{-1}$ was reported for these samples at 290 K, but the authors did not express complete confidence in this measurement.

F2   AlGaN Properties

A continuous range of $Al_xGa_{1-x}N$ alloys can be synthesized as first demonstrated by Hagen et al [52] and Baranov et al [53]. Several groups discuss electrical transport properties for $Al_xGa_{1-x}N$ [53-55]. Yoshida et al [54] found that the electron concentration for unintentionally doped samples dropped from $10^{20} \, cm^{-3}$ to $10^{17} \, cm^{-3}$ as the AlN mole fraction was increased from x = 0 to x = 0.3, with larger Al concentrations (x ≥ 0.35) resulting in insulating material. Mobilities varied from 10 to $30 \, cm^2 \, V^{-1} \, s^{-1}$ but did not change appreciably with Al mole fraction. On the other hand, both Baranov et al [53] and Khan et al [55] found that the 300 K electron carrier concentration dropped from $10^{20} \, cm^{-3}$ at x = 0 to $10^{17} \, cm^{-3}$ at x = 0.40 [55] or to $10^{19} \, cm^{-3}$ at x = 0.35 [54], while the mobility decreased from ~ $100 \, cm^2 \, V^{-1} \, s^{-1}$ for x = 0 to roughly $10 \, cm^2 \, V^{-1} \, s^{-1}$ for x = 0.35 or 0.40 for both groups.

Recently, two groups have concentrated on compositions near x = 0.10. For $Al_{0.09}Ga_{0.91}N$, 300 K Hall measurements yielded a mobility of $35 \, cm^2 \, V^{-1} \, s^{-1}$ for an electron concentration of $5 \times 10^{18} \, cm^{-3}$ [56]. Successful n-type doping using Si has been reported for $Al_{0.1}Ga_{0.9}N$ [57]. Undoped material had electron carrier concentrations of $1 \times 10^{17} \, cm^{-3}$ and two Si-doped films had free electron concentrations of approximately 6×10^{17} and $2 \times 10^{18} \, cm^{-3}$ with corresponding mobilities (calculated using resistivities provided in the text) of 40 and $180 \, cm^2 \, V^{-1} \, s^{-1}$.

G   MAGNETOTRANSPORT RESULTS

The temperature dependence of the Hall scattering factor for GaN has been calculated using iterative solutions of the Boltzmann equation with Fermi statistics and is shown in FIGURE 11 [38] (also included in FIGURE 11 are the scattering and magnetoresistance factors for GaAs). The scattering factor for hexagonal GaN is actually anisotropic (± 1 %), but the anisotropy is neglected in FIGURE 11 where the average r_H is plotted. The scattering factor is maximum just above 300 K and is about 1.25. Rode points out r_H is maximum at approximately 1/2 the Debye temperature because polar scattering is a rapidly varying function at this temperature. No experimental measurements of r_H for GaN have been reported to date.

A two-dimensional electron gas (2DEG) has been observed in GaN/AlGaN heterostructures by Khan and co-workers [56,58]. Approximately 3000 Å of unintentionally doped GaN was deposited on an AlN buffer layer and capped by 500 Å of unintentionally doped $Al_{0.09}Ga_{0.91}N$. Presumably the 2DEG was formed from electrons donated from the cap which contained

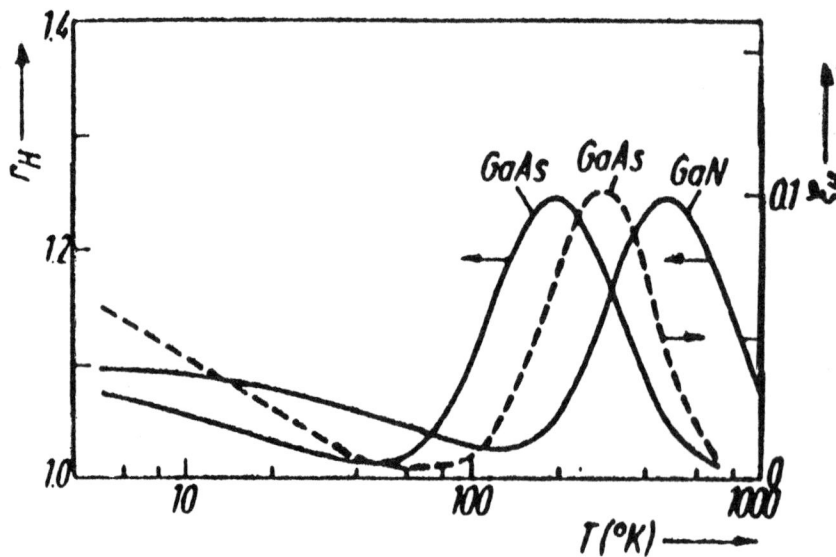

FIGURE 11 The Hall scattering factor, r_H, for GaN as calculated by iterative solutions of the Boltzmann equation with Fermi statistics. Also included are the Hall scattering and magnetoresistance factors for GaAs (from [38]).

$5 \times 10^{18} \, cm^{-3}$ free electrons. The mobility of this structure was enhanced to 620 and $1600 \, cm^2 \, V^{-1} \, s^{-1}$ at 300 and 77 K, respectively. The sheet carrier density was estimated using a parallel conduction model to be $5 \times 10^{12} \, cm^{-2}$. An 18-period multiple heterojunction was also formed and mobilities were further enhanced to 860 and $1980 \, cm^2 \, V^{-1} \, s^{-1}$ at 300 and 77 K, respectively.

H   TRANSPORT STUDIES OF ZINC BLENDE GaN

Some believe that zinc blende GaN may be more amenable to doping since all III-V semiconductors that can be efficiently doped n- or p-type are cubic [59,60]. In addition, calculations suggest that zinc blende GaN may exhibit higher electron drift velocities and reduced phonon scattering since the lattice is isotropic [61]. Zinc blende GaN has been grown on (100) Si [62], (100) GaAs [34,63,64], (100) 3C-SiC [61], and (100) MgO [65] by MBE.

On (100) Si, a single-crystal GaN buffer was needed in order to obtain the zinc blende structure. The resistivity was approximately $170 \, \Omega \, cm$ at 300 K and had an activation energy, approximately 80 meV, that was smaller than that of the wurtzite structure, 110 meV [62]. On (100) MgO, n-type zinc blende GaN was obtained with 300 K resistivities ranging from approximately 10^2 to $10^5 \, \Omega \, cm$ with mobilities $< 10 \, cm^2 \, V^{-1} \, s^{-1}$ [65].

Using (100) GaAs as the substrate, p-type zinc blende GaN has been grown. Van der Pauw Hall measurements show 300 K hole concentrations ranging from $8 \times 10^{16} \, cm^{-3}$ to $8 \times 10^{18} \, cm^{-3}$ with corresponding mobilities ranging from 39 to $3 \, cm^2 \, V^{-1} \, s^{-1}$, respectively [64].

I CONCLUSION

The electrical properties of AlN, GaN and AlGaN, measured mainly by Hall effect, have been reviewed. Most reports have focused on 300 K mobilities and carrier concentrations, and the better values are summarized in FIGURES 2 and 7. Limited data are available at reduced temperatures for n-type GaN and none for p-type GaN or, for that matter, the alloys. Of the low temperature data that are available, impurity band conduction dominates the transport properties and the onset of this conduction mechanism decreases to lower temperatures as the overall purity of GaN is improved. This raises the question of the nature of the impurity conduction band. Two-band conductivity analysis, similar to the work of Molnar et al [33], performed on high mobility material may shed light on the nature of the impurity conduction band and how it may be affected by growth conditions. Variable temperature Hall measurements on high quality n- and p-type material are also desirable and may yield more information on the transport scattering mechanisms.

To date, theoretical calculations have been limited mainly by poor or questionable material parameters which are the foundation of these calculations. As more reliable material parameters become available, comparison of the calculations with transport data from better quality samples will probably yield information on the degree of compensation in GaN. To make these comparisons quantitatively meaningful, calculated and measured values of the Hall scattering factor as a function of temperature are necessary. Also, Hall measurement data at 77 K may be useful for evaluating contributions from impurity band conduction and perhaps for determining the magnitude of compensation, as was found for the case of GaAs several years ago [66].

As in all semiconductor systems, the motivating force is the promise of new or improved device performance. Using AlGaN alloys as important elements of devices, or exploiting the properties of the cubic form of this semiconductor system, awaits further improvements in growth and the subsequent investigations on the properties of these materials. Still, the rate of evolutionary, and sometimes revolutionary, improvements in the growth of this semiconductor system strongly suggests that many of the needs and questions raised in this study will be answered, and a new set will arise.

ACKNOWLEDGMENTS

The authors gratefully acknowledge the data contributions of Dr. Wickenden, Prof. Abernathy, Prof. Morkoç and Prof. Amano. Illuminating discussions with Prof. Rode are also happily acknowledged. The contributions of Prof. Skowronski, Dr. Giordana and Mr. Bailey are greatly appreciated. One of us (LBR) acknowledges a National Research Council/Naval Research Laboratory Cooperative Research Associateship.

REFERENCES

[1]   I. Akasaki, H. Amano, M. Kito, K. Hiramatsu [*J. Lumin. (Netherlands)* vol.48-49 (1991) p.666-70]

[2]   D.A. Anderson, N. Apsley [*Semicond. Sci. Technol. (UK)* vol.1 (1986) p.187-202]

[3]   D.L. Rode, C.M. Wolfe, G. E. Stillman [*J. Appl. Phys. (USA)* vol.54 (1983) p.10-3]

[4]   See for example D.L. Rode [*Semicond. Semimet. (USA)* vol.10 (1975) p.1-89]

[5]   A.W.R. Leitch. [*J. Appl. Phys. (USA)* vol.65 (1989) p.2357-60]

[6]   N.F. Mott, W.D. Twose [*Adv. Phys. (UK)* vol.10 (1961) p.107-63]

[7]   L.J. van der Pauw [*Philips Res. Rep. (Netherlands)* vol.13 (1958) p.1-9]

[8]   H. Wieder [*Laboratory Notes on Electrical and Galvano Magnetic Measurements* (Elsevier, 1979)]

[9]   D.W. Koon, C.J. Knickerbocker [*Rev. Sci. Instrum. (USA)* vol.64 (1993) p.1-4]

[10]   R.L. Peterson [*Semicond. Semimet. (USA)* vol.10 (1975) p.221-89]

[11]   S.C. Binari, H.B. Dietrich, G. Kelner, L.B. Rowland, K. Doverspike, D.K. Gaskill [*Electron. Lett. (UK)* in press]

[12]   M. Ilegems, H.C. Montgomery [*J. Phys. Chem. Solids (UK)* vol.34 (1973) p.885-95]

[13]   B.B. Kosicki, D. Kahng [*J. Vac. Sci. Technol. (USA)* vol.6 (1969) p.593-6]

[14]   D.L. Rode [unpublished results (1973)]

[15]   J.I. Pankove, E.A. Miller, J.E. Berkeyheiser [*RCA Rev. (USA)* vol.32 (1971) p.383]

[16]   H.P. Maruska, J.J. Tietjen [*Appl. Phys. Lett. (USA)* vol.15 (1969) p.327-9]

[17]   R.K. Crouch, W.J. Debnam, A.L. Fripp [*J. Mater. Sci. (UK)* vol.13 (1978) p.2358-64]

[18]   H. Amano, K. Hiromatsu, M. Kitto, N. Sawaki, I. Akasaki [*J. Cryst. Growth (Netherlands)* vol.93 (1988) p.79-82]

[19]   S. Yoshida, S. Misawa, S. Gonda [*Appl. Phys. Lett. (USA)* vol.42 (1983) p.427-9]

[20]   M.A. Khan, R.A. Skogman, R.G. Schulze, M. Gershenzon [*Appl. Phys. Lett. (USA)* vol.42 (1983) p.430-2]

[21]   H. Amano, N. Sawaki, I. Akasaki, Y. Toyoda [*Appl. Phys. Lett. (USA)* vol.48 (1986) p.353-5]

[22]   S. Nakamura [*Jpn. J. Appl. Phys. (Japan)* vol.30 (1991) p.L1705-7]

[23]   D.K. Wickenden, W.A. Bryden [*Proc. Int. Conf. on Silicon Carbide and Related Materials*, Washington, DC, 1-3 Nov 1993 (IOP Publishing, Bristol, England) in press]

[24]   H. Morkoç [private communication]

[25]   L.B. Rowland, K. Doverspike, D.K. Gaskill, J.A. Freitas Jr. [*Mater. Res. Soc. Symp. Proc. (USA)* San Francisco, CA, 4-8 April 1994 (Materials Research Society, Pittsburgh, PA, 1994) in press]

[26]   S. Nakamura, T. Mukai, M. Senoh [*J. Appl. Phys. (USA)* vol.71 (1992) p.5543-9]

[27]   S. Nakamura, T. Mukai, M. Senoh [*Jpn. J. Appl. Phys. (Japan)* vol.31 (1992) p.2883-8]

[28]   M.A. Khan, R.A. Skogman, R.G. Schulze, M. Gershenzon [*Appl. Phys. Lett. (USA)* vol.42 (1983) p.430-2]

[29]   M. Asif Khan, J.N. Kuznia, J.M. Van Hove, D.T. Olson, S. Krishnankutty, R.M. Kolbas [*Appl. Phys. Lett. (USA)* vol.58 (1991) p.526-7]

[30]   S. Nakamura, T. Makui, M. Senoh [*J. Appl. Phys. (USA)* vol.71 (1992) p.5543-9]

[31]   I. Akasaki, H. Amano [*Mater. Res. Soc. Symp. Proc. (USA)* vol.242 (1992) p.383-94]

[32]   M. Rubin, N. Newman, J.S. Chan, T.C. Fu, J.T. Ross [*Appl. Phys. Lett. (USA)* vol.64 (1994) p.64-6]

[33]   R.J. Molnar, T. Lei, T.D. Moustakas [*Appl. Phys Lett. (USA)* vol.62 (1993) p.72-4]

[34]   C.R. Abernathy, S.J. Pearton, F.Ren, P.W. Wisk [to be published]

[35]   H. Amano [private communication]

[36] S. Nakamura, M. Senoh, T. Mukai [*Jpn. J. Appl. Phys. (Japan)* vol.30 (1991) p.L1708-11]

[37] B. Goldenberg, J.D. Zook, R. J. Ulmer [*Appl. Phys. Lett. (USA)* vol.62 (1993) p.381-3]

[38] D.L. Rode [*Phys. Status Solidi B (Germany)* vol.55 (1973) p.687-96]

[39] D.L. Rode [unpublished work]

[40] The relevant material parameters can be found in Table I of [4]

[41] D.L. Camphausen, G.A.N. Connell, W. Paul [*Phys. Rev. Lett. (USA)* vol.26 (1971) p.184-8]

[42] M.A. Littlejohn, J.R. Hauser, T.H. Glisson [*Appl. Phys. Lett. (USA)* vol.26 (1975) p.625-7]

[43] B. Gelmont, K. Kim, M. Shur [*J. Appl. Phys. (USA)* vol.74 (1993) p.1818-21]

[44] M. Shur, B. Gelmont, C. Saavedra-Munoz, G. Kelner [*Proc. Int. Conf. on Silicon Carbide and Related Materials*, Washington, DC, 1-3 Nov 1993 (IOP Publishing, Bristol, England) in press]

[45] R.P. Joshi [*Appl. Phys. Lett. (USA)* vol.64 (1994) p.223-5]

[46] J. Edwards, K. Kawabe, G. Stevens, R.H. Tredgold [*Solid State Commun. (USA)* vol.3 (1965) p.99-100]

[47] G.A. Cox, D.O. Cummins, K. Kawabe, R.H. Tredgold [*J. Phys. Chem. Solids (UK)* vol.28 (1967) p.543-8]

[48] K. Kawabe, R.H. Tredgold, Y. Inuishi [*Electr. Eng. Jpn. (USA)* vol.87 (1967) p.62-70]

[49] R.F. Rutz [*Appl. Phys. Lett. (USA)* vol.28 (1976) p.379-81]

[50] T.L. Chu, D.W. Ing, A.J. Norieka [*Solid-State Electron. (UK)* vol.10 (1967) p.1023-7]

[51] A.G. Gorbatov, V.M. Kamyshov [*Sov. Powder Metall. Met. Ceram. (USA)* vol.9 (1970) p.917-20]

[52] J. Hagen, R.D. Metcalfe, D. Wickenden, W. Clark [*J. Phys C (UK)* vol.11 (1978) p.L143-6]

[53] B. Baranov, L. Däweritz, V.B. Gutan, G. Jungk, H. Neumann, H. Raidt [*Phys. Status Solidi A (Germany)* vol.29 (1978) p.629-36]

[54] S. Yoshida, S. Misawa, S. Gonda [*J. Appl. Phys. (USA)* vol.53 (1982) p.6844-7]

[55] M.A. Khan, R.A. Skogman, R.G. Schulze, M. Gershenzon [*Appl. Phys. Lett. (USA)* vol.43 (1983) p.492-4]

[56] M. Asif Khan, J.N. Kuznia, J.M. Van Hove, N. Pan, J. Carter [*Appl. Phys. Lett. (USA)* vol.60 (1992) p.3027-9]

[57] H. Murakami, T. Asahi, H. Amano, K. Hiramatsu, N. Sawaki, I. Akasaki [*J. Cryst. Growth (Netherlands)* vol.115 (1991) p.648-51]

[58] M. Asif Khan, J.M. Van Hove, J.N. Kuznia, D.T. Olson [*Appl. Phys. Lett. (USA)* vol.58 (1991) p.2408-10]

[59] J. Pankove [*Mater. Res. Soc. Symp. Proc. (USA)* vol.162 (1990) p.515-24]

[60] J.P. Dismukes, M.W. Yim, J.J. Tietjen, R.E. Novak [*RCA Rev. (USA)* vol.31 (1970) p.680-91]

[61] M.J. Paisley, Z. Sitar, J.B. Posthill, R.F. Davis [*J. Vac. Sci. Technol. A (USA)* vol.3 (1989) p.701-5]

[62] T. Lei, T.D. Moustakas, R.J. Graham, Y. He, S.J. Berkowitz [*J. Appl. Phys. (USA)* vol.71 (1992) p.4933-43]

[63]   H. Okumura, S. Misawa, S. Yoshida [*Appl. Phys. Lett. (USA)* vol.59 (1991) p.1058-60]

[64]   M.E. Lin, G. Xue, G.L. Zhou, J.E. Greene, H. Morkoç [*Appl. Phys. Lett. (USA)* vol.74 (1993) p.932-3]

[65]   R.C. Powell, G.A. Tomasch, Y.W. Kim, J.A. Thornton, J.E. Greene [*Mater. Res. Soc. Symp. Proc. (USA)* vol.162 (1990) p.525-30]

[66]   W. Walukiewicz, J. Lagowski, H.C. Gatos [*J. Appl. Phys. (USA)* vol.53 (1982) p.769-70]

3.3 Electrical transport properties of InN, GaInN and AlInN

W.A. Bryden and T.J. Kistenmacher

May 1994

A INTRODUCTION

Of the group IIIA nitride semiconductors, indium nitride (InN) is by far the least studied. The two principal reasons for the relative scarcity of data on InN are 1) the difficulty in preparation of high quality samples and, 2) the relative abundance of other, more easily produced, semiconductors with similar bandgaps (for InN, $E_g = 2\,eV$). However, band structure calculations [1,2], indicating a direct bandgap and low carrier effective masses, predict the potential for high speed photonic devices. Furthermore, as the only member of the group IIIA nitride family with a band edge in the visible regime, InN is a natural choice for visible light photonic devices based on alloys with the other, wider bandgap, group IIIA nitrides, gallium nitride (GaN, $E_g = 3.4\,eV$) and aluminium nitride (AlN, $E_g = 6.0\,eV$). Broad reviews of the physical properties of these materials can be found in recent publications by Davis [3], Strite and Morkoç [4], and Matsuoka [5]. We focus here on the electrical transport properties of InN and its alloys with GaN and AlN.

B ELECTRICAL TRANSPORT PROPERTIES OF InN

B1 Studies on Bulk InN

The first reported study of InN was by Fischer and Schröter in 1910 [6]. They reacted indium metal with nitrogen in a cathodic discharge. A thermal chemical synthesis of InN, using the decomposition of ammonium hexafluoroindate $[(NH_4)_3InF_6]$ in an ammonia atmosphere, was carried out by Juza and Hahn [7] in 1940. Further research was in part stimulated by the development of the injection laser in the 1960s and an extensive search for wide, direct bandgap semiconductors was carried out. As a result of this search, the group IIIA nitrides were identified as candidates for visible light emitters. Some of the basic thermophysical properties of these nitrides were investigated at that time, with the goal of determining conditions for growth of single crystals. Material formed using the procedure of Juza and Hahn [7] was the subject of study by MacChesney et al [8] on the thermal stability of InN at elevated temperatures. They concluded that InN rapidly dissociates at higher temperatures and that extraordinarily high nitrogen overpressure would be required to stabilize the material up to the melting point, realizations that would make single crystal growth practically impossible. No electrical transport properties of chemically synthesized InN were reported in these early works but a fundamental problem, nonstoichiometry (or nitrogen vacancies), that would haunt succeeding generations of researchers was identified even at this time.

B2   Thin Films of InN

Given then that growth of bulk InN crystals using equilibrium techniques was unlikely, attention turned to deposition of thin films using several nonequilibrium techniques. These investigations began in 1972 with the work of Hovel and Cuomo [9] and continue until the present day. A table of the electrical properties of InN films deposited using various techniques follows here.

TABLE 1 Electrical properties of InN thin films.

n-type carrier concentration (cm^{-3})	Carrier mobility $(cm^2\,V^{-1}\,s^{-1})$	Substrate	Deposition technique	Year	Ref
$5 - 8 \times 10^{18}$	250 ± 50	sapphire, silicon, various metals	reactive sputtering	1972	[9]
10^{20}	20	sapphire	reactive evaporation	1974	[10]
$3 - 10 \times 10^{18}$	20 - 50	glass, fused quartz	reactive sputtering	1976	[11]
$1 - 200 \times 10^{18}$	3	fused quartz	reactive sputtering	1977	[12]
$2 - 80 \times 10^{20}$	35 - 50	sapphire	CVD	1977	[13]
5×10^{18}	20	glass, NaCl	reactive sputtering	1980	[14-16]
6×10^{20}	2	fused quartz	cathodic sputtering	1980	[17]
6×10^{16}	2	glass, silicon, 304 stainless steel	RF ion plating	1982	[18]
$7 - 70 \times 10^{16}$	730 - 3980	glass, silicon	reactive sputtering	1984	[19]
3×10^{16} at 150 K	5000 at 150 K	glass, silicon	reactive sputtering	1984	[20]
10^{20}	10	glass	reactive DC magnetron sputtering	1988	[21]
4.8×10^{20}	38	sapphire	magnetron sputtering	1989	[22]
$1 - 8 \times 10^{20}$	50	sapphire	plasma assisted MOVPE	1990	[23]
$1 - 10 \times 10^{20}$	50	sapphire	MOVPE	1990	[24]
$2 - 3 \times 10^{20}$	20 - 60	sapphire, silicon, mica	reactive RF magnetron sputtering	1990-1993	[25-30]
2×10^{20}	~100	GaAs	ECR-assisted MOMBE	1993	[31]
10^{20}*	220*	GaAs	plasma-assisted MBE	1993	[32]

* Zinc blende polytype

C   ELECTRICAL TRANSPORT PROPERTIES OF InN ALLOYS

C1   Ga$_x$In$_{1-x}$N Alloys

An electron-beam plasma technique was used to prepare Ga$_x$In$_{1-x}$N alloys by Osamura and co-workers in 1972 [33]. They reported stable growth across the entire compositional range (as evidenced by the predictable change in optical energy gap) but did not report on the electrical properties of the materials. In a follow-up study in 1975 [34], they further confirmed the stability of these alloys by detailed X-ray scattering and infrared spectroscopic techniques. This alloy system was also grown, using MOVPE, by Nagatoma et al in 1989 [35]. They used a fixed flow of reactant gases and monitored the physical properties of the films produced as a function of deposition temperature. The films were n-type as determined by thermoelectric power measurements and the resistivity of the films decreased from $10^{10}\,\Omega$ cm to $10^{-2}\,\Omega$ cm as the growth temperature increased from 500 °C to > 600 °C. In the following year, Matsouka et al reported the MOVPE growth of the complete alloy system [24]; in 1991, the detailed electrical properties of these films were first reported by Yoshimoto et al [36]. For a particular alloy composition (x ~ 0.2) grown on sapphire, they found a reduction in carrier concentration from $10^{20}\,\text{cm}^{-3}$ to $10^{18}\,\text{cm}^{-3}$ and an increase in mobility from $< 10\,\text{cm}^2\text{V}^{-1}\text{s}^{-1}$ to $100\,\text{cm}^2\,\text{V}^{-1}\,\text{s}^{-1}$ as the deposition temperature was increased from 500 °C to 900 °C. They reported similar results in 1992 for films grown onto ZnO substrates [37]. Finally, in 1992 and 1993, Nakamura and co-workers reported [38,39] the growth of the alloy system using the two-flow MOCVD system. They have produced high quality specimens, as determined by X-ray scattering and photoluminescence measurements, but have not reported on the electrical properties of the films.

C2   Al$_x$In$_{1-x}$N Alloys

The first report of Al$_x$In$_{1-x}$N alloy growth, by RF sputtering, was by Starosta in 1981 [40] and X-ray scattering, RHEED measurements and infrared spectroscopy were employed to confirm the miscibility of solid solutions across the entire compositional range. Later, Starosta and Marsik [41] used Auger spectroscopy to further confirm the formation of solid solutions for the entire range of x. In 1989, Kubota et al reported on the preparation, structural and optical properties of the alloy system using RF magnetron sputtering [22]. Finally, Kistenmacher et al have reported on the growth of these alloys using RF reactive magnetron sputtering from a composite metal target [42]. At a growth temperature near 300 °C, they report a mobility of $35\,\text{cm}^2\text{V}^{-1}\text{s}^{-1}$ and a carrier concentration of $2 \times 10^{20}\,\text{cm}^{-3}$ for a film with x = 0.04 and a mobility of $2\,\text{cm}^2\,\text{V}^{-1}\,\text{s}^{-1}$ and a carrier concentration of $8 \times 10^{19}\,\text{cm}^{-3}$ for x = 0.25.

D   CONCLUSION

The electrical transport properties of InN are, in general, characterized by high carrier concentrations (~ 10^{20} cm^{-3}) and low mobilities (~ $100\,\text{cm}^2\,\text{V}^{-1}\,\text{s}^{-1}$). The high carrier concentration is generally believed to derive from a high density of nitrogen vacancies produced during growth. Given the high carrier concentration, a low mobility is expected and even lower values are realized due to the intergrain scattering which is present even in the best samples. An exception to this general behaviour is found in the work of Tansley and

Foley [19,20], where an RF sputtering technique (with extremely long pre-sputtering times) was used to produce films with carrier concentrations near $10^{16}\,cm^{-3}$ and with mobilities near $4000\,cm^2\,V^{-1}\,s^{-1}$. To date, unfortunately, all attempts to produce device quality films with similar electrical transport properties (using techniques more amenable to standard semiconductor processes) have been unsuccessful. Initial studies on the electrical transport properties of alloys of InN with GaN and AlN have been published and seem to indicate films with intermediate properties, but no more definitive conclusions can be drawn from the work published at this time.

REFERENCES

[1]   C.P. Foley, T.L. Tansley [*Phys. Rev. B (USA)* vol.33 (1986) p.1430-3]

[2]   D.W. Jenkins, R.-D. Hong, J.D. Dow [*Superlattices Microstruct. (UK)* vol.3 (1987) p.365-9]

[3]   R.F. Davis [in *The Physics and Chemistry of Carbides, Nitrides and Borides* Ed R. Freer (Kluwer Academic Publishers, 1990) p.653-69]

[4]   S. Strite, H. Morkoç [*J. Vac. Sci. Technol. B (USA)* vol.10 (1992) p.1237-66]

[5]   T. Matsuoka [*J. Cryst. Growth (Netherlands)* vol.124 (1992) p.433-8]

[6]   F. Fischer, F. Schröter [*Ber. Dtsch. Chem. Ges. (Germany)* vol.43 (1910) p.1465-79]

[7]   R. Juza, H. Hahn [*Z. Anorg. Allg. Chem. (Germany)* vol.244 (1940) p.133-8]

[8]   J.B. MacChesney, P.M. Bridenbaugh, P.B. O'Conner [*Mater. Res. Bull. (USA)* vol.5 (1970) p.783-92]

[9]   H.J. Hovel, J.J. Cuomo [*Appl. Phys. Lett. (USA)* vol.20 (1972) p.71-3]

[10]   J.W. Trainor, K. Rose [*J. Electron. Mater. (USA)* vol.3 (1974) p.821-8]

[11]   N. Puchevrier, M. Menoret [*Thin Solid Films (Switzerland)* vol.36 (1976) p.141-5]

[12]   H. Takeda, T. Hada [*Quarterly of the Toyama Industrial-Technical School (Japan)* vol.11 (1977) p.73-5]

[13]   L.A. Marasina, I.G. Pichugin, M. Tlaczala [*Krist. Tech. (Germany)* vol.12 (1977) p.541-5]

[14]   B.R. Natarajan, A.H. Eltoukhy, J.E. Greene, T.L. Barr [*Thin Solid Films (Switzerland)* vol.69 (1980) p.201-16]

[15]   B.R. Natarajan, A.H. Eltoukhy, J.E. Greene, T.L. Barr [*Thin Solid Films (Switzerland)* vol.69 (1980) p.217-27]

[16]   B.R. Natarajan, A.H. Eltoukhy, J.E. Greene, T.L. Barr [*Thin Solid Films (Switzerland)* vol.69 (1980) p.229-35]

[17]   G.V. Samsonov, A.F. Andreeva [*Sci. Sinter. (Serbia)* vol.12 (1980) p.155-61]

[18]   O. Takai, J. Ebisawa, Y. Hisamatsu [*Proc. 7th Int. Conf. Vacuum Metallurgy*, Tokyo, Japan, 1982 (Societe Francaise du Vide) p.137-44]

[19]   T.L. Tansley, C.P. Foley [*Proc. 3rd Int. Conf. on Semi-Insulating III-V Materials*, Warm Springs, OR, Eds D.C. Look, J.S. Blakemore (Shiva Publishing, Nantwich, England, 1984) p.497-500]

[20]   T.L. Tansley, C.P. Foley [*Electron. Lett. (UK)* vol.20 (1984) p.1066-8]

[21]   K.L. Westra, R.P.W. Lawson, M.J. Brett [*J. Vac. Sci. Technol. A (USA)* vol.6 (1988) p.1730-2]

[22]   K. Kubota, Y. Kobayashi, K. Fujimoto [*J. Appl. Phys. (USA)* vol.66 (1989) p.2984-8]

[23]   A. Wakahara, T. Tsuchiya, A. Yoshida [*J. Cryst. Growth (Netherlands)* vol.99 (1990) p.385-9]

[24]   T. Matsouka, T. Sasaki, A. Katsui [*Optoelectron., Devices Technol. (Japan)* vol.5 (1990) p.53-64]

[25]   W.A. Bryden, J.S. Morgan, T.J. Kistenmacher, D. Dayan, R. Fainchtein, T.O. Poehler [*Mater. Res. Soc. Symp. Proc. (USA)* vol.162 (1990) p.567-72]

[26]   J.S. Morgan, T.J. Kistenmacher, W.A. Bryden, S.A. Ecelberger [*Mater. Res. Soc. Symp. Proc. (USA)* vol.202 (1991) p.383-8]

[27]   T.J. Kistenmacher, W.A. Bryden, J.S. Morgan, D. Dayan, R. Fainchtein, T.O. Poehler [*J. Mater. Res. (USA)* vol.6 (1991) p.1300-7]

[28]   T.J. Kistenmacher, W.A. Bryden [*Appl. Phys. Lett. (USA)* vol.59 (1991) p.1844-6]

[29]   W.A. Bryden, J.S. Morgan, R. Fainchtein, T.J. Kistenmacher [*Thin Solid Films (Switzerland)* vol.213 (1992) p.86-93]

[30]   T.J. Kistenmacher, W.A. Bryden [*Appl. Phys. Lett. (USA)* vol.62 (1993) p.1221-3]

[31]   C.R. Abernathy, S.J. Pearton, F. Ren, P.W. Wisk [*J. Vac. Sci. Technol. B (USA)* vol.11 (1993) p.179-82]

[32]   S. Strite et al [*J. Cryst. Growth (Netherlands)* vol.127 (1993) p.204-9]

[33]   K. Osamura, K. Nakajima, Y. Murakami [*Solid State Commun. (USA)* vol.11 (1972) p.617-21]

[34]   K. Osamura, S. Naka, Y. Murakami [*J. Appl. Phys. (USA)* vol.46 (1975) p.3432-7]

[35]   T. Nagatoma, T. Kuboyama, H. Minamino, O. Omoto [*Jpn. J. Appl. Phys. (Japan)* vol.28 (1989) p.L1334-6]

[36]   N. Yoshimoto, T. Matsouka, T. Sasaki, A. Katsui [*Appl. Phys. Lett. (USA)* vol.59 (1991) p.2251-3]

[37]   T. Matsouka, N. Yoshimoto, T. Sasaki, A. Katsui [*J. Electron. Mater. (USA)* vol.21 (1992) p.157-63]

[38]   S. Nakamura, T. Mukai [*Jpn. J. Appl. Phys. (Japan)* vol.31 (1992) p.L1457-9]

[39]   S. Nakamura, T. Mukai, M. Senoh [*Jpn. J. Appl. Phys. (Japan)* vol.32 (1993) p.L16-9]

[40]   K. Starosta [*Phys. Status Solidi A (Germany)* vol.68 (1981) p.K55-7]

[41]   K. Starosta, J. Marsik [*Thin Solid Films (Switzerland)* vol.128 (1985) p.L41-3]

[42]   T.J. Kistenmacher, S.A. Ecelberger, W.A. Bryden [*Mater. Res. Soc. Symp. Proc. (USA)* vol.280 (1993) p.509-12]

CHAPTER 4

BAND STRUCTURE OF PURE GROUP III NITRIDES

4.1   General remarks and notations on the band structure of pure group III nitrides

W.R.L. Lambrecht and B. Segall

August 1994

In this chapter, we present the band structures of each of the nitrides for several crystal structures. This introduction contains some general comments and a brief description of the format in which the data are presented.

The most important structures for each of the nitrides are the wurtzite and zinc blende. The zinc blende and wurtzite structures (figures of which can be found elsewhere in this book) are both tetrahedrally coordinated and hence are closely related. The {0001} basal plane of wurtzite corresponds to one of the {111} planes of zinc blende. In the latter plane, the structure is formed by staggered (chair-shaped) hexagonal rings. The rings between the planes are chair-shaped (or staggered) for zinc blende and are boat-shaped or eclipsed for the wurtzite. When the in-plane hexagons are lined up in wurtzite and zinc blende, the following directions are parallel:

$$\text{wurtzite } [0001] \parallel \text{zinc blende } [111]$$
$$\text{wurtzite } [11\bar{2}0] \parallel \text{zinc blende } [10\bar{1}]$$
$$\text{wurtzite } [1\bar{1}00] \parallel \text{zinc blende } [1\bar{2}1]$$

These in turn lead to relations between the symmetry directions and k-points in the reciprocal lattices for the two structures. The unit cell of wurtzite, however, has twice the number of atoms as the zinc blende unit cell. As a result of the above, it is useful to compare the bands of wurtzite with those of a doubled zinc blende unit cell displayed in the conventional Brillouin zone of the wurtzite phase. The band structures of zinc blende and wurtzite are, of course, not identical. Nevertheless, this unconventional way of presenting the zinc blende bands brings out more clearly the similarity between the two band structures. While we have here chosen to represent the bands in their conventional way, the alternative representation of the zinc blende band structures can be found in [1] for AlN and GaN. TABLE 1 lists the equivalences between some of the symmetry k-points, shown in FIGURE 1. We hope this will help the reader to compare the results for wurtzite and zinc blende given in the tables for each material below.

TABLE 1   Equivalence between wurtzite and zinc blende symmetry k-points, assuming the ideal c/a ratio for wurtzite.

Zinc blende	Wurtzite
$L_\perp$	Γ
$L_\parallel$	U_1 at 2/3 of M - L
X	U_1 at 2/3 of M - L
W	T_1 at 3/4 of Γ - K
U, K	Σ_1 at 3/4 of Γ - M

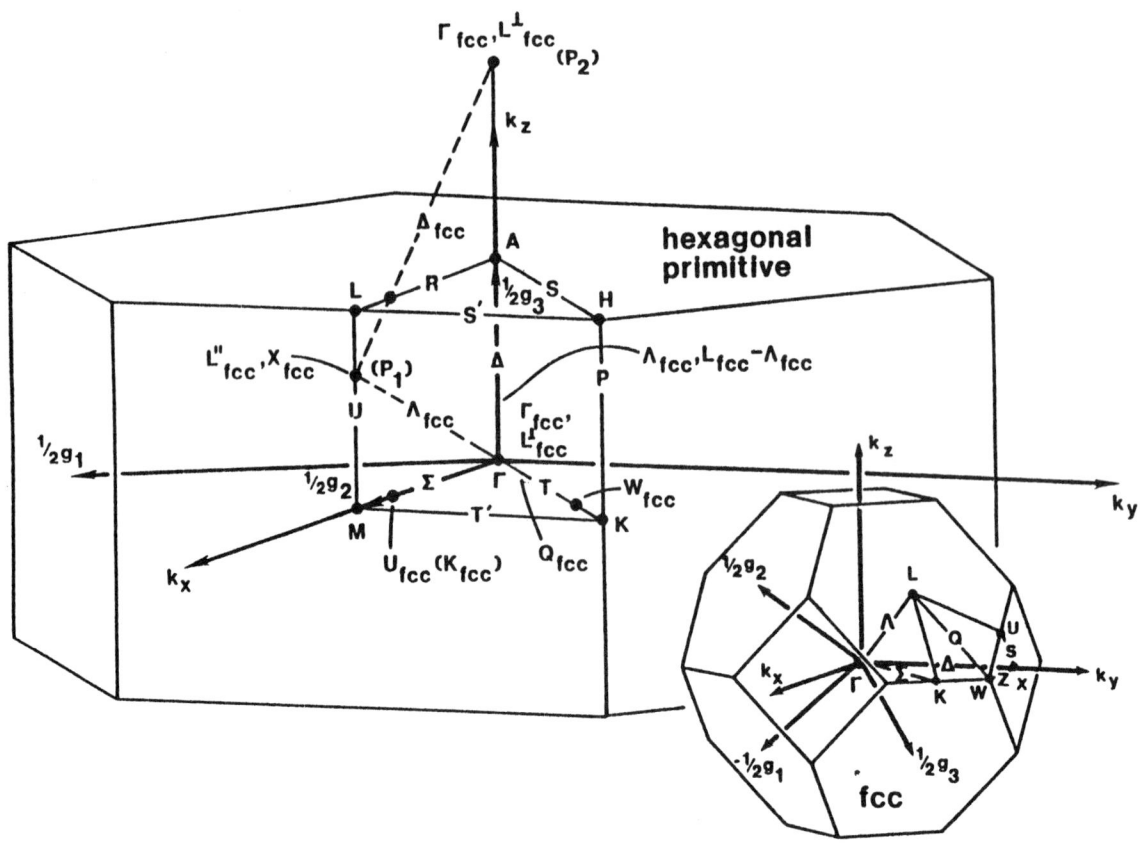

FIGURE 1 Zinc blende and wurtzite Brillouin zones, indicating equivalent k-points (from [10]).

Of the two tetrahedral structures, the lowest energy phase is zinc blende for BN and wurtzite for AlN, GaN and InN. The rocksalt structure is a high pressure phase observed for AlN, GaN and InN. BN has a layered hexagonal phase (similar to graphite) which is the stable phase at 1 atm.

In the following, we present for each nitride for the wurtzite and zinc blende structures:

• a brief literature overview and discussion

• a figure of the band structure along symmetry lines of the Brillouin zone

• a table of selected eigenvalues

— at Γ, K and M for wurtzite
— at Γ, X and L for zinc blende

• a summary table of band widths, bandgaps, bandgap pressure coefficients and deformation potentials.

In addition to the band structure data, we give the experimental lattice constants at which the calculations were performed and the calculated bulk moduli for use in converting pressure coefficients to deformation potentials, using $dE_g/dp = -(dE_g/d\ln V)/B$ with B the bulk modulus.

The tables of selected eigenvalues and the figures use our own LMTO-ASA calculations because, in most cases, we could not find detailed tabular information on the band structure in the literature. Our calculations were performed at the experimental lattice parameters (except as otherwise indicated). These calculations are scalar-relativistic [2], and use Hedin-Lundqvist exchange and correlation [3] in the local density approximation (LDA) to density functional theory [4]. They use well-converged special k-point sets for the Brillouin zone integrations in the determination of the self-consistent potential. In order to describe the open structures by a system of closely packed spheres, empty spheres are included in the interstitial region and the combined correction terms are included [2].

The eigenvalues are indicated by the k-point, a subscript indicating the irreducible symmetry representation and superscript indicating valence or conduction band state plus an additional number to distinguish different eigenvalues of the same symmetry where necessary. For wurtzite, we use the symmetry representation notations of Rashba [5]. For zinc blende, we use the standard notation of Bouckaert et al [6].

The summary tables, on the other hand, include comparisons to other calculations and to experimental values. For the rocksalt structures, we provide data on the transition pressure p_t and minimum bandgaps at the transition pressure, $E_g(p_t)$, and at the theoretical equilibrium lattice constant, $E_g(p=0)$, of this structure. The corresponding cubic lattice constants are also given. For BN, we also include some data on the low-pressure h-BN phase.

We have made no attempt at being complete in the comparisons to other calculations. The main purpose is to give some idea of the uncertainties arising from the use of different methods and the choice of certain parameters. We have generally selected the most recent calculations based on density functional theory in the LDA. Because of the uncertainties involved in the semi-empirical pseudopotential and tight-binding methods, we believe these to be less reliable. This is especially the case for the present materials, because these methods generally depend on fitting parameters to optical data. Because of the relatively poor quality of the crystals that were available at the time of these measurements, the data themselves had significant uncertainties.

The LDA on the other hand has the disadvantage of underestimating the bandgap. This is a well understood problem and calculations going beyond the LDA, for example using Hedin's GW-approximation [7-9] to the quasiparticle self-energy, are currently becoming available for a number of semiconductors. Unfortunately, no such calculations are available at present for the nitrides, except for zinc blende GaN. The GW calculations available for other semiconductors [8,9] indicate that the correction is constant to within a few 0.1 eV. The recommended values are thus the LDA values plus a constant bandgap correction adjusted to the minimum gap obtained from experiment. In the tables of selected eigenvalues, we have indicated these recommended values in parentheses for the conduction-band states in addition to the LDA values. For convenience, a list of frequently used acronyms of computational methods is provided in TABLE 2.

TABLE 2 Frequently used acronyms of computational methods.

Acronym	Method	Ref
ASA	atomic sphere approximation (version of LMTO)	[2]
DFT	density functional theory	[4]
FP	full-potential (version of LMTO)	[11,12]
FLAPW	full-potential linearized augmented plane-wave method	[2,13]
LDA	local density approximation	[4]
LMTO	linear muffin-tin orbital	[2]
OLCAO	orthogonalized linear combination of atomic orbitals	[14]
PP-PW	norm-conserving pseudopotential plane-wave	[15]

ACKNOWLEDGEMENTS

It is a pleasure to thank Dr. S. Satpathy for providing us with FIGURE 1, and N.E. Christensen and C.O. Bertoni for allowing us to quote their latest results prior to publication. We acknowledge financial support by the NSF under grant No. DMR-92-22387.

REFERENCES

[1]   W.R.L. Lambrecht, B. Segall [*Mater. Res. Soc. Symp. Proc. (USA)* vol.242 (1992) p.367]

[2]   O.K. Andersen [*Phys. Rev. B (USA)* vol.12 (1975) p.3060]; O.K. Andersen, O. Jepsen, M. Sob [*Electronic Band Structure and its Applications*, Ed. M. Yussouff (Springer, Heidelberg, 1987) p.1]

[3]   L. Hedin, B.I. Lundqvist [*J. Phys. C (UK)* vol.4 (1971) p.2064]

[4]   P. Hohenberg, W. Kohn [*Phys. Rev. B (USA)* vol.136 (1964) p.864]; W. Kohn, L.J. Sham [*Phys. Rev. A (USA)* vol.140 (1965) p.1133]

[5]   E.I. Rashba [*Sov. Phys.-Solid State (USA)* vol.1 (1959) p.368]

[6]   L.P. Bouckaert, R. Smoluchowski, E. Wigner [*Phys. Rev. (USA)* vol.50 (1936) p.58]

[7]   L. Hedin [*Phys. Rev. A (USA)* vol.139 (1965) p.796]

[8]   R.W. Godby, M. Schluter, L.J. Sham [*Phys. Rev. B (USA)* vol.37 (1988) p.10159]

[9]   M.S. Hybertsen, S.G. Louie [*Phys. Rev. B (USA)* vol.34 (1986) p.5390]

[10]   M.R. Salehpour, S. Satpathy [*Phys. Rev. B (USA)* vol.41 (1990) p.3048]

[11]   M. Methfessel [*Phys. Rev. B (USA)* vol.38 (1988) p.1537]

[12]   K.H. Weyrich [*Phys. Rev. B (USA)* vol.37 (1988) p.10269]

[13]   E. Wimmer, H. Krakauer, M. Weinert, A.J. Freeman [*Phys. Rev. B (USA)* vol.24 (1981) p.864]

[14]   J. Ihm, A. Zunger, M.L. Cohen [*J. Phys. C (UK)* vol.12 (1979) p.4409]

[15]   W.Y. Ching [*J. Am. Ceram. Soc. (USA)* vol.71 (1990) p.3135]

4.2 Band structure of pure BN

W.R.L. Lambrecht and B. Segall

August 1994

A INTRODUCTION

For general comments and notations the reader is referred to Datareview 4.1.

B HEXAGONAL BN

The graphite-like hexagonal BN structure is the groundstate of BN at ambient pressure. Recent band structure calculations were presented by Xu and Ching [2] who included references to earlier work. Their main results are summarized in TABLE 1.

TABLE 1 Band structure of hexagonal BN (values in eV).

Minimum bandgap (indirect)	H — M	4.07
Direct gap at	H	4.2
	K	4.5
	M	4.6
	L	5.6
	Γ	8.9
	A	10.5
Band width	N 2s	4.02
	N 2p	10.40
	N 2s - N 2p gap	4.42
	total valence band	18.8

C ZINC BLENDE BN

Zinc blende BN is normally referred to as c-BN, or cubic BN. It is the stable form at high-pressure and has a lattice constant of 3.616 Å [1].

FIGURE 1 shows the band structure of the zinc blende polytype. Recent LDA band-structure results were obtained by Xu and Ching [2] using the OLCAO method, by Wentzcovitch et al [3] using the PP-PW method, by Park et al [4] using the FLAPW method and by Lambrecht and Segall [5] using LMTO-ASA. The results given in TABLE 2 deviate slightly from those of [5] because they were recalculated including the combined correction and at the experimental instead of theoretical equilibrium lattice constant. A comparison between the results obtained by the various methods is given for some relevant energy differences in TABLE 3.

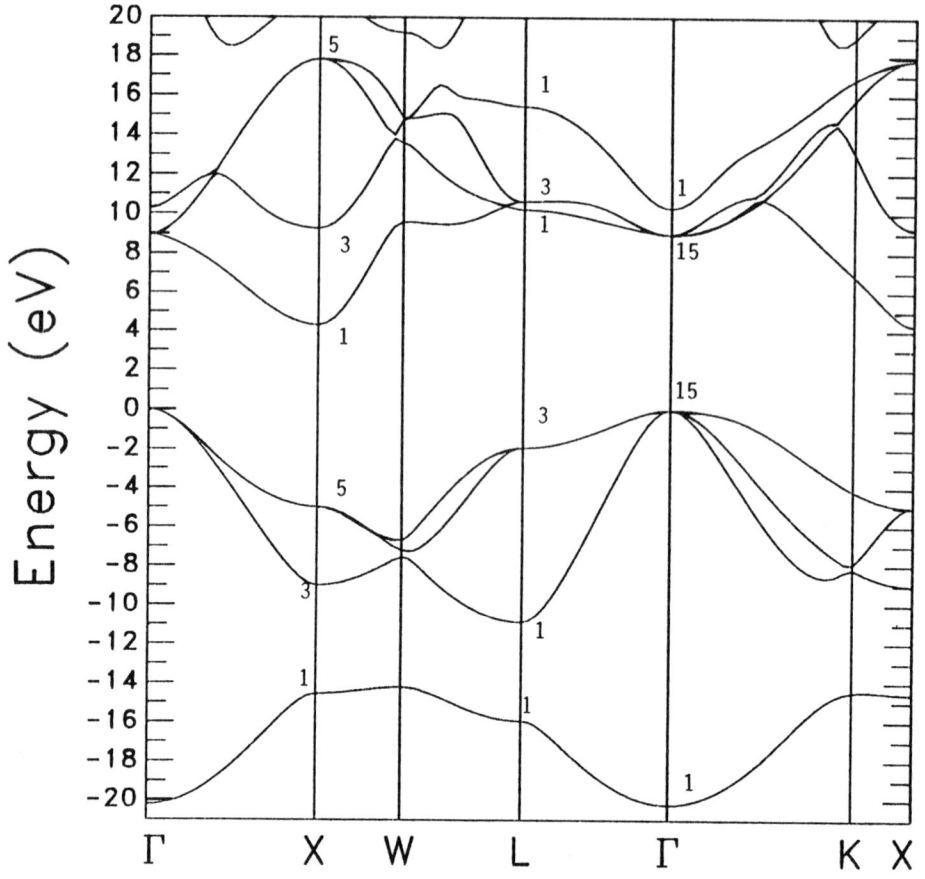

FIGURE 1 Band structure of zinc blende BN.

TABLE 2 Selected eigenvalues (in eV) for zinc blende BN with respect to valence-band maximum (Γ_{15}^{v}).

Γ_1^v	-20.23
Γ_{15}^v	0.00
Γ_{15}^c	8.94 (11.0)[a]
Γ_1^c	10.28 (12.4)
X_1^v	-14.56
X_3^v	-8.99
X_5^v	-4.99
X_1^c	4.30 (6.4)
X_3^c	9.24 (11.3)
L_1^{v1}	-15.94
L_1^{v2}	-10.87
L_3^v	-1.94
L_1^{c1}	10.21 (12.3)
L_3^c	10.64 (12.7)
L_1^{c2}	15.44 (17.5)

[a] Values in parentheses include bandgap correction.

TABLE 3 Bandgaps, widths and deformation potentials (in eV) for zinc blende BN.

		LTMO[a]	OLCAO[b]	PP-PW[c]	FLAPW[d]	Expt.
Bandgap (indirect)	$\Gamma - X$	4.30	5.18	4.2	4.4	6.4[e]
						6.1 ± 0.2[f]
Bandgap (direct)	Γ	8.94	8.7	8.6	8.8	
	X	9.29	10.3			
	L	12.15	12.4			
	K	10.99	11.8			
Band width	N 2s ($\Gamma - W$)	6.00	6.92	5.9	5.9	
	N 2s-N 2p gap ($\Gamma - W$)	3.35	3.28	3.6	3.5	
	N 2p (L $- \Gamma$)	10.87	10.94	10.8	10.7	
	total valence band	20.23	21.1	20.3	20.1	
$dE_g/d\ln V$				-2.2[c]		
dE_g/dp[g]	(meV/GPa)			6[c]		

[a] Lambrecht and Segall (present calculation) at a = 3.616 Å.
[b] Xu and Ching [2].
[c] Wentzcovitch et al [3].
[d] Park et al [4].
[e] Chrenko [12].
[f] Miyata et al [13].
[g] $dE_g/dp = -(dE_g/d\ln V)/B$ with bulk modulus B = 367 GPa [3].

D   WURTZITE BN

Wurtzite BN is a metastable high-pressure phase which can be produced by shock compression of h-BN and transforms to zinc blende at high temperature [6]. Unlike the other nitrides, the wurtzite form of this compound is predicted to have an indirect bandgap, namely one at $\Gamma - K$. Its band structure is similar to that of lonsdaleite diamond in this respect (ref [10] of Datareview 4.1 and see FIGURE 2). Calculations of wurtzite BN have been reported by Park et al [4], Lam et al [7], Xu and Ching [2] and Gorczyca and Christensen [8], and are summarised in TABLES 4 and 5. We here report our own results for this compound for the first time.

The lattice constants used here assume the same volume per atom in the zinc blende phase and the ideal wurtzite structure:

a = 2.557 Å, c = 4.176 Å (c/a = $\sqrt{(8/3)} \sim 1.633$), u = 3/8

E   ROCKSALT BN

Rocksalt BN has not been observed experimentally. Theoretically a phase transition from zinc blende to rocksalt BN was predicted by Wentzcovitch et al [10]. Gorczyca and Christensen [8] also predict a transition from wurtzite. Experimentally, it was verified by Ueno et al [11] that no transition from zinc blende takes place up to 106 GPa, which is still an order of magnitude below the predicted transition pressures.

TABLE 4 Selected eigenvalues (in eV) for wurtzite BN[a] with respect to valence-band maximum (Γ_1^{v2}).

Γ_1^{v1}	-20.20
Γ_3^{v1}	-16.00
Γ_3^{v2}	-10.72
Γ_5^{v}	-2.05
Γ_1^{v}	-0.08
Γ_1^{v2}	0.00
Γ_6^{c}	8.89 (11.0)[b]
Γ_3^{c1}	9.41 (11.5)
Γ_1^{c1}	9.46 (11.55)
Γ_1^{c2}	10.83 (13.0)
K_3^{v1}	-14.85
K_1^{v1}	-9.39
K_3^{v2}	-7.02
K_2^{v2}	-5.50
K_3^{v3}	-5.20
K_2^{c1}	5.70 (7.8)
K_1^{c2}	13.88 (16.0)
K_3^{c1}	14.02 (16.1)
K_3^{c2}	14.54 (16.7)
M_3^{v1}	-16.08
M_1^{v1}	-14.81
M_1^{v2}	-9.42
M_3^{v2}	-7.74
M_1^{v3}	-6.68
M_2^{v}	-4.98
M_3^{v3}	-4.2
M_4^{v}	-1.95
M_1^{c1}	6.65 (8.7)
M_3^{c1}	8.28 (10.4)
M_4^{c2}	11.19 (13.3)
M_3^{c2}	11.18 (13.9)

[a] At same volume as zinc blende and ideal c/a and u.
[b] Values in parentheses include the same gap correction as for zinc blende.

The bandgaps given in TABLE 6 are LDA values at the theoretical equilibrium lattice constant of the rocksalt structure (E_g (p = 0)) and at the transition pressure p_t ($E_g(p_t)$). They were calculated by Gorczyca and Christensen [9].

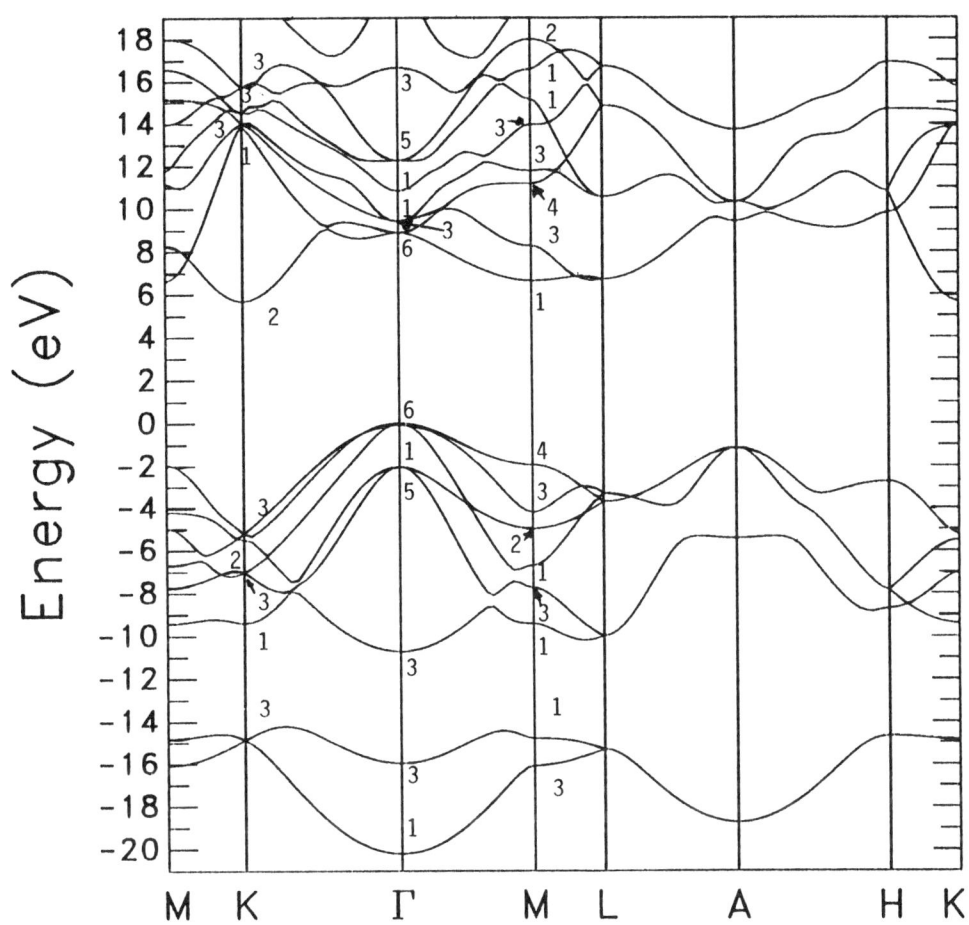

FIGURE 2 Band structure of wurtzite BN.

TABLE 5 Bandgaps and widths (in eV) for wurtzite BN.

		LMTO[a]	OLCAO[b]	PP-PW[c]	FLAPW[d]
Bandgap (indirect)	Γ — K	5.7	5.81	4.9	4.9
Bandgap (direct)	Γ	8.89	8.0	8.2	
	M	8.6	9.3		
	L	10.03	10.7		
	A	10.56	10.6		
	H	12.58	12.8		
	K	10.90	11.7		
Band width	N 2s	5.99	6.28		6.0
	N 2s-N 2p gap	3.49	2.93		3.3
	N 2p	10.72	11.76		11.0
	total valence band	20.20	21.0		20.3

[a] Present calculation using ideal u, c/a and volume/atom of zinc blende BN.
[b] Xu and Ching [2].
[c] Lam et al [7].
[d] Park et al [4].

TABLE 6 Properties of rocksalt BN, transition pressure p_t, and minimum bandgaps at p_t and theoretical equilibrium.

	p_t (GPa)	E_g (p = 0) (eV)	E_g (p_t) (eV)		Ref
zinc blende → rocksalt	1110				[10]
wurtzite → rocksalt	1025, 930[c]	1.3[a]	6.4[a]	$\Sigma \to X^b$	[8,9]

[a] Lattice constants $a(p_0)$ = 3.47 Å, $a(p_t)$ = 2.79 Å [9].
[b] The valence-band maximum is located at ~ 0.5 (¾,¾,0)$2\pi/a$ along Σ.
[c] From a recent calculation with more k-points: recommended value [9].

REFERENCES

[1]   R.W.G. Wyckoff [*Crystal Structures, Second Edition*, vol.1 (John Wiley & Sons, New York, 1989) ch.II p.108-12]

[2]   Y.-N. Xu, W.Y. Ching [*Phys. Rev. B (USA)* vol.44 (1991)]

[3]   R.M. Wentzcovitch, K.J. Chang, M.L. Cohen [*Phys. Rev. B (USA)* vol.34 (1986) p.1071]

[4]   K.T. Park, K. Terakura, N. Hamada [*J. Phys. C (UK)* vol.20 (1987) p.1241]

[5]   W.R.L. Lambrecht, B. Segall [*Phys. Rev. B (USA)* vol.40 (1989) p.9909 and vol.41 (1990) p.5409]

[6]   A. Onodera, H. Miyazaki, N. Fujimoto [*J. Chem. Phys. (USA)* vol.74 (1981) p.5814]

[7]   P.K. Lam, R.M. Wentzcovitch, M.L. Cohen [*Mater. Sci. Forum (Switzerland)* vol.54&55 (1990) p.165]

[8]   I. Gorczyca, N.E. Christensen [*Physica B (Netherlands)* vol.185 (1993) p.410]

[9]   I. Gorczyca, N.E. Christensen [private communication]

[10]   R.M. Wentzcovitch, M.L. Cohen, P.K. Lam [*Phys. Rev. B (USA)* vol.36 (1987) p.6058]

[11]   M. Ueno et al [*Jpn. J. Appl. Phys. (Japan)* vol.32 suppl.32-1 (1993) p.42]

[12]   R.M. Chrenko [*Solid State Commun. (USA)* vol.14 (1974) p.511]

[13]   N. Miyata, K. Moriki, O. Mishima, M. Fujisawa, T. Hattori [*Phys. Rev. B (USA)* vol.40 (1989) p.12028]

4.3   Band structure of pure AlN

W.R.L. Lambrecht and B. Segall

August 1994

A   INTRODUCTION

For general comments and notations the reader is referred to Datareview 4.1.

B   WURTZITE AlN

The wurtzite structure is the stable crystal structure at ambient pressure and has lattice constants (see ref [1] of Datareview 4.2):

a = 3.111 Å, c = 4.978 Å   (c/a = 1.601), u = 0.385

Recent band structure calculations were reported by Ching and Harmon [1], Xu and Ching [2] using the OLCAO method, Lambrecht and Segall (ref [1] of Datareview 4.1) and Gorczyca et al [3,5] both using the LMTO-ASA method, Christensen and Gorczyca [6] using the FP-LMTO method, and Van Camp et al [7] and Munoz and Kunc [8] using the norm-conserving pseudopotential plane wave (PP-PW) method. All of these used the LDA and agree fairly well among each other except for the calculation by van Camp et al [7] which reports a significantly smaller bandgap. The older results of semi-empirical pseudopotential calculations by Bloom [9], semi-empirical tight-binding calculations by Kobayashi et al [10] and semi-ab-initio calculations by Huang and Ching [11] deviate significantly from the LDA results. The results are summarised in TABLES 1 and 2 and FIGURE 1 shows the band structure of the wurtzite polytype.

C   ZINC BLENDE AlN

The zinc blende phase could potentially be stabilized by epitaxial growth and has been reported to exist in AlN precipitates produced by ion bombardment in FCC Al by Schwabe and Mader [12]. Assuming the same volume per atom as in the closely related wurtzite structure its lattice constant would be a = 4.37 Å. Yeh et al [13] have calculated its energy difference from wurtzite to be 18 meV/atom, while Christensen and Gorczyca [6] report 43 meV/atom.

Band structure calculations for zinc blende AlN were reported by Lambrecht and Segall [14]. Details of this calculation are here reported for the first time. Since no experimental value for the gap is available, we estimate it on the basis of the wurtzite experimental value. One estimate for the bandgap correction is to take it to be the same as for wurtzite (1.72 eV). If one assumes instead that the bandgap correction is proportional to the LDA gap, one would obtain 1.27 eV. Both assumptions seem plausible. Thus our estimate of the correction is the average, 1.5 ± 0.2 eV.

TABLE 1 Selected eigenvalues (in eV) for wurtzite AlN[a] with respect to valence-band maximum (Γ_1^{v2}).

	LMTO-LDA	PPPW-LDA[c]	GW[c]
Γ_1^{v1}	-15.07	-15.2	-17.4
Γ_3^{v1}	-12.93	-13.0	-15.2
Γ_3^{v2}	-5.97	-6.1	-6.9
Γ_5^{v}	-0.85	-0.9	-1.1
Γ_6^{v}	-0.25	-0.3	-0.2
Γ_1^{v2}	0.00	0.0	0.0
Γ_1^{c1}	4.56 (6.28)	3.9	5.8
Γ_3^{c1}	6.64 (8.36)	6.2	8.3
Γ_1^{c2}	11.13 (12.85)	10.5	12.4
Γ_6^{c2}	11.65 (13.37)	10.6	12.9
K_3^{v1}	-12.51	-12.7	-14.8
K_1^{v1}	-5.01	-4.9	-5.6
K_3^{v2}	-3.73	-3.9	-4.5
K_2^{v2}	-2.68	-2.5	-3.0
K_3^{v3}	-2.34	-2.5	-2.9
K_2^{c1}	5.58 (7.3)	4.8	6.7
K_1^{c2}	8.53 (10.25)	8.1	10.7
K_3^{c1}	9.8 (11.52)	9.4	12.0
K_3^{c2}	12.09 (13.82)		
M_3^{v1}	-13.07	-13.2	-15.4
M_1^{v1}	-12.35	-12.5	-14.6
M_1^{v2}	-5.53	-5.6	-6.3
M_3^{v2}	-4.28	-4.3	-4.9
M_1^{v3}	-3.34	-3.4	-4.0
M_2^{v}	-2.11	-2.2	-2.6
M_3^{v3}	-1.71	-1.8	-2.1
M_4^{v}	-0.67	-0.7	-0.9
M_1^{c1}	5.9 (7.62)	5.5	7.4
M_3^{c1}	6.15 (7.87)	5.6	7.6
M_3^{c2}	8.47 (10.18)	7.8	10.1
M_1^{c2}	10.53 (12.25)	10.0	12.9

[a] Present calculation at u = 0.375, expt. a and c/a.
[b] Values in parentheses include gap correction.
[c] Rubio et al [19].

D   ROCKSALT AlN

The rocksalt structure is a high-pressure phase, metastable at ambient pressure [3,4]. This appears to be different from the other nitrides where the phase transition was found to be reversible. For other possible high pressure phases see [6]. Calculated and experimental values for the transition pressure are given in TABLE 5.

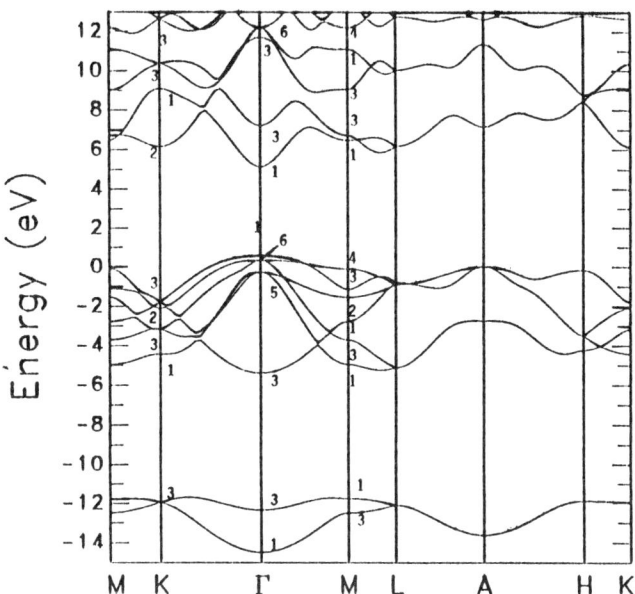

FIGURE 1   Band structure of wurtzite AlN.

TABLE 2   Band widths, bandgaps, bandgap deformation potential (in eV), and bandgap pressure coefficient
of wurtzite AlN.

	Theory	Expt.
N 2s band width	2.78[a], 2.89[b], 2.8[h,i]	3[f]
N 2p band width	5.97[a], 6.16[b]	6[f]
	6.1[h], 6.9[i]	
N 2s-N 2p gap	6.32[a], 6.07[b]	6[f]
	6.3[h], 7.7[i]	
Total valence band width	15.07[a], 15.11[b]	16[f]
	15.2[h], 17.4[i]	
Minimum bandgap (direct at Γ)	4.56[a], 4.64[b]	6.28[g]
	4.52[c], 3.09[d]	
	3.9[h], 5.8[i]	
Bandgap pressure coefficient* (dE_g/dp) (meV/GPa)	43[e], 37[c], 36[d]	
Bandgap deformation potential* ($dE_g/d\ln V$)	-9.5[e], -7.6[c], -7.1[d]	
$\partial E_g/\partial(c/a)$	3.9[c]	
$\partial E_g/\partial u$	-8.3[c]	

* (dE_g/dp) = -($dE_g/d\ln V$)/B with bulk modulus B = 205 GPa [6].
[a] Lambrecht and Segall, present calculation, LMTO-ASA at expt. a, c/a, u = 0.375.
[b] Xu and Ching [2] OLCAO expt. structure parameters.
[c] Christensen and Gorczyca [6] FP-LMTO; E_g at theoretical equilibrium c/a = 1.596, u = 0.382; dE_g/dp and
   $dE_g/d\ln V$ using optimised c/a and u, $\partial E_g/\partial(c/a)$ at fixed expt. a, ideal u, c/a, $\partial E_g/\partial u$ at fixed expt. a, ideal u, c/a.
[d] Van Camp et al [7] (norm-conserving PP-PW).
[e] Perlin et al [5] (LMTO-ASA calc. at expt. a, ideal c/a and u).
[f] Estimated from photoemission and X-ray emission spectroscopic data of Gautier et al [16].
[g] Perry and Rutz [15] absorption at 300 K, from excitonic edge assuming exciton binding energy of 75 meV.
[h] Rubio et al [19] LDA value.
[i] Rubio et al [19] GW value.

TABLE 3 Selected eigenvalues (in eV) for zinc blende AlN with respect to valence-band maximum (Γ_{15}^v).

	LMTO-LDA	PPPW-LDA[b]	GW
Γ_1^v	-14.86	-15.1	-17.0
Γ_{15}^v	0.00	0.0	0.0
Γ_1^c	4.52 (6.0)[a]	4.2	6.0
Γ_{15}^c	13.18 (14.7)	12.3	14.6
X_1^v	-12.12	-12.3	-14.3
X_3^v	-4.96	-5.0	-5.6
X_5^v	-1.79	-1.8	-2.1
X_1^c	3.36 (4.9)	3.2	4.9
X_3^c	8.75 (10.25)	8.4	10.5
L_1^{v1}	-12.77	-12.9	-14.9
L_1^{v2}	-5.93	-6.0	-6.7
L_3^v	-0.48	-0.5	-0.6
L_1^{c1}	7.65 (9.15)	7.3	9.3
L_1^{c2}	10.20 (11.7)	10.0	12.6
L_3^c	11.33 (12.8)	11.0	13.2

[a] Values in parentheses include bandgap correction estimated as described in text.
[b] Rubio et al [19].

TABLE 4 Band widths and bandgaps (in eV) for zinc blende AlN.

	LMTO-LDA	PPPW-LDA[b]	GW[b]
N 2s band width	2.74	2.8	2.7
N 2p band width	5.93	6.0	6.7
N 2s-N 2p gap	6.19	6.3	7.6
total valence band	14.86	15.1	17.0
minimum gap (indirect Γ_{15}^v — X_1^c)	3.36(4.9 $\pm$ 0.2)[a]	3.2	4.9
minimum direct gap at Γ	4.52 (6.24)	4.2	6.0

[a] Gap correction estimated as described in text.
[b] Rubio et al [19].

TABLE 5 Properties of rocksalt AlN, transition pressure p_t, and minimum bandgap at p_t and theoretical equilibrium.

	p_t (GPa)	E_g (p = 0) (eV)	E_g (p_t) (eV)		Ref
expt. (visual)	16 - 17				[3]
expt. (X-ray, 300 K)	22.9				[4]
expt. (quenching 1400 - 1600 K)	16.5				[17]
LMTO-ASA	16.6	4.5[a]	4.7[a]	$\Gamma \to X$	[3,18]
LMTO-FP	12.5				[6]
PP-PW	12.9	4.04[a]		$\Gamma \to X$	[7]

[a] Lattice constants a(p = 0) = 4.06 Å, a(p_t) = 3.99 Å [18].
[b] Lattice constants a(p = 0) = 4.06 Å [7].

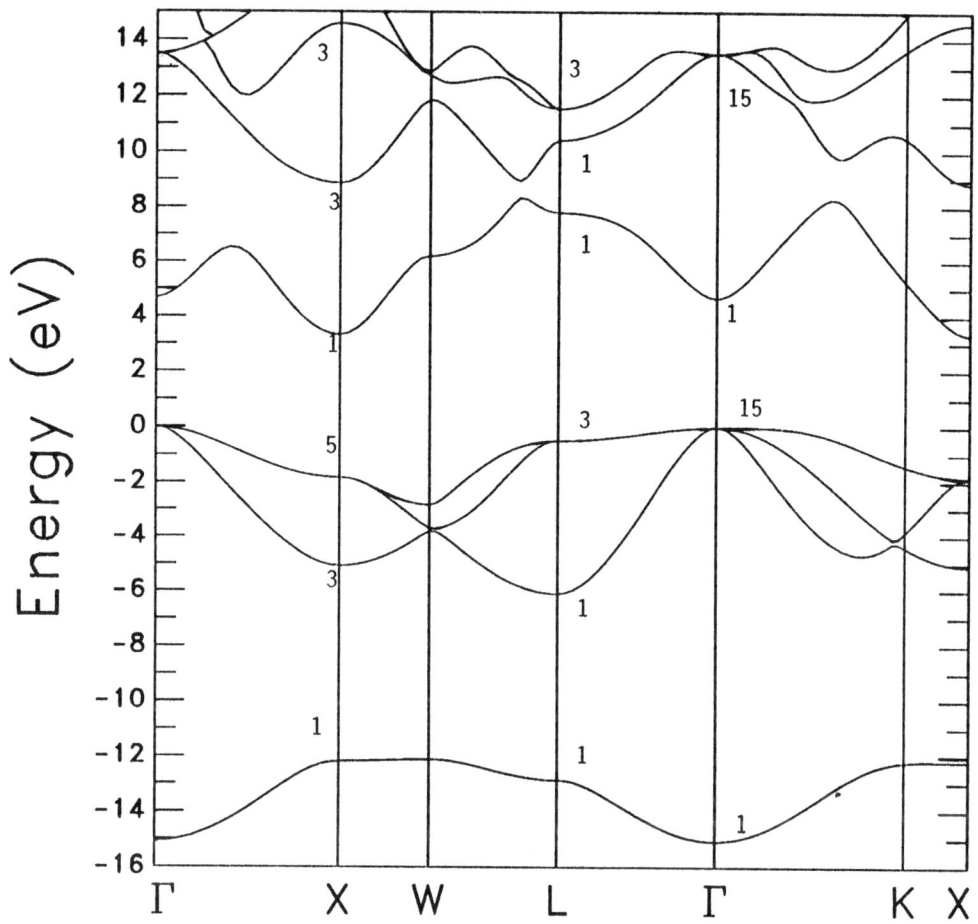

FIGURE 2 Band structure of zinc blende AlN.

NOTE ADDED WHILE IN PRINT

After completion of our Datareview, Rubio et al [19] published complete band structure results for wurtzite and zinc blende AlN calculated using the GW approach. These results include self-energy quasiparticle corrections not included in any of the LDA results. Both their LDA and GW results are included in the tables. Additional data on the eigenvalues at the L, A and H points of the hexagonal Brillouin zone can by found in [19]. We note that the quasiparticle self-energy corrections are not only important for the gap but also for the N 2s band. They also have an effect on the band widths and show a non-negligible k and state dependence. There are a few inconsistencies in the symmetry labelling of the states. We note that our symmetry labelling was checked explicitly by inspection of the eigenvectors. The PPPW-LDA results are in good agreement with the LMTO results except for an overall ~ 0.5 eV lower conduction band for wurtzite which may be due to unconverged basis sets or different use of lattice constants.

REFERENCES

[1]   W.Y. Ching, B.N. Harmon [*Phys. Rev. B (USA)* vol.34 (1986) p.5305]
[2]   T.-N. Xu, W.Y. Ching [*Phys. Rev. B (USA)* vol.48 (1993) in press]
[3]   I. Gorczyca, N.E. Christensen, P. Perlin, I. Grzegory, J. Jun, M. Bockowski [*Solid State Commun. (USA)* vol.79 (1991) p.1033]
[4]   M. Ueno, A. Onodera, O. Shimomura, K. Takemura [*Phys. Rev. B (USA)* vol.45 (1992) p.10123]
[5]   P. Perlin, I. Gorczyca, S. Porowski, T. Suski, N.E. Christensen, A. Polian [*Jpn. J. Appl. Phys. (Japan)* vol.32 suppl.32-1 (1993) p.334]
[6]   N.E. Christensen, I. Gorczyca [*Phys. Rev. B (USA)* vol.47 (1993) p.4307]
[7]   P.E. Van Camp, V.F. Van Doren, J.T. Devreese [*Phys. Rev. B (USA)* vol.44 (1991) p.9056]
[8]   A. Munoz, K. Kunc [*Physica B (Netherlands)* vol.185 (1993) p.422]
[9]   S. Bloom [*J. Phys. Chem. Solids (UK)* vol.32 (1971) p.2027]
[10]   A. Kobayashi, O.F. Sankey, S.M. Volz, J.D. Dow [*Phys. Rev. B (USA)* vol.28 (1983) p.935]
[11]   M.Z. Huang, W.Y. Ching [*J. Phys. Chem. Solids (UK)* vol.46 (1985) p.977]
[12]   D. Schwabe, W. Mader [*Proc. 1ˢᵗ Eur. Conf. Adv. Mater. Processes* Eds H.E. Exner, V. Schumacher (DGM Informationsges., Oberursel, Germany, 1990) vol.2 p.1267-72]
[13]   C.-Y. Yeh, Z.W. Lu, S. Froyen, A. Zunger [*Phys. Rev. B (USA)* vol.46 (1992) p.10086]
[14]   W.R.L. Lambrecht, B. Segall [*Phys. Rev. B (USA)* vol.43 (1991) p.7070]
[15]   P.B. Perry, R.F. Rutz [*Appl. Phys. Lett. (USA)* vol.33 (1978) p.39]
[16]   M. Gautier, J.P. Duraud, C. Le Gressus [*J. Appl. Phys. (USA)* vol.61 (1987) p.574]
[17]   H.Volstadt, E. Ito, M. Akaishi, S. Akimoto, O. Fukunara [*Proc. Jpn. Acad. B (Japan)* vol.66 (1991) p.7]
[18]   I. Gorczyca, N.E. Christensen [private communication]
[19]   A. Rubio, J.L. Corkhill, M.L. Cohen, E. Shirley, S.G. Louie [*Phys. Rev. B (USA)* vol.48 (1993) p.11810-5]

4.4 Band structure of pure GaN

W.R.L. Lambrecht and B. Segall

August 1994

A   INTRODUCTION

For general comments and notations the reader is referred to Datareview 4.1.

B   WURTZITE GaN

The wurtzite structure is the stable phase at ambient pressure and has lattice parameters (see ref [1] of Datareview 4.2):

a = 3.180 Å, c = 5.166 Å (c/a = 1.624), u ~ 0.375

Recent band structure calculations were reported by Gorczyca and Christensen [1] and Lambrecht and Segall (see ref [1] of Datareview 4.1) both using the LMTO-ASA method in the LDA and Xu and Ching [25] using the OLCAO-LDA method. Norm-conserving pseudopotential calculations were reported by several groups [2-5] emphasizing total energy properties. Only Palummo et al [5] give details of the band structure in tabular form. The older semi-empirical calculations [26] were judged to be less reliable.

Some discussion is necessary of the effects of the Ga 3d semi-core bands. Using the LDA 'all-electron' calculations, the latter overlap with the N 2s bands forming two separate sets of bands with energy ranges - 15.8 to - 14.75 eV and - 13.44 to - 11.05 eV. In pseudopotential calculations, the Ga 3d bands are treated as core states and thus do not show up in the band structures. We find the inclusion of Ga 3d band-dispersion important for an accurate description of the self-consistent potential and the equilibrium properties. In particular, the usual core-charge density renormalization effects lead to errors when treating the Ga 3d states as core states because of the small lattice constant and hence small Ga Wigner-Seitz sphere.

However, photoemission data of Hunt et al [6] indicate that the Ga 3d quasiparticle band occurs below the N 2s band at ~ 18 eV below the valence band maximum. This shift of the semi-localized state can be explained (and calculated) as a combination of a self-interaction correction and final state screening effects [7], i.e. effects beyond the LDA single particle picture. The position of the Ga 3d band calculated including these effects is given in TABLE 2.

Due to the above mentioned effects beyond LDA, the Ga 3d states become decoupled from the N 2s band. Thus, for comparison to photoemission and related spectroscopies probing the N 2s band region, it is desirable to remove the coupling effect between the Ga 3d and N 2s states. One should clearly note here that these spectroscopies probe the quasiparticle band structure while the ground state properties are based on the Kohn-Sham eigenvalues. In the former the Ga 3d and N 2s appear as separate bands while in the latter they overlap.

In the LMTO method, one can easily accomplish the decoupling by choosing a high 'linearization energy' (E_v in LMTO jargon) for the Ga d partial waves such that the latter become essentially 4d-like. The resulting bands are then similar to the ones of the PP-PW calculation [5]. We note, however, that in our treatment the Ga 3d band dispersion is properly included in the determination of the potential.

TABLE 1   Selected eigenvalues (in eV) for wurtzite GaN with respect to valence-band maximum (Γ_1^{v2}).

	LMTO-LDA[a]	PP-PW[b]	PP-PW[d]	GW[d]
Γ_1^{v1}	-15.80	-15.98	-16.3	-18.2
Γ_3^{v1}	-13.35	-13.57	-13.8	-15.7
Γ_3^{v2}	-7.00	-7.18	-7.4	-8.0
Γ_5^{v}	-0.97	-0.95	-1.1	-1.2
Γ_6^{v}	-0.02	-0.08	-1.1	-1.2
Γ_1^{v2}	0.00	0.00	0.0	0.0
Γ_1^{c1}	2.42 (3.65)	2.76	2.3	3.5
Γ_3^{c1}	4.30 (5.52)	4.76	4.6	5.9
Γ_1^{c2}	9.80 (11.03)	9.91	9.5	12.1
Γ_6^{c2}	10.97 (12.20)	10.04	10.1	11.9
K_3^{v1}	-12.73	-13.04	-13.2	-15.2
K_1^{v1}	-5.47	-5.37	-5.6	-6.1
K_3^{v2}	-5.24	-5.29	-5.5	-6.1
K_3^{v3}	-2.96	-3.02	-3.2	-3.5
K_2^{v2}	-2.85	-2.80	-3.0	-3.2
K_2^{c1}	5.74 (6.97)	4.93	4.9	6.6
K_3^{c1}	8.14 (9.36)	8.63	8.3	10.6
K_1^{c2}	8.62 (9.85)	8.64	8.6	10.8
K_3^{c2}	10.44 (11.67)	10.10		
M_3^{v1}	-13.42	-13.68	-13.9	-15.8
M_1^{v1}	-12.55	-12.84	-13.0	-15.0
M_1^{v1}	-6.49	-6.56	-6.8	-7.4
M_1^{v2}	-5.32	-5.35	-5.6	-6.1
M_1^{v3}	-4.18	-4.25	-4.4	-4.9
M_2^{v}	-2.75	-2.57	-2.8	-2.6
M_3^{v3}	-2.14	-2.29	-2.4	-2.6
M_4^{v}	-0.93	-0.88	-1.0	-1.1
M_1^{c1}	5.22 (6.45)	5.02	5.1	6.5
M_3^{c1}	5.79 (7.02)	5.73	5.7	7.4
M_3^{c2}	6.63 (7.86)	6.47	6.2	8.1
M_1^{c2}	8.98 (10.20)	9.93	9.1	11.5

[a] Present calculation expt. c/a., a, u = 0.375; N 2s bands (below - 10.0 eV) calculated without Ga 3d hybridization.
[b] Palummo et al [5].
[c] Values in parentheses include gap correction.
[d] Rubio et al [32].

The N 2s bands (below - 10 eV) given in TABLE 1 were calculated using the above procedure, i.e. without explicit hybridization to Ga 3d in the final band calculation, but with hybridization included in construction of the potential. The upper valence band was calculated including the coupling to Ga 3d. In FIGURES 1(a) and 1(b), the upper valence bands as calculated respectively with or without Ga 3d hybridization are indistinguishable. The labels of the valence-band eigenvalues do not include the Ga 3d bands in the numbering so as to be compatible with those of AlN and BN. In TABLE 2, we give separately the N 2s and Ga 3d 'quasiparticle' band positions which are comparable to photoemission and the hybridized (Kohn-Sham) bands as resulting from LDA (for comparison to other calculations addressing the ground state properties).

TABLE 2 Band widths, bandgaps, bandgap deformation potential (in eV) and bandgap pressure coefficient of wurtzite GaN.

	Theory	Expt.
Ga 3d band centre	-18[a]	-18[d]
N 2s band minimum	-15.8[a], -15.98[g], -16.3[l], -18.2[m]	-15.5[d]
N 2s band width	3.4[a], 3.14[g], 3.5[l], 3.2[m]	3[d]
N 2s-Ga 3d LDA lower band width	1.08[b], 0.98[c]	
N 2s-Ga 3d LDA gap	1.34[b], 1.39[c]	
N 2s-Ga 3d LDA upper band	2.42[b], 2.65[c]	
N 2p band width	7.00[b], 7.26[c], 7.18[g], 7.4[l], 8.0[m]	7[d]
N 2s-N 2p gap	5.4[a], 5.66[g]	
	4.04[b], 3.76[c]	4.5[d]
Total valence band width	15.8[b], 16.03[c], 15.98[g], 16.3[l], 18.2[m]	16[d]
Minimum bandgap	2.42[b], 2.71[c]	
	1.63[f], 2.76[g], 2.3[l], 3.5[m]	3.65[h], 3.44[e]
Bandgap pressure coefficient (dE_g/dp) (meV/GPa)[k]	41[j]	47[e], 42[i]
Bandgap deformation potential $(dE_g/d\ln V)$[k]	-9.8[j]	

[a] ΔSCF calculation of Ga 3d, N 2s band calculated without Ga 3d hybridization [7].

[b] LDA LMTO-ASA bands including Ga 3d hybridization [7]; Christensen (private communication) obtained 2.48 eV also using ASA-LMTO but averaging the interaction with Ga 3d and Ga 4d and using a two-panel calculation for the potential.

[c] Xu and Ching [25].

[d] Band widths and gaps are estimates based on UPS spectrum measured by Hunt et al [6].

[e] Perlin et al [30] measured value at room temperature.

[f] Van Camp et al [2].

[g] Palummo et al [5].

[h] Dingle et al [22] optical absorption and reflection at 2 K.

[i] Camphausen et al [9].

[j] Perlin et al [8].

[k] $dE_g/dp = -(dE_g/d\ln V)/B$ with bulk modulus B = 239 GPa [8].

[l] Rubio et al [32] LDA.

[m] Rubio et al [32] GW.

Perlin et al [8] reported experimental results for the pressure dependence of the minimum bandgap,

$$E_g(P) = E_g(P = 0) + 0.047\,P - 0.0018\,P^2$$

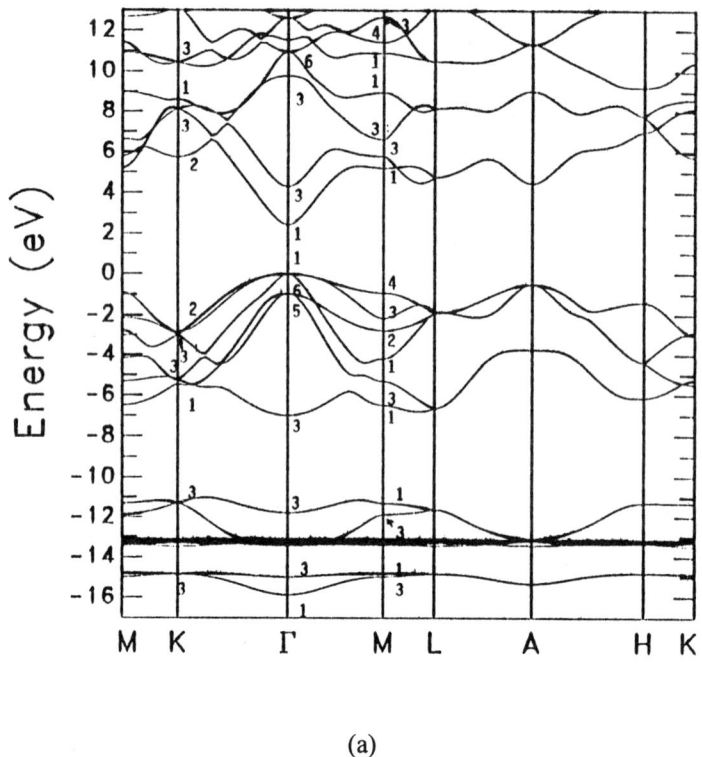

(a)

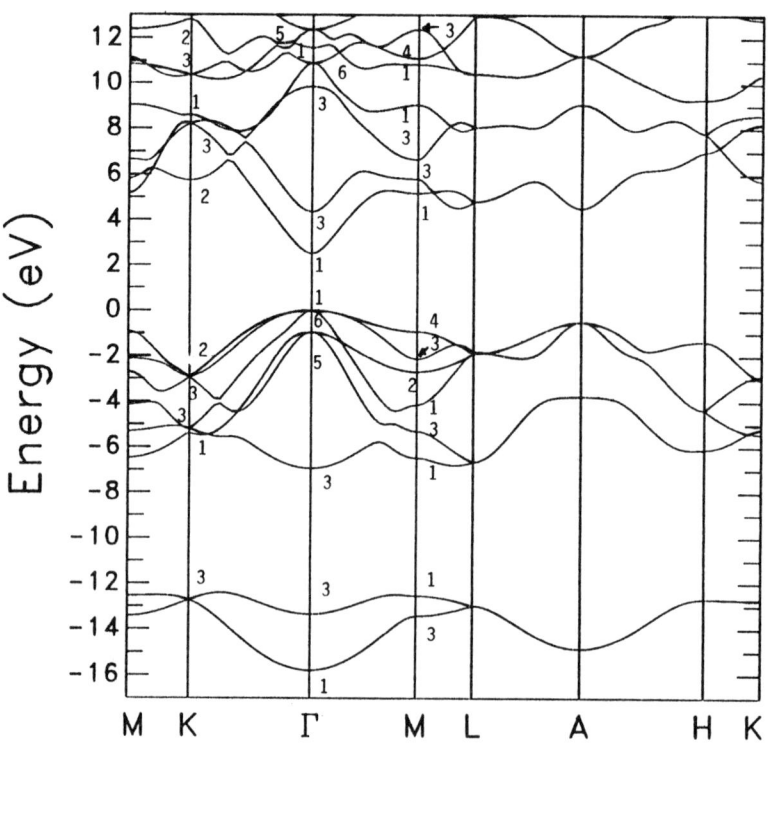

(b)

FIGURE 1 Band structure of wurtzite GaN: (a) including Ga 3d hybridization, (b) without Ga 3d hybridization. Note that both use the same potential and lattice constant.

with the gap in eV and the pressure in GPa, extending the previous determinations of the bandgap pressure coefficient by Camphausen et al [9]. $\partial E_g/\partial P = 42\,\text{meV GPa}^{-1}$ and is in good agreement with the calculated value of Gorczyca and Christensen [1] ($41\,\text{meV GPa}^{-1}$).

Calculated pressure coefficients for other bandgaps can be found in Perlin et al [8].

C   ZINC BLENDE GaN

The zinc blende phase (see FIGURE 2 for the band structure) is very close in total energy to the wurtzite phase and has been stabilized by epitaxial growth on GaAs {001} substrates [10-12], on β-SiC {001} substrates [13] and on MgO [14]. The total energy difference between zinc blende and wurtzite was calculated to be 15 meV/atom by Munoz and Kunc [3], and 10 meV/atom by Yeh et al [27] in favour of the wurtzite structure. Min et al [4] and Palummo et al [5] obtained the opposite ordering with energy differences 9 and 35 meV/atom respectively. Experimentally, wurtzite is known to be the groundstate structure.

The value for the lattice constant used below for our own calculations assumes the same volume per atom as for the wurtzite and is a = 4.49 Å.

LMTO-ASA calculations were reported by Lambrecht and Segall (ref [1] of Datareview 4.1), and Fiorentini et al [18] reported FP-LMTO calculations. These authors give contour plots of the wave functions at certain k-points. The bands are in excellent agreement with ASA-LMTO given below. Remarks about the Ga 3d band similar to those for wurtzite are applicable here.

Previous semi-empirical calculations [19] predicted a very small change in bandgap from wurtzite but these calculations used the same pseudopotential atomic form factors for wurtzite and zinc blende. Also, Palummo et al [5] found the gap of zinc blende to be only 0.11 eV lower than that of wurtzite while our ASA-LMTO calculations predict a difference of 0.41 eV. (Christensen and Gorczyca obtain a difference of 0.3 eV using similar LMTO-ASA calculations with a slightly different treatment of the Ga 3d's.)

Palummo et al [20] reported a GW-calculation of zinc blende GaN. The bandgap obtained was 3.60 eV, while their LDA value was 2.65 eV, indicating a 0.95 eV self-energy correction. Very recently, using a somewhat different pseudopotential, converged calculations with a 100 Ry cut-off and including non-linear core-corrections, they found the LDA gap to be decreased to 2.21 eV (in closer agreement to the LMTO all-electron calculations), while the self-energy correction, 0.97 eV, was almost unchanged from their previous result. This gives a final gap of 3.18 eV calculated from GW and PP-PW calculations [21].

Experimental data on the bandgap of zinc blende GaN have been determined on thin films. Strite et al [12] concluded that the gap at 53 K would be slightly larger than 3.52 eV. This is only ~ 0.1 eV lower than for wurtzite. Strite et al [12] give data on the temperature dependence of cathodoluminescence peaks. At temperatures above 200 K, a linear dependence with slope - $3.5 \times 10^{-4}\,\text{eV K}^{-1}$ is obtained, leading to a room temperature value of the gap of 3.45 eV. It should be noted that this value and the 3.52 eV value at low temperature are based

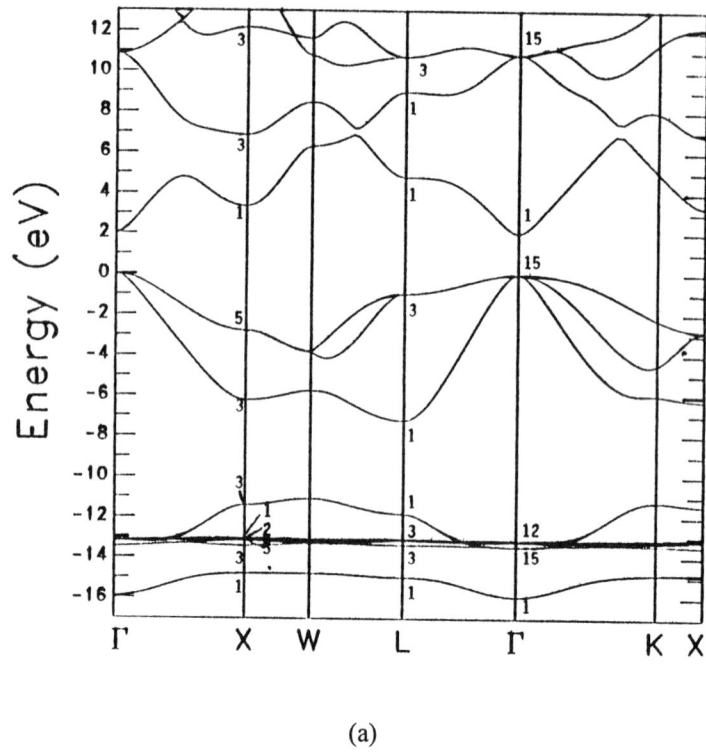

(a)

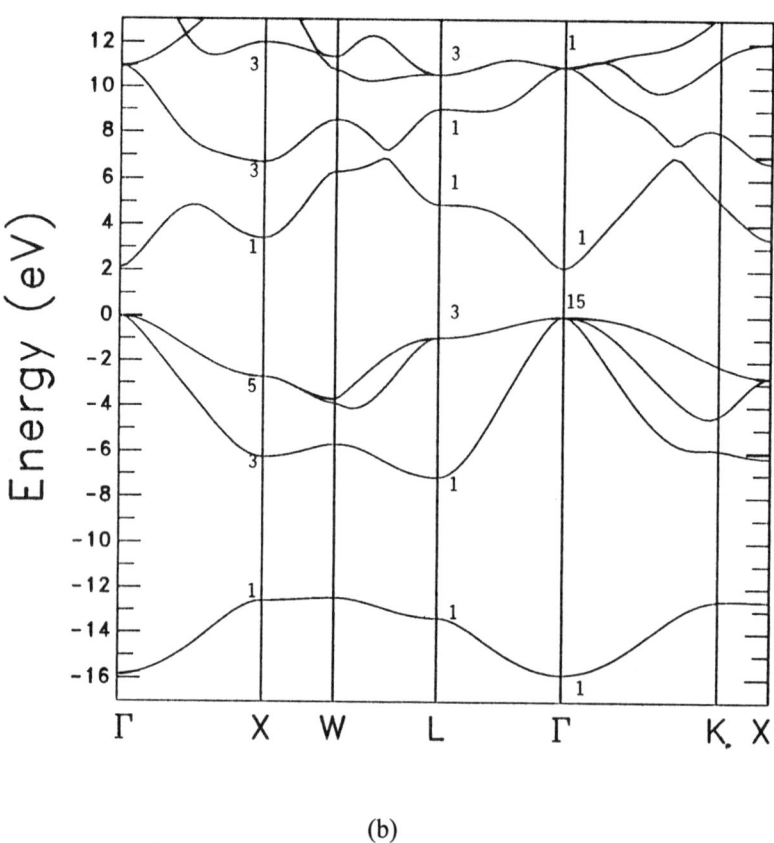

(b)

FIGURE 2 Band structure of zinc blende GaN: (a) including Ga 3d hybridization, (b) without Ga 3d hybridization. Note that both use the same potential and lattice constant.

on interpreting the a'-peak as being closest to the bandgap. An alternative interpretation [16], however, is to assign this peak to small wurtzitic crystallites in the film and the main α-peak to the corresponding bandgap exciton emission of zinc blende GaN. In that case, the zero temperature exciton bandgap E_{gx} would be 3.266 eV and the room temperature gap would be 3.20 eV. This value is in better agreement with measurements by Powell et al who gave a value of 3.3 eV for the room temperature gap of GaN grown on MgO and by Lei et al who found 3.2 eV for GaN grown on Si. It thus seems likely that the zinc blende gap is about 0.3 eV lower than the wurtzite gap, in agreement with the all electron LMTO-ASA calculations. Data on single crystals are not yet available. TABLES 3 and 4 summarise the eigenvalues, band widths and bandgaps of zinc blende GaN.

TABLE 3 Selected eigenvalues (in eV) for zinc blende[a] GaN with respect to valence-band maximum (Γ_{15}^v).

	LMTO-LDA	PPPW-LDA[d]	GW[d]
Γ_1^v	-15.82[b]	-16.3	-17.8
Γ_{15}^v	0.00	0.0	0.0
Γ_1^c	2.01 (3.24)[c]	2.1	3.1
Γ_{15}^c	10.86 (12.10)	10.6	12.2
X_1^v	-12.58	-13.0	-14.8
X_3^v	-6.23	-6.5	-6.9
X_5^v	-2.79	-2.8	-3.0
X_1^c	3.34 (4.57)	3.2	4.7
X_3^c	6.86 (8.09)	6.9	8.4
L_1^{v1}	-13.35	-13.8	-15.5
L_1^{v2}	-7.24	-7.4	-7.8
L_3^v	-0.98	-1.0	-1.1
L_1^{c1}	4.81 (6.04)	5.0	6.2
L_1^{c2}	8.99 (10.22)	9.1	11.2
L_3^c	10.75 (11.98)	10.6	12.3

[a] At same volume/atom as wurtzite.
[b] N 2s band (below - 10 eV) calculated without Ga 3d hybridization.
[c] Values in parentheses include same gap correction as for wurtzite.
[d] Rubio et al [32].

D   ROCKSALT GaN

Rocksalt is a high-pressure phase of GaN [23,24]. Other possible high-pressure phases, e.g. the NiAs structure, were discussed by Munoz and Kunc [3,28] and Gorczyca and Christensen [1,29].

TABLE 4 Band widths and bandgaps (in eV) for zinc blende GaN.

	Theory	Expt.
N 2s band width	3.4, 3.3[i], 3.0[j]	
N 2p band width	6.2, 7.4[i], 7.8[j]	
N 2s-N 2p gap	5.3, 5.6[i], 7.0[j]	
Total valence band	15.8, 16.3[i], 17.8[j]	
Minimum gap (direct $\Gamma_{15}^v - \Gamma_1^{c1}$)	2.01[a], (3.24)[b]	3.52[f]
	1.48[c], 2.65 (2.21)[d]	3.3[g]
	3.60 (3.18)[e], 2.1[i], 3.1[j]	3.2[h]

[a] LDA-ASA-LMTO value with coupling to Ga 3d at a = 4.50 Å; using an LMTO two-panel calculation and averaging Ga 3d and Ga 4d interaction, Christensen (private communication) obtains 2.17 eV at a = 4.47 Å.
[b] Assuming that the bandgap correction is the same as for wurtzite (namely 1.23 eV).
[c] Van Camp et al [2].
[d] Palummo et al [5] (PP-PW-LDA). The value in parenthesis is a recently obtained refined value [21].
[e] Palummo et al [20] (GW). The value in parenthesis is a recently obtained refined value [21].
[f] Strite et al [12] at 53 K from cathodoluminescence on epitaxial films; however, see text for an alternative interpretation of the spectra which gives E_{gx} = 3.266 eV at 0 K.
[g] Powell et al [14], room temperature.
[h] Lei et al [15], room temperature.
[i] Rubio et al [32] LDA.
[j] Rubio et al [32] GW.

TABLE 5 Properties of rocksalt GaN: transition pressure p_t, and minimum bandgaps at p_t and theoretical equilibrium.

Comment	p_t (GPa)	E_g (p = 0) (eV)	E_g (p_t) (eV)		Ref
expt.	47 - 50				[30]
LMTO-ASA	65	0.6[a]	1.7[a]	$\Sigma \to X^b$	[30,31]
PP-PW (cut-off 70 Ry)	55	0.5[c]		$\Gamma \to X$	[3]
PP-PW (cut-off 34 Ry)	55	1.01[d]		$\Gamma \to X$	[2]

[a] Lattice constants a(p = 0) = 4.17 Å, a(p_t) = 3.94 Å [31].
[b] Valence-band maximum at ~ 0.5(¾,¾,0)2π/a.
[c] At theoretical equilibrium lattice constant 4.22 Å.
[d] At theoretical equilibrium lattice constant 4.098 Å.

NOTE ADDED WHILE IN PRINT

After completion of our Datareview, Rubio et al [32] published complete band structure results for wurtzite and zinc blende GaN calculated using the GW approach. These results include self-energy quasiparticle corrections not included in any of the LDA results. Both their LDA and GW results are included in the tables. Additional data on the eigenvalues at the L, A and H points of the hexagonal Brillouin zone can be found in [32]. The LDA results are in good agreement with the LMTO results given here, but do not include the Ga 3d states. We note that the quasiparticle self-energy corrections are not only important for the gap but also for the N 2s band. They also have an effect on the band widths and show a non-negligible k and state dependence.

REFERENCES

[1]   I. Gorczyca, N.E. Christensen [*Solid State Commun. (USA)* vol.80 (1991) p.335]

[2]   P.E. Van Camp, V.E. Van Doren, J.T. Devreese [*Solid State Commun. (USA)* vol.81 (1992) p.23]

[3]   A. Munoz, K. Kunc [*Phys. Rev. B (USA)* vol.44 (1991) p.10372]

[4]   B.J. Min, C.T. Chan, K.M. Ho [*Phys. Rev. B (USA)* vol.45 (1992) p.1159]

[5]   M. Palummo, C.M. Bertoni, L. Reining, F. Finochi [*Physica B (Netherlands)* vol.185 (1993) p.404]

[6]   R.W. Hunt et al [*Physica B (Netherlands)* vol.185 (1993) p.415-21]

[7]   W.R.L. Lambrecht, B. Segall [*Bull. Am. Phys. Soc. (USA)* vol.38 (1993) p.622, and paper in preparation]

[8]   P. Perlin, I. Gorczyca, N.E. Christensen, I. Grzegory, H. Teisseyre, T. Suski [*Phys. Rev. B (USA)* vol.45 (1992) p.13307-13]

[9]   D.L. Camphausen, G.A. Neville Conell, W. Paul [*Phys. Rev. Lett. (USA)* vol.26 (1971) p.184]

[10]   M. Mizuta, S. Fujieda, Y. Matsumoto, T. Kawamura [*Jpn. J. Appl. Phys. (Japan)* vol.25 (1986) p.L945]

[11]   G. Martin, S. Strite, J. Thornton, H. Morkoc [*Appl. Phys. Lett. (USA)* vol.58 (1991) p.21]

[12]   S. Strite et al [*J. Vac. Sci. Technol. B (USA)* vol.9 (1991) p.192]

[13]   M.J. Paisley, Z. Sitar, J.B. Posthill, R.F. Davis [*J. Vac. Sci. Technol. A (USA)* vol.7 (1989) p.701]

[14]   R.C. Powell, G.A. Tomasch, Y.-W. Kim, J.A. Thornton, J.E. Greene [*Mater. Res. Soc. Symp. Proc. (USA)* vol.162 (1990) p.525]

[15]   T. Lei, T.D. Moustakas, R.J. Graham, Y. He, S.J. Berkowitz [*J. Appl. Phys. (USA)* vol.71 (1992) p.4933]

[16]   S. Strite [private communication]

[17]   H. Volstadt, E. Ito, M. Akaishi, S. Akimoto, O. Fukunara [*Proc. Jpn. Acad. B (Japan)* vol.66 (1991) p.7]

[18]   V. Fiorentini, M. Methfessel, M. Scheffler [*Phys. Rev. B (USA)* vol.48 (1993) p.13353-62]

[19]   S. Bloom, G. Harbeke, E. Meier, I.B. Ortenburger [*Phys. Status Solidi B (Germany)* vol.66 (1974) p.161]

[20]   M. Palummo, L. Reining, R.W. Godby, C.M. Bertoni [in *Proc. 21ˢᵗ Int. Conf. on the Physics of Semiconductors*, Beijing, Aug.1992, Eds Ping Jiang, Hou-Zhi Zheng (World Scientific, Singapore, 1993) p.89]

[21]   C.M. Bertoni [private communication]

[22]   R. Dingle, D.D. Sell, S.E. Stokowski, P.J. Dean, R.B. Zetterstrom [*Phys. Rev. B (USA)* vol.3 (1971) p.497]

[23]   P. Perlin, C. Jauberthie-Carillon, J.P. Ithie, A. San Miguel, I. Grzegory, A. Polian [*Phys. Rev. B (USA)* vol.45 (1992) p.83]

[24]   H. Xia, Q. Xia, A.L. Ruoff [*Phys. Rev. B (USA)* vol.47 (1993) p.12925]

[25]   T.-N. Xu, W.Y. Ching [*Phys. Rev. B (USA)* vol.48 (1993) in press]

[26]   S. Bloom [*J. Phys. Chem. Solids (UK)* vol.32 (1971) p.2027]
[27]   C.-Y. Yeh, Z.W. Lu, S. Froyen, A. Zunger [*Phys. Rev. B (USA)* vol.46 (1992) p.10086]
[28]   A. Munoz, K. Kunc [*Physica B (Netherlands)* vol.185 (1993) p.422]
[29]   I. Gorczyca, N.E. Christensen [*Physica B (Netherlands)* vol.185 (1993) p.410]
[30]   P. Perlin, I. Gorczyca, S. Porowski, T. Suski, N.E. Christensen, A. Polian [*Jpn. J. Appl. Phys. (Japan)* vol.32 suppl.32-1 (1993) p.334]
[31]   I. Gorczyca, N.E. Christensen [private communication]
[32]   A. Rubio, J.L. Corkhill, M.L. Cohen, E. Shirley, S.G. Louie [*Phys. Rev. B (USA)* vol.48 (1993) p.11810-5]

4.5 Band structure of pure InN

W.R.L. Lambrecht and B. Segall

August 1994

A INTRODUCTION

For general comments and notations the reader is referred to Datareview 4.1.

B WURTZITE InN

The wurtzite structure (see FIGURE 1 for the band structure of wurtzite InN) is the stable phase at ambient pressure and has lattice parameters (see ref [1] of Datareview 4.2):

$$a = 3.533\,\text{Å}, \ c = 5.693\,\text{Å} \ (c/a = 1.611), \ u \sim 0.375$$

The In 4d bands hybridize with the N 2s band in the LDA and give rise to two separate sets of bands in the ranges - 14.59 to - 14.44 eV and - 13.06 to - 12.65 eV. Undoubtedly these bands will be found at a few eV lower energy in photoemission spectroscopy if screening and self-interaction corrections (beyond LDA) are included (as discussed for the Ga 3d bands in GaN). Calculations of the shift have not yet been performed. Photoemission measurements are not available either to our knowledge. In TABLE 1, we give the N 2s bands without hybridization to the In 4d and the upper valence band including it. We note that including the hybridization interchanges the position of the Γ_6^v and Γ_1^{v2} levels, making the Γ_6^v the valence-band maximum. For other calculations of InN and related alloys, see Datareview 5.1 by D. Jenkins in this book.

C ZINC BLENDE InN

The zinc blende structure (see FIGURE 2) has been observed by Strite et al [1] but optical or electronic properties have not been reported. Yeh et al [2] calculated the energy difference between zinc blende and wurtzite to be 9.88 meV/atom. We performed ASA-LMTO calculations assuming the lattice constant obtained by assuming the same volume per atom as for wurtzite, i.e. a = 4.97 Å. Due to the slight bandgap reduction at Γ from the wurtzite phase and the LDA bandgap underestimate the band structure becomes semimetallic. TABLES 3 and 4 summarise the eigenvalues, band widths and bandgaps for the zinc blende polytype.

D ROCKSALT InN

A phase transition to rocksalt has been predicted by Gorczyca and Christensen [3] and was observed by Perlin and by Ueno et al [4]. TABLE 5 summarises the relevant properties for this phase.

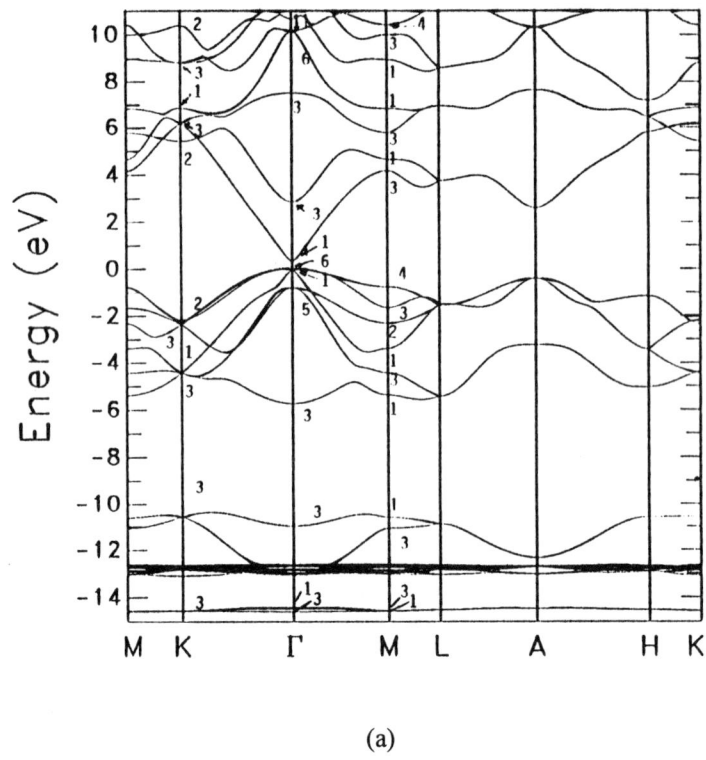

(a)

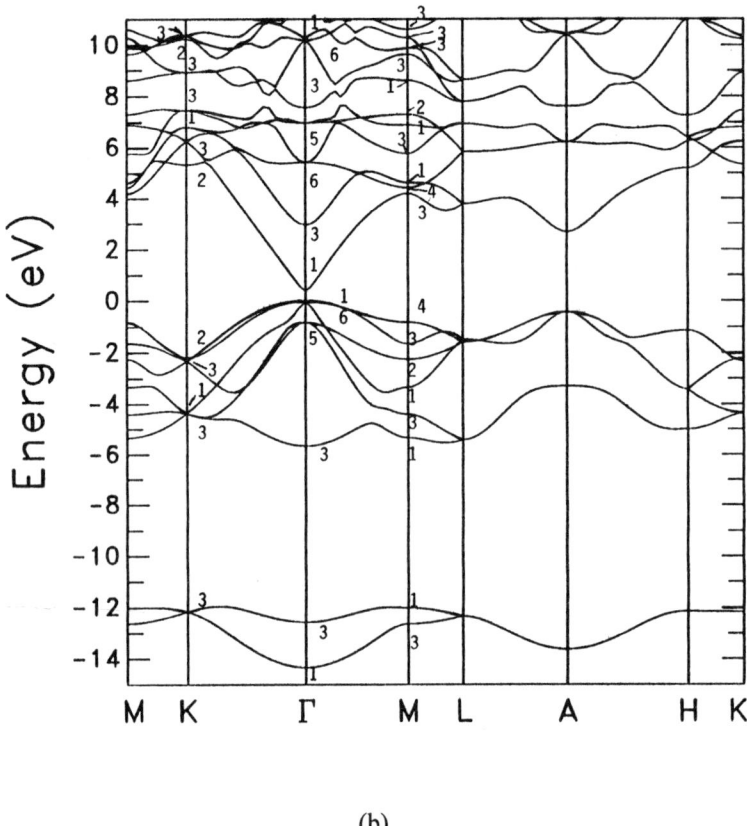

(b)

FIGURE 1 Band structure of wurtzite InN: (a) including In 4d hybridization, (b) without In 4d hybridization. Note that both use the same potential and lattice constant.

TABLE 1 Selected eigenvalues (in eV) for wurtzite InN with respect to valence-band maximum (Γ_1^{v2}).

Γ_1^{v1}	-14.34[a]
Γ_3^{v1}	-12.56
Γ_3^{v2}	-5.73
Γ_5^{v}	-0.80
Γ_1^{v}	-0.01
Γ_1^{v2}	0.00
Γ_1^{c1}	0.30 (2.05)[b]
Γ_3^{c1}	2.86 (4.61)
Γ_3^{c2}	7.51 (9.26)
Γ_6^{c}	10.19 (11.94)
K_3^{v1}	-12.16
K_3^{v2}	-4.45
K_1^{v1}	-4.42
K_3^{v3}	-2.35
K_2^{v}	-2.18
K_2^{c}	5.41 (7.16)
K_3^{c1}	6.18 (7.93)
K_1^{c1}	6.85 (8.6)
K_3^{c2}	8.83 (10.58)
M_3^{v1}	-12.62
M_1^{v1}	-12.01
M_1^{v2}	-5.38
M_3^{v2}	-4.44
M_1^{v3}	-3.39
M_2^{v}	-2.31
M_3^{v3}	-1.66
M_4^{v}	-0.76
M_3^{c1}	4.18 (5.93)
M_1^{c1}	4.65 (6.40)
M_3^{c2}	5.78 (7.53)
M_1^{c2}	6.83 (8.58)

[a] N 2s bands (below - 10 eV) calculated without In 4d hybridization.
[b] Values in parentheses include gap correction.

TABLE 2 Band widths, bandgaps, bandgap deformation potential (in eV) and bandgap pressure coefficient of wurtzite InN.

N 2s band width	2.39
N 2p band width	5.73
N 2s-N 2p gap	6.29
Total valence band width	14.34
Minimum gap (direct $\Gamma_{15}^{v} - \Gamma_1^{c1}$)	0.30[a], 0.65[b], 2.05[c]
Bandgap pressure coefficient (dE_g/dp) (meV/GPa)	25[b]
Bandgap deformation potential ($dE_g/d\ln V$)	-4.25[b]

[a] Present calculation LMTO-ASA at expt. a, c/a, u = 0.375.
[b] Gorczyca and Christensen [3], $dE_g/dp = -(dE_g/d\ln V)/B$ with bulk modulus B = 165 GPa.
[c] Tyagai et al [5].

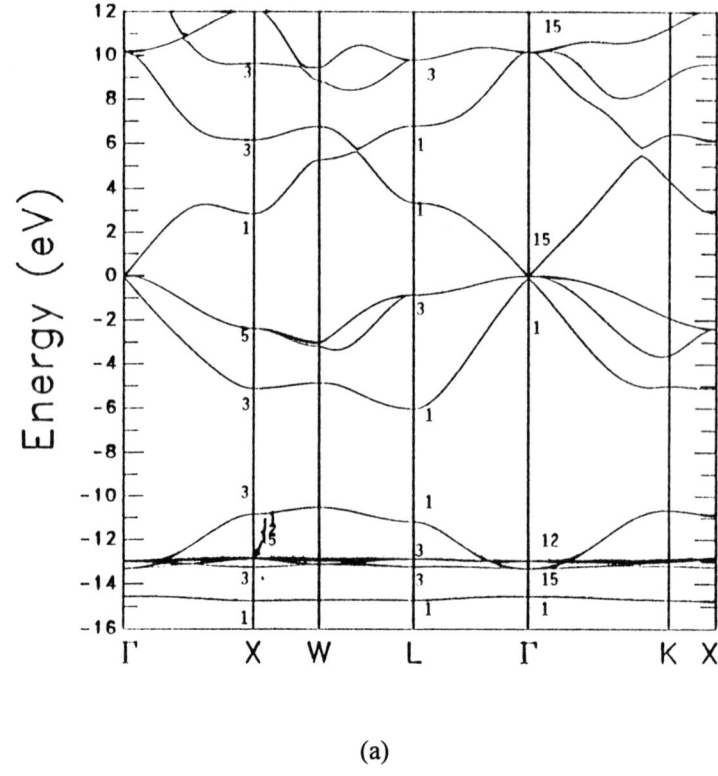

(a)

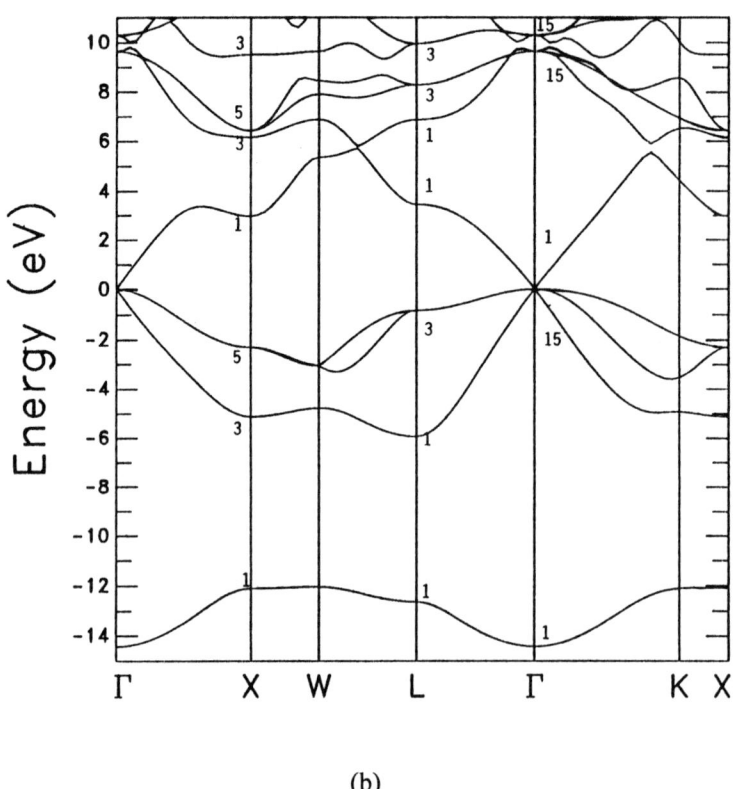

(b)

FIGURE 2 Band structure of zinc blende InN: (a) including In 4d hybridization, (b) without In 4d hybridization. Note that both use the same potential and lattice constant.

TABLE 3 Selected eigenvalues (in eV) for zinc blende[a] InN with respect to valence-band maximum.

Γ_1^v	-14.43[b]
Γ_{15}^v	0.00
Γ_1^c	-0.09 (1.66)[c]
Γ_{15}^c	10.18 (11.93)
X_1^v	-12.10
X_3^v	-5.12
X_5^v	-2.38
X_1^c	2.85 (4.60)
X_3^c	6.15 (7.90)
L_1^{v1}	-12.63
L_1^{v2}	-6.04
L_3^v	-0.86
L_1^{c1}	3.37 (5.12)
L_1^{c2}	6.79 (8.54)
L_3^c	9.81 (11.56)

[a] At same volume/atom as wurtzite.
[b] N 2s band (below - 10 eV) calculated without In 4d hybridization.
[c] Values in parentheses include same gap correction as for wurtzite.

TABLE 4 Band widths and bandgaps (in eV) for zinc blende InN.

N 2s band width	2.33[a]
N 2p band width	6.04
N 2s-N 2p gap	6.16
Total valence band width	14.43
Minimum gap (direct Γ_{15}^v — Γ_1^{c1})	-0.09, (1.66)[b]

[a] N 2s band calculated without In 4d hybridization.
[b] Value in parentheses assumes same gap correction as for wurtzite.

TABLE 5 Properties of rocksalt InN: transition pressure p_t and minimum bandgaps at p_t and theoretical equilibrium.

Comment	p_t (GPa)	E_g (p = 0) (eV)	E_g (p_t) (eV)		Ref
expt.	23.0				[3]
expt.	12.0				[4]
LMTO-ASA	25.4, 17.7[d]	0.7[a]	1.0[a]	$\Sigma \to \Gamma$[b]	[3,6]
PP-PW	5.0	0.7[c]			[7]

[a] Lattice constants a(p = 0) = 4.58 Å, a(p_t) = 4.47 Å [6].
[b] Valence-band maximum at $\sim 0.5(\frac{3}{4},\frac{3}{4},0)2\pi/a$.
[c] At lattice constant 4.57 Å.
[d] From a recent calculation with more k-points: recommended value [6].

REFERENCES

[1]   S. Strite et al [*J. Cryst. Growth (Netherlands)* vol.127 (1993) p.204-8]

[2]   C.-Y. Yeh, Z.W. Lu, S. Froyen, A. Zunger [*Phys. Rev. B (USA)* vol.46 (1992) p.10086]

[3]   I. Gorczyca, N.E. Christensen [*Physica B (Netherlands)* vol.185 (1993) p.410]

[4]   M. Ueno et al [*Jpn. J. Appl. Phys. (Japan)* vol.32 suppl.32-1 (1993) p.42]

[5]   V.A. Tyagai, A.M. Evstigneev, A.N. Krasiko, A.F. Andreeva, V.Ya. Malakhov [*Sov. Phys.-Semicond (USA)* vol.11 (1977) p.1257]

[6]   I. Gorczyca, N.E. Christensen [private communication]

[7]   A. Munoz, K. Kunc [*Physica B (Netherlands)* vol.185 (1993) p.422]

CHAPTER 5

BAND STRUCTURE OF GROUP III NITRIDE ALLOYS

5.1   Band structure of InN, GaInN and AlInN

D. Jenkins

June 1994

A   INTRODUCTION

InN is a material which has not been studied extensively and few features of the band structure have been investigated experimentally. Only recently has the semiconductor been grown in a crystalline wurtzite bulk form [1,2]. Previous growths resulted in polycrystalline samples which retain evidence of hexagonal wurtzite crystal structure. Polycrystalline samples have been grown which have high mobility ($1800 \, \text{cm}^2 \, \text{V}^{-1} \, \text{s}^{-1}$) and low electron concentration ($n = 5 \times 10^{16} \, \text{cm}^{-3}$) [3].

At present there is little experimental data on the band structure of wurtzite InN as a function of wave vector k. Several theoretical calculations have been made: two empirical pseudopotential models [4,5]; an empirical tight-binding model [6]; and an ab initio pseudofunction model [7]. Each of these models suffers some deficiencies. The empirical models are interpolations from other materials and must be adjusted to account for the gap and the ab initio model suffers from the well known underestimation of the bandgap. The ab initio model is believed to give a good representation of the valence bands and, if shifted to give the correct gap, a good representation of the conduction band. It is likely that it is the best band structure published to date. The pseudopotential models have been used to identify measured optical transitions. The tight binding model has been used to model the deep defect levels with modest success [8].

The pseudofunction band structure is shown in FIGURE 1. The band structure of InN has several features which are common to all models. The bandgap is direct and located at the centre of the Brillouin zone (Γ point, k = [000]). The experimentally determined value of the gap, $E_g = 1.89 \, \text{eV}$, is several electron volts lower than all other optical transitions. At the Γ point, there is a crystal field splitting of the valence band maximum and a level approximately 1 - 2 eV below the valence band maximum. The conduction band has no pronounced local minima other than at Γ and the valence band has no pronounced local maxima other than at Γ. This is common among wurtzite semiconductors, but differs from other cubic III-V materials.

B   REFLECTIVITY

Despite the lack of good material several facts about InN are known. The energy bandgap has been reported anywhere from 1.7 eV to 3.1 eV. Most material grown has significant electron concentrations on the order of $n \sim 10^{19} - 10^{20} \, \text{cm}^{-3}$ and exhibits a significant Moss-Burstein shift. Tansley and Foley report high mobility polycrystalline material [3] with $E_g = 1.89 \, \text{eV}$ correcting for the concentration effects [9].

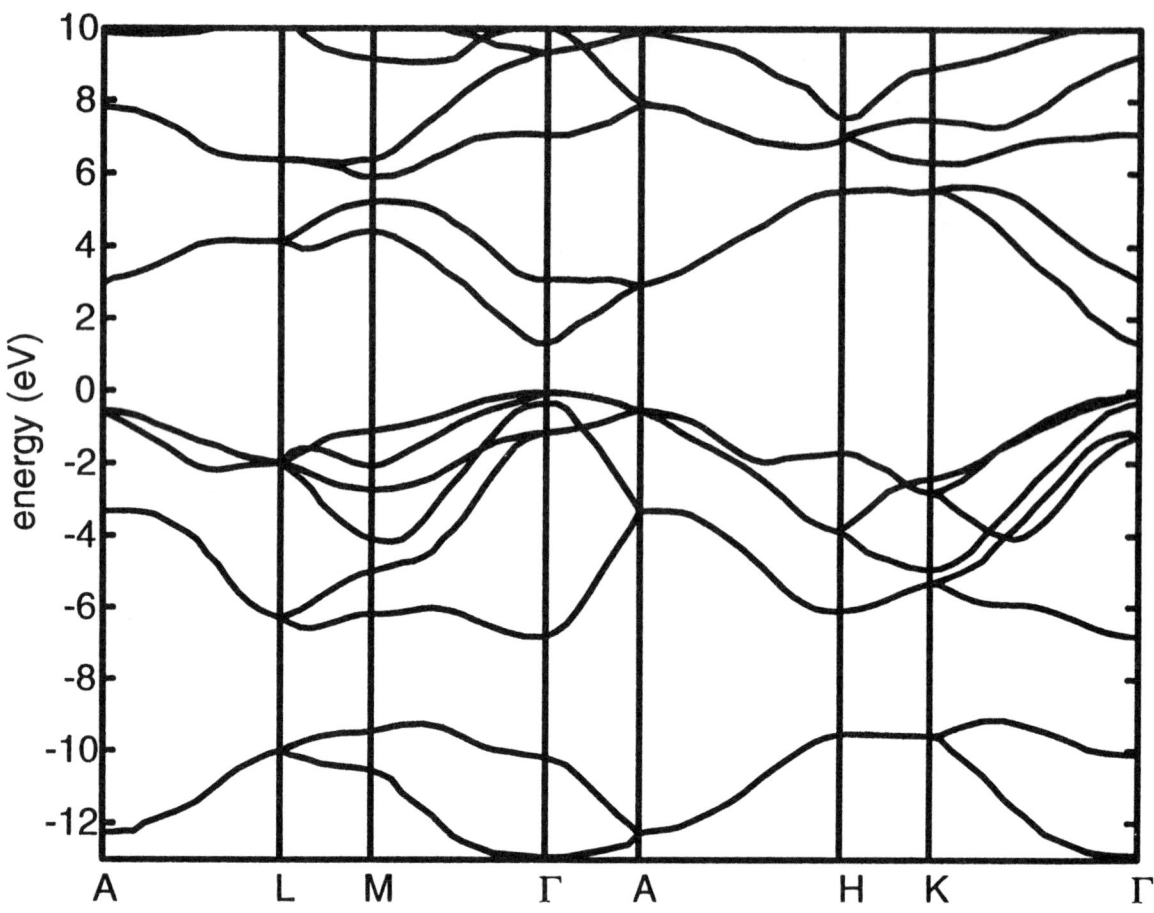

FIGURE 1   The band structure of InN computed with the ab initio pseudofunction method.   The model underestimates the gap because of the local density approximation but gives a good representation of the valence band and the general structure of the conduction band.

Sobolev et al report optical transitions for energies 5.0, 5.5, 5.8, 7.3 and 8.8 eV in reflectivity data and make transition assignments based on similar assignments made for GaN [10]. Their measured transition energies in polycrystalline material differ from transitions measured by Guo et al of 2.3, 5.3, 7.9, 8.9, 10.1 and 11.2 eV measured in crystalline material [11]. Listed in TABLE 1 is the Sobolev data with assignments taken from the pseudopotential band structure calculations.

TABLE 1 Identification of optical transitions in InN. Two pseudopotential band structure calculations have been used to assign specific band transitions to the observed transitions. The reader is cautioned that the data was taken on polycrystalline material and only for the range of 5 - 12.5 eV. Unless noted the data is taken from [5].

Exp (eV)	Theory [4]	Transition	Theory [5]	Transition
1.89 [9]	1.89	$\Gamma_6 - \Gamma_1$	2.04	$\Gamma_6 - \Gamma_1$
5.0	4.70	$\Gamma_5 - \Gamma_3$	4.5	$\Gamma_5 - \Gamma_3$
5.5	4.95	$M_2 - M_1$	6.4	$M_2 - M_1$
5.8	5.40	$H_3 - H_3$	5.4	$U_3 - U_1$
7.3	7.20	$K_3 - K_2$	7.5	$K_3 - K_2$
8.8	8.90	$\Gamma_5 - \Gamma_6$	8.6, 8.9	$\Gamma_6 - \Gamma_1, \Gamma_5 - \Gamma_6$

C   EFFECTIVE MASSES

The effective mass of InN has been measured but not confirmed. A value of $m_e = 0.11\,m_0$ was inferred from the free carrier absorption measurements by Tyagai et al [12], for low mobility, highly doped material. However, their estimate from the Moss-Burstein shifts suggests $m_e = 1.6\,m_0$. Other estimates of effective band masses have been inferred from band structure calculations and used to interpret optical absorption by shallow impurity states [10]. The masses from the band calculations are listed in TABLE 2 with the corresponding estimates of the shallow levels.

TABLE 2 Effective band masses for InN and shallow impurity energy levels. All energies are in meV. The methods of band structure calculation are: (EP) empirical pseudopotential method; (PF) ab initio pseudofunction method; (TB) semi-empirical tight binding method.

	m_e/m	E_e	m_{lh}/m	$E_\perp$	m_{hh}/m	$E_{\parallel}$
Exp.	0.11 [12]	40 [8]		155 [8]		
EP [5]	0.12	23	0.50		0.17	
PF [7]	0.34	67	1.60	316	1.70	316
TB [6]	0.59	118 [7]	2.70	533 [7]	2.70	553 [7]

D   ALLOYS OF InN: $Ga_xIn_{1-x}N$ AND $Al_xIn_{1-x}N$

The solid solution of $Ga_xIn_{1-x}N$ has been fabricated by a number of groups which report the bandgap as a function of composition x [13-15]. Osamura and co-workers [13] measured the optical absorption of polycrystalline $Ga_xIn_{1-x}N$ and report that the bandgap varies quadratically as a function of composition x:

$$E_g(x) = (1-x)\,E_g(InN) + x\,E_g(GaN) - Cx(1-x) \tag{1}$$

where $E_g(GaN) = 3.40$ eV, $E_g(InN) = 2.07$ eV and $C = 1.0$ eV. They also report the temperature shift in the gap as a function of composition to be

$$\frac{\partial E_g}{\partial T}(x) = 0.27 + 0.09x \qquad \text{meV K}^{-1} \tag{2}$$

by absorption measurements at 78 K and room temperature assuming the linear variation over the temperature range. Pressure coefficients have not been reported.

Crystalline $Al_xIn_{1-x}N$ has been grown by Kubota and co-workers who find the bandgap from absorption measurements [16]. The data varies from 2 eV to 6 eV, with a value of 3.34 eV for x = 0.83. Their bandgap data does not fit to a quadratic form as well as data for most III-V semiconductors.

E   CUBIC InN

At this time there is only one report of InN fabricated in the zinc blende form. The full details of this work have not been published [17]. One report of an ab initio pseudopotential computation of the band structure of InN yields pressure coefficients for each of the minima at Γ, X and L [18]. They find:

$$
\begin{aligned}
E_\Gamma(p) - E_\Gamma(0) &= (2.54p - 1.15p^2) \quad \text{eV} \\
E_X(p) - E_X(0) &= (0.59p - 0.53p^2) \quad \text{eV} \\
E_L(p) - E_L(0) &= (3.35p - 1.94p^2) \quad \text{eV}
\end{aligned}
\tag{3}
$$

if the pressure p is measured in Mbar.

REFERENCES

[1]   A. Wakahara, A. Yoshida [*Appl. Phys. Lett. (USA)* vol.54 no.8 (1989) p.709-11]

[2]   O. Igarashi [*Jpn. J. Appl. Phys. (Japan)* vol.31 (1992) p.2665-8]

[3]   T.L. Tansley, C.P. Foley [*Electron. Lett. (UK)* vol.20 no.25/26 (1984) p.1066-8]

[4]   S.N. Grinyeav, V.Ya. Malakhov, V.A. Chaldyshev [*Sov. Phys. J. (USA)* vol.29 (1986) p.311-4]

[5]   C.P. Foley, T.L. Tansley [*Phys. Rev. B (USA)* vol.33 no.2 (1986) p.1430-3]

[6]   D.W. Jenkins, R.-D. Hong, J.D. Dow [*Superlattices Microstruct. (UK)* vol.3 no.4 (1987) p.365-9]

[7]   M.-H. Tsai, D.W. Jenkins, J.D. Dow, R.V. Kasowski [*Phys. Rev. B (USA)* vol.38 no.2 (1988) p.1541-3]

[8]   T.L. Tansley, R.J. Egan [*Phys. Rev. B (USA)* vol.45 no.19 (1992) p.10942-50]

[9]   T.L. Tansley, C.P. Foley [*J. Appl. Phys. (USA)* vol.59 no.9 (1986) p.3241-4]

[10]   V.V. Sobolev, S.G. Kroitoru, A.F. Andreeva, V.Ya. Malakhov [*Sov. Phys.-Semicond. (USA)* vol.13 no.3 (1979) p.485-6]

[11]   Q. Guo, O. Kato, M. Fujisawa, A. Yoshida [*Solid State Commun. (USA)* vol.83 no.9 (1992) p.721-3]

[12]   V.A. Tyagai, A.M. Evstigneev, A.N. Krasiko, A.F. Andreeva, V.Ya. Malakhov [*Sov. Phys.-Semicond. (USA)* vol.11 no.11 (1977) p.1257-9]; V.A. Tyagai, O.V. Snitko, A.M. Evstigneev, A.N. Krasiko [*Phys. Status Solidi B (Germany)* vol.103 no.2 (1977) p.589-94]

[13]   K. Osamura, K. Nakajima, Y. Murakami [*Solid State Commun. (USA)* vol.11 no.5 (1972) p.617-21]; K. Osamura, S. Naka, Y. Murakami [*J. Appl. Phys. (USA)* vol.46 no.8 (1975) p.3432-7]

[14]   T. Nagatomo, T. Kuboyama, H. Minamino, O. Omoto [*Jpn. J. Appl. Phys. (Japan)* vol.28 no.8 (1989) p.L1334-L1336]

[15]   N. Yoshimoto, T. Matsuoka, A. Katsui [*Appl. Phys. Lett. (USA)* vol.59 no.18 (1991) p.2251-3]

[16]   K. Kubota, Y. Kobayashi, K. Fujimoto [*J. Appl. Phys. (USA)* vol.66 no.7 (1989) p.2984-8]

[17]   S. Strite, H. Morkoc [*J. Vac. Sci. Technol. B (USA)* vol.10 no.4 (1992) p.1237-66]

[18]   P.E. Van Camp, V.E. Van Doren, J.T. Devreese [*Phys. Rev. B (USA)* vol.41 no.3 (1990) p.1598-602]

5.2 Band structure of AlGaN, $(AlN)_x(SiC)_{1-x}$ and diamond/c-BN alloys

W.R.L. Lambrecht and B. Segall

June 1994

A   INTRODUCTION

The bandgap $E_g(x)$ of an A_xB_{1-x} alloy as a function of concentration x can be approximated by

$$E_g(x) = \bar{E}_g + \Delta E_g(x - \tfrac{1}{2}) - bx(1 - x) \tag{1}$$

with $\bar{E}_g = [E_g(0) + E_g(1)]/2$ the average gap and $\Delta E_g = E_g(1) - E_g(0)$ the gap difference between the endpoints A and B. The parameter b is called the bowing parameter. Note that a downward bending corresponds to $b > 0$ and an upward bending to $b < 0$. The above equation can be applied to several gaps of interest for a given system, e.g. the minimum indirect and minimum direct gap. A subscript will be used to indicate the type of gap the parameters refer to.

B $Al_xGa_{1-x}N$

The band structure of $Al_xGa_{1-x}N$ alloys has been studied theoretically by Albanesi et al [1] using the cluster expansion method combined with LDA-LMTO-ASA calculations of zinc blende derived structures with specific cation ordering. For details on the computational method, see [1] and Chapter 4 on band structure calculations of the binary compounds (BN, AlN, GaN and InN). Because zinc blende GaN has a minimum direct gap at Γ and AlN has a minimum indirect gap ($\Gamma \rightarrow X$), both these gaps were studied. The results are summarized by the bowing parameter b_Γ^{zb} for the direct gap $\Gamma_{15}^v \rightarrow \Gamma_1^c$ and b_X^{zb} for the indirect $\Gamma_{15}^v \rightarrow X_1^c$ gap given in TABLE 1. The superscript indicates that these refer to the zinc blende structure.

TABLE 1   Bandgap bowing parameters (in eV) for $Al_xGa_{1-x}N$.

	b_X^{zb}	b_Γ^{zb}	b_Γ^{wz}	Ref
No relaxation	0	0		[1]
With relaxation	-0.92	-0.40		[1]
Absorption edge			-0.6 ± 0.2	[2]
Photoluminescence			0.98	[3]
Absorption edge			1 ± 3	[4]
Photoluminescence			0	[5]

The bowing parameters were found to be negligible when relaxation of the bond lengths was not included, i.e. the average bond length d_{av} (determined by the average lattice constant a_{av} of the alloy as $d_{av} = a_{av}\sqrt{3}/4$) of the alloy was assumed for each bond. A small negative bowing was obtained under the assumption that the bond lengths relax to their 'ideal' value,

- 163 -

i.e. the value they have in the binary compounds GaN or AlN. The latter gives an upper limit on the bandgap bowing. A cross-over from indirect to direct minimum gap takes place at $x_{Ga} \sim 0.4$.

Since the Γ conduction band is very similar in wurtzite and zinc blende the behaviour of the direct gap curve as a function of concentration can also be considered as a good approximation for the behaviour of the gap in the wurtzite structure. In that case, the minimum gap is direct over the full concentration range. The bandgap bowing parameter b_Γ^{wz} for wurtzite $Ga_xAl_{1-x}N$, which should thus be comparable to b_Γ^{zb}, has been studied by photoluminescence [3,5] and optical absorption [2,4]. The results of Khan et al [3,4] cover only a small concentration range ($0 \leq x \leq 0.24$). We note that there is considerable spread in the results among the various experimental determinations and the theory (for b_Γ) and experiment. This suggests that the results may be sample preparation dependent. Further work on high-quality samples would seem desirable.

The calculations of the energies of formation [1] (which are of order 15 - 20 meV/atom) indicate full mutual solubility at growth temperatures $\geq 600\,K$.

C $(AlN)_x(SiC)_{1-x}$

The band structure of $(AlN)_x(SiC)_{1-x}N$ alloys has been studied theoretically by Lambrecht [6] using the cluster expansion method combined with LDA-LMTO-ASA calculations of zinc blende derived structures with specific cation and anion ordering in a pseudobinary model.

The bandgap is indirect ($\Gamma_{15}^v \rightarrow X_1^c$) [8] over the full concentration range and varies from 2.417 eV to 4.9 ± 0.2 eV. The former is the experimental value for 3C-SiC at zero temperature [7] and the latter is the calculated value including an estimated gap correction for zinc blende AlN (see Chapter 4). The downward bowing parameter is 3.1 ± 0.3 eV.

The calculations indicate that there is very little mutual solubility up to the decomposition or melting temperatures, suggesting that the solid solutions are metastable.

Experimental data are available for the wurtzite, or 2H, polytype of the alloys [9,10]. There is a transition from the indirect gap $\Gamma_6^v \rightarrow K_2^c$ in SiC to the direct gap $\Gamma_1^v \rightarrow \Gamma_1^c$ in AlN at $x \sim 0.75$.

The indirect gap varies from 3.3 eV to 7.2 eV with a bowing parameter of 5.2 eV.

The direct gap varies from 4.6 eV to 6.3 eV with a bowing parameter of 4.4 eV.

TABLE 2 gives an overview of the gap parameters.

TABLE 2 Bandgaps and bowing coefficients (in eV) in (AlN)$_x$(SiC)$_{1-x}$ alloys.

Polytype	$E_g(0)$	$E_g(1)$	b	Ref
Zinc blende (3C) $\Gamma \to X$	2.4	4.9	3.1	[6]
Wurtzite (2H) $\Gamma \to K$	3.3	6.3	5.2	[9]
$\Gamma \to \Gamma$	4.6	7.2	4.4	[9]

D   TETRAHEDRAL C$_x$(BN)$_{1-x}$

Alloys of diamond and c-BN have been reported by Badzian [11]. The electronic structure and miscibility were studied by Lambrecht and Segall [12]. The bandgaps of diamond and c-BN are respectively 5.5 eV and 6.4 eV. The bandgap bowing in this system is large and calculated to be b = 6.8 eV. This is related to the type-II offset at the corresponding heterojunction. It implies that alloys would have gaps that are smaller than that of diamond up to about 80% BN. The calculations predict that the system has almost no miscibility up to the melting point. This indicates that solid solutions must be considered to be thermodynamically metastable. They may nevertheless be stabilized kinetically when quenched from a gas or liquid phase.

E   CONCLUSION

The main conclusion from the alloy studies so far is that there is a strong difference in behaviour between heterovalent systems such as diamond/c-BN or SiC/AlN and the homovalent systems, such as Al$_x$Ga$_{1-x}$N. The former have very low miscibility and are basically metastable only while true solid solutions are possible among the group III nitrides. The bandgap bowing is also very different. In the heterovalent case it is very large, which is due to charge transfer effects, while it is small in the Al$_x$Ga$_{1-x}$N case. Effects of lattice relaxation are expected to be somewhat larger in the case of Al and Ga alloys with In and need further study (see other chapters).

ACKNOWLEDGEMENTS

We acknowledge financial support by the NSF under grant No. DMR-92-22387 and partial support from NASA Lewis Research Center.

REFERENCES

[1]   E.A. Albanesi, W.R.L. Lambrecht, B. Segall [*Phys. Rev. B (USA)* vol.48 (1993) submitted]
[2]   S. Yoshida, S. Misawa, S. Gonda [*J. Appl. Phys. (USA)* vol.53 (1982) p.6844-8]
[3]   M.A. Khan, R. Skogman, R. Schulge, M. Gerschenzon [*Appl. Phys. Lett. (USA)* vol.43 (1983) p.492]

[4]   M.R. Khan, Y. Koide, H. Itoh, N. Sawaki, I. Akasaki [*Solid State Commun. (USA)* vol.60 (1986) p.509-12]

[5]   B. Goldenberg [private communication]

[6]   W.R.L. Lambrecht [unpublished]

[7]   R.G. Humphreys, D. Bimberg, W.J. Choyke [*Solid State Commun. (USA)* vol.39 (1981) p.163]

[8]   Some of the ordered compounds used in the cluster expansion method have the valence-band maximum not at Γ. Because we do not expect such behaviour for the random alloys, we have used the $\Gamma_{15}^v \rightarrow X_1^c$ gaps even for these compounds. If we use the actual minimum gaps for those cases, the bowing coefficient becomes 3.6 eV.

[9]   G.K. Safaraliev, Yu.M. Tairov, V.F. Tsvetkov [*Sov. Phys.-Semicond. (USA)* vol.25 (1991) p.865]

[10]   Sh.A. Nurmagomedov, A.N. Pikhtin, V.N. Razbegaev, G.K. Safaraliev, Yu.M. Tairov, V.F. Tsvetkov [*Sov. Phys.-Semicond. (USA)* vol.23 (1989) p.100]

[11]   A.R. Badzian [*Mater. Res. Bull. (USA)* vol.16 (1981) p.1285]; A.R. Badzian [*Advances in X-ray Analysis*, vol.31, Eds C.S. Barrett, J.V. Gilfrich, R. Jenkins, J.C. Russ, J.W. Richardson Jr., P.K. Predcki (Plenum, New York, 1988) p.113]

[12]   W.R.L. Lambrecht, B. Segall [*Phys. Rev. B (USA)* vol.47 (1993) p.9289]

CHAPTER 6

OPTICAL FUNCTIONS

6.1 Optical functions of BN

G.L. Doll

January 1994

A INTRODUCTION

The optical functions of the hexagonal (h-BN), cubic (c-BN), wurtzitic (w-BN) and rhombohedral (r-BN) phases of boron nitride are not nearly as well defined as those of the other group III nitrides. This is due in large part to the absence of high quality, single crystal samples. Consequently, most optical studies on BN have been performed on polycrystalline samples with various amounts of impurities. As would be expected, the results of optical studies performed on these materials vary greatly. Values of the h-BN optical bandgap energies range from 3.2 to 5.8 eV (experimental) [1] and 2.4 to 12.7 eV (calculated) [2]. The c-BN bandgap energies range from 5.5 [3] to 6.4 eV [4,5] (experimental) and 3.0 to 11.3 eV (calculated) [6]. Although there is disagreement as to whether the bandgaps in c-BN and h-BN are direct or indirect, a consensus seems to be forming that the bandgap is direct in the case of h-BN, and indirect in the case of c-BN.

In this Datareview, we discuss two experiments that investigate the optical properties of h-BN and c-BN. These experiments are singled out because they were performed on a high purity, preferentially oriented sample in the case of h-BN, and on a reasonably high purity, single crystal sample in the case of c-BN. While measurements of the optical functions of w-BN have never been reported, a calculation will also be discussed. The optical functions of r-BN have yet to be measured or calculated.

B HEXAGONAL BORON NITRIDE

A recent optical study on h-BN was performed by Hoffman et al [1] who examined the optical reflectivity of high purity, c-axis oriented, polycrystalline h-BN in the energy range 0.05 to 10.0 eV. The reflectivity measurement was performed with $E \perp c$ polarization on a sample with an approximate 10° mosaic spread of the c-axis. The reflectance spectrum is plotted versus the photon energy in FIGURE 1. A Kramers-Krönig transform of the reflectance data yields the refractive index (n), extinction coefficient (k) and the absorption coefficient (α) (FIGURES 2(a), 2(b) and 2(c)). The structure at low energies in n and k is due to IR-active phonons. Electronic absorption in k and α starts near 1 eV and a broad continuum extends to beyond 5 eV. The authors attribute this extrinsic absorption to impurity and defect states distributed throughout the bandgap. Above 5 eV, the absorption rises rapidly due to intrinsic π-π transitions corresponding to the direct bandgap. The authors define the bandgap energy (5.2 ± 0.2 eV) to be that energy at the transition from the continuum absorption to the intrinsic absorption. Band structure calculations [2] suggest that the direct bandgap in h-BN is associated with the H point ($H_3 \rightarrow H_2$). Optical transitions corresponding to different locations in the Brillouin zone such as $Q_{2g}^- \rightarrow Q_{2u}^-$ (6.10 eV) and $Q_{2u}^- \rightarrow Q_{2g}^-$ (6.85 eV) are also observed in the data [2]. A more recent study [7] of the optical properties of h-BN reports a bandgap energy of 5.89 eV.

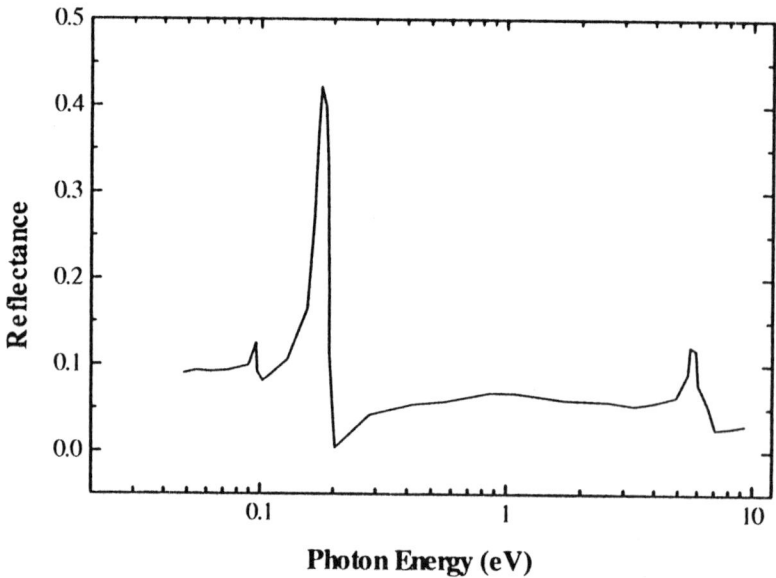

FIGURE 1 Reflectance spectrum of h-BN taken at near-normal incidence to the c-face (from [1]).

C   CUBIC BORON NITRIDE

The most complete study of the optical constants of c-BN has been reported by Miyata et al [8]. They performed reflectance measurements on a yellow single crystal of c-BN (5 mm² area x 0.16 mm thick) over the photon energy range from 2 to 23 eV, and transmittance measurements from 2 to 7 eV. The spectra are shown in FIGURES 3(a) and 3(b). The optical constants n and k were determined directly from the reflectance and transmittance spectra in the 2 to 7 eV photon energy region by the equations

$$k = \frac{\lambda}{4\pi d} \ln \frac{(1-R)^2}{T} \tag{1}$$

$$n = \frac{R+1+[(R+1)^2-(R-1)^2(1+k^2)]^{1/2}}{1-R} \tag{2}$$

At higher photon energies, n and k were determined by a self consistent Kramers-Krönig analysis of the reflectance spectrum, where the n and k values determined directly from 2 to 7 eV were used in the evaluation of the phase integral from 2 to 23 eV. The results for n and k, as well as the absorption coefficient (α) are shown in FIGURES 4(a), 4(b) and 4(c). The absorption coefficient in FIGURE 4(c) shows an onset near 6.1 ± 0.2 eV which is identified as the bandgap energy. Several other absorption edges are observed at higher energies, but were not assigned to specific transitions. Band structure calculations [6] suggest that the 6.1 ± 0.2 eV indirect bandgap in c-BN is associated with the $\Gamma_{15}^v \rightarrow X_1^c$ transition. Other reports have determined the bandgap energy of c-BN to be > 5.5 eV [3], $\leq 6.0 \pm 0.5$ eV [4], and 6.4 ± 0.5 eV [5].

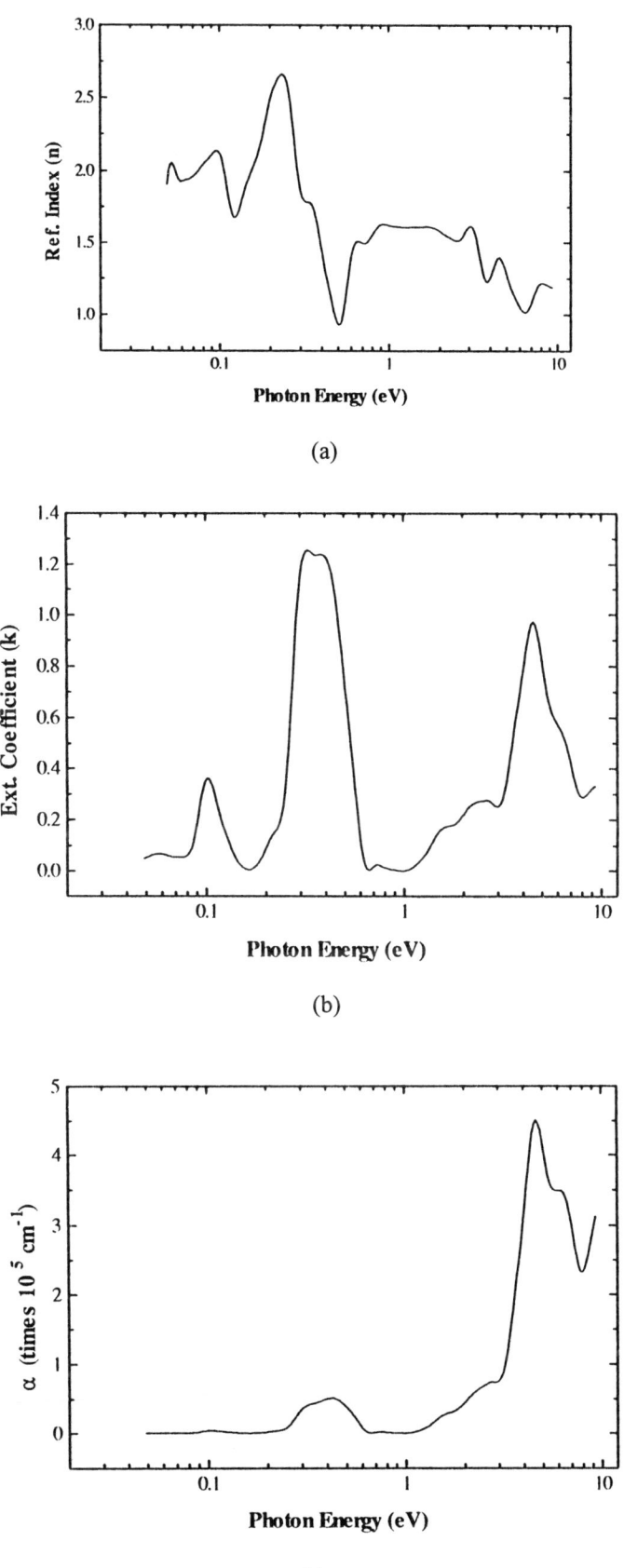

(a)

(b)

(c)

FIGURE 2   Optical constants (n and k) and the absorption coefficient (α) of h-BN determined by a
Kramers-Krönig analysis of the reflectance data in FIGURE 1 (from [1]).

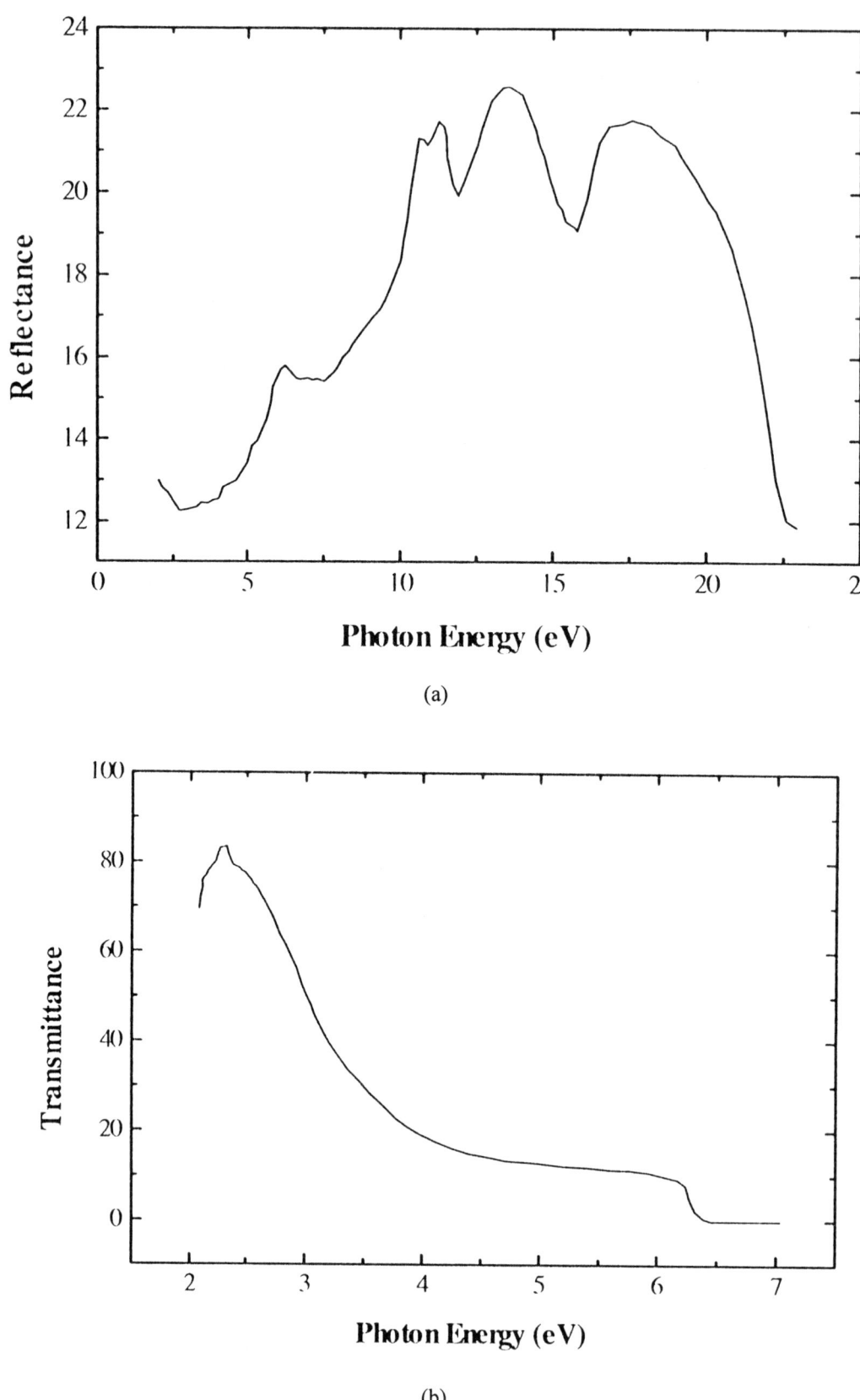

(a)

(b)

FIGURE 3 (a) Reflectance and (b) transmittance spectra of a single crystal c-BN sample (from [8]).

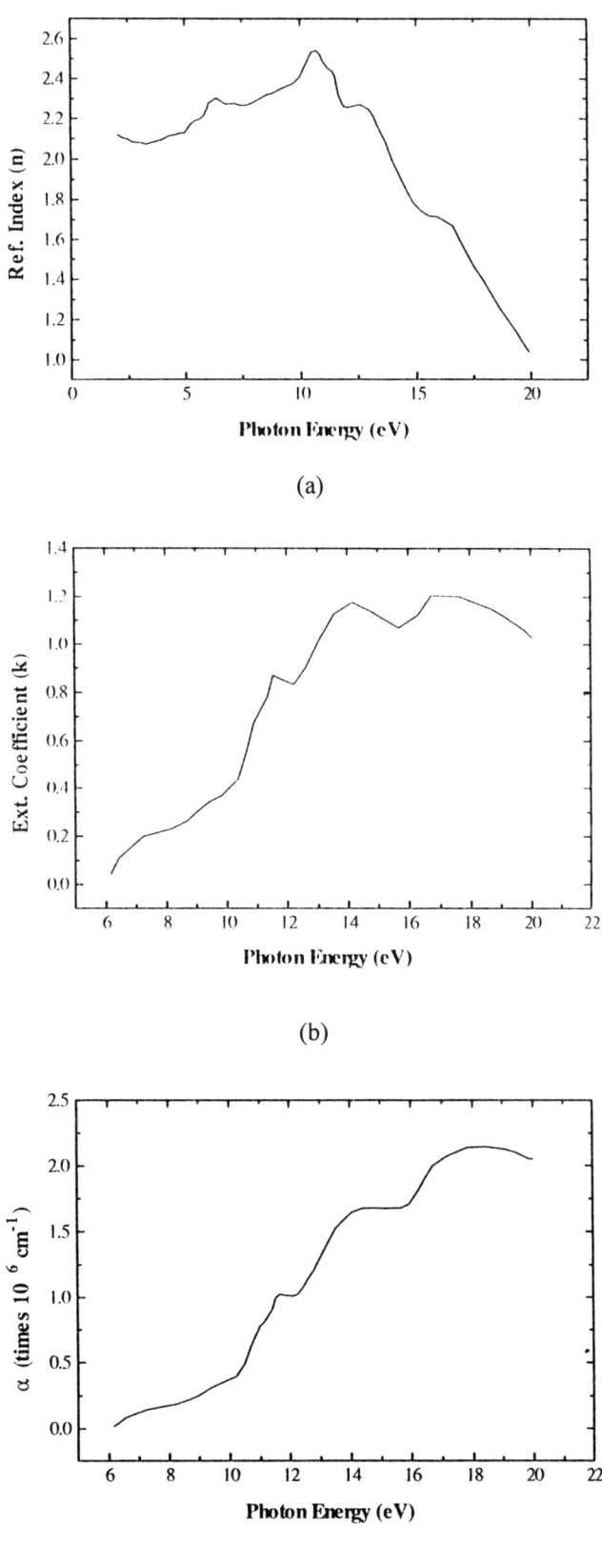

(a)

(b)

(c)

FIGURE 4 (a) Refractive index n, (b) extinction coefficient k, and (c) absorption coefficient α for c-BN obtained from the data in FIGURES 3(a) and 3(b). From [8].

D   WURTZITIC BORON NITRIDE

The measured optical properties of w-BN have not been reported, but have been calculated by Li et al [9] for E ∥ c and E⊥c polarizations. Local density approximations of the band structure of w-BN indicate that the indirect bandgap ($\Gamma \rightarrow$K) energy is 4.9 eV [10], while others [11] estimate the energy to be higher (5.81 eV).

REFERENCES

[1]   D.M. Hoffman, G.L. Doll, P.C. Eklund [*Phys. Rev B (USA)* vol.30 (1984) p.6051 and references 5-13 therein]

[2]   References 8,10,14-20 in [1]

[3]   H.R. Phillip, E.A. Taft [*Phys. Rev. (USA)* vol.127 (1962) p.159]

[4]   P.V.A. Fomichev, M.A. Rumsh [*J. Phys. Chem. Solids (UK)* vol.29 (1968) p.1015]

[5]   R.M. Chrenko [*Solid State Commun. (USA)* vol.14 (1974) p.511]

[6]   R.M. Wentzcovitch, K.J. Chang, M.L. Cohen [*Phys. Rev. B (USA)* vol.34 (1986) p.1071]

[7]   A.J. Lukomskii, V.B. Shipilo, L.M. Gameza [*J. Appl. Spectrosc. (USA)* vol.57 (1993) p.607]

[8]   N. Miyata, K. Moriki, O. Mishima, M. Fujisawa, T. Hattori [*Phys. Rev. B (USA)* vol.40 (1989) p.12028]

[9]   D. Li, Y.-N. Xu, W.Y. Ching [*Phys. Rev. B (USA)* vol.45 (1991) p.5895]

[10]   P.K. Lam, R.M. Wentzcovitch, M.L. Cohen [in *Synthesis and Properties of Boron Nitride* Eds J.J. Pouch, S.A. Alterovitz (Trans Tech Publications, Brookfield, Vermont, 1990) vol.54 & 55, p.165]

[11]   Y.-N. Xu, W.Y. Ching [*Phys. Rev. B (USA)* in press]

6.2 Optical functions of AlN

S. Loughin and R.H. French

August 1994

A INTRODUCTION

Aluminium nitride is a uniaxial crystal with a wurtzite structure. Its symmetry is described by the space group P6$_3$mc or in the Schoenfleis notation, C$_{6v}^4$. The optical properties of AlN, with a bandgap of 6.2 eV, have been the subject of a number of investigations since the pioneering work of Slack et al [1-3]. A limited amount of data exists on the optical constants of AlN below the band edge and on the optical absorption near the band edge. Several researchers have probed the optical response of AlN above the bandgap by employing either vacuum ultraviolet (VUV) reflectance or electron energy loss spectroscopy (EELS). This Datareview reviews the literature values of optical constants, the band edge optical absorption and the VUV reflectance of AlN. From the reflectance, the important optical functions of AlN are calculated. Finally, a critical point analysis of the optical function, J$_{cv}$, (interband transition strength) is presented to relate the optical response to the band structure.

AlN has a high oxygen solubility [4,5], and all material prepared for commercial application contains some oxygen. Previous investigators [6] have related oxygen impurities to shifts in the energy of peak luminescence intensity, suggesting that oxygen defects affect the electronic structure. Investigation of AlN is complicated by the difficulty of growing high-purity AlN and the difficulty of obtaining oxygen-free AlN surfaces. Mroz [7] notes that hydrolysis of AlN occurs easily, forming ammonia and bohmite, and hence the best optical results are obtained on non-aqueously polished samples.

B OPTICAL CONSTANTS

Optical constants have been reported by several investigators, usually for films prepared by sputtering or chemical vapour deposition. At visible wavelengths, an index of 2.1 - 2.2 is generally accepted, although thin films often give lower values [8]. TABLE 1 summarizes these results while FIGURE 1 presents the index, n, and extinction coefficient, k, as a function of wavelength, λ. AlN is a birefringent crystal, and Geidur and Yaskov [9] have calculated the birefringence of the index, $\Delta n \equiv n_{E \parallel c} - n_{E \perp c}$, the value of which is about 0.075 at 633 nm.

C BAND EDGE ABSORPTION

Slack and McNelly [2] investigated the optical absorption of sublimation-grown crystals of AlN over the range 0.12 - 6.2 eV as shown in FIGURE 2. While pure AlN should transmit light for $0.31 \leq h\nu < 5.33$ eV, their sublimation-grown crystals have a broad background absorption in this region which is attributed to nitrogen vacancies, and an absorption feature at 2.86 eV which is attributed to oxygen.

TABLE 1 Optical constants for AlN films and polycrystals at selected wavelengths. Some samples were prepared by plasma-enhanced CVD.

λ (nm)	n	k	Notes	Ref
250	1.98	-	PCVD	Bauer et al [11]
250	2.17 - 2.34	-	CVD	
300	1.89	-	PCVD	
300	2.04 - 2.18	-	CVD	
300	2.08	0.018	LA-CVD	Demiryont et al [10]
250	2.2	-	Polycrystalline	Kutolin et al [12]
400	2.3	-		
620	2.4	-		
633	2.14	-	Calculated	Geidur and Yaskov [9]
-	1.99	-	CVD	Chu and Kelm [13]
633	1.9	0.0	Sputtered	Yin and Harding [14]
633	2.11	-	Sputtered (800 W)	Kubiak et al [15]

D   UV REFLECTANCE

The UV reflectance of AlN was measured by several investigators, most recently by the authors [16], over the range 4 - 40 eV for a high-purity single crystal prepared by Slack [17] with a non-aqueous polish. FIGURE 3 shows the reflectance of single crystal AlN (Λ = 275 W m^{-1} K^{-1}) which was found to agree, feature for feature, with that of non-aqueously polished polycrystalline AlN substrates with thermal conductivities Λ = 70 and 170 W m^{-1} K^{-1}.

AlN exhibits a sharp increase in reflectance just above 6 eV. A shoulder just above the edge at 6.4 eV was previously attributed to excitonic behaviour [18,19]. However, it is more properly understood as the correct shape for a two-dimensional critical point at the bandgap [20]. FIGURE 4 compares the previous literature with this result and also with a spectrum taken from a hydrated sample with an aqueous polish. Above the fundamental edge Yamashita et al [18] show a second edge which rises rapidly to a prominent feature at 9.5 eV labelled 'A'. Meleshkin et al [21] reported a feature at 9 eV but did not include it in their graphical results. They also observed a doublet at 13 - 15 eV as well as a shoulder at 22 eV labelled 'C'. The feature labelled 'A' is observed for the aqueously polished specimen, but not for the non-aqueously polished, single-crystal spectrum. The higher energy peak in the doublet matches a feature labelled 'B' in both the aqueous and non-aqueous spectra. The feature at 'C' was attributed by Meleshkin et al to plasma oscillations. In Section E the bulk plasmon for the high-purity single crystal is seen to occur at 24 eV. However, oxygen content will lower the energy of the bulk plasmon [20,22]. All of this suggests that the AlN examined in this earlier work was not sufficiently free of oxygen. Yamashita et al [18] do not report oxygen content for their thin film specimens while Meleshkin et al [21] report oxygen contents in the range of 0.7 - 1.35 atom %.

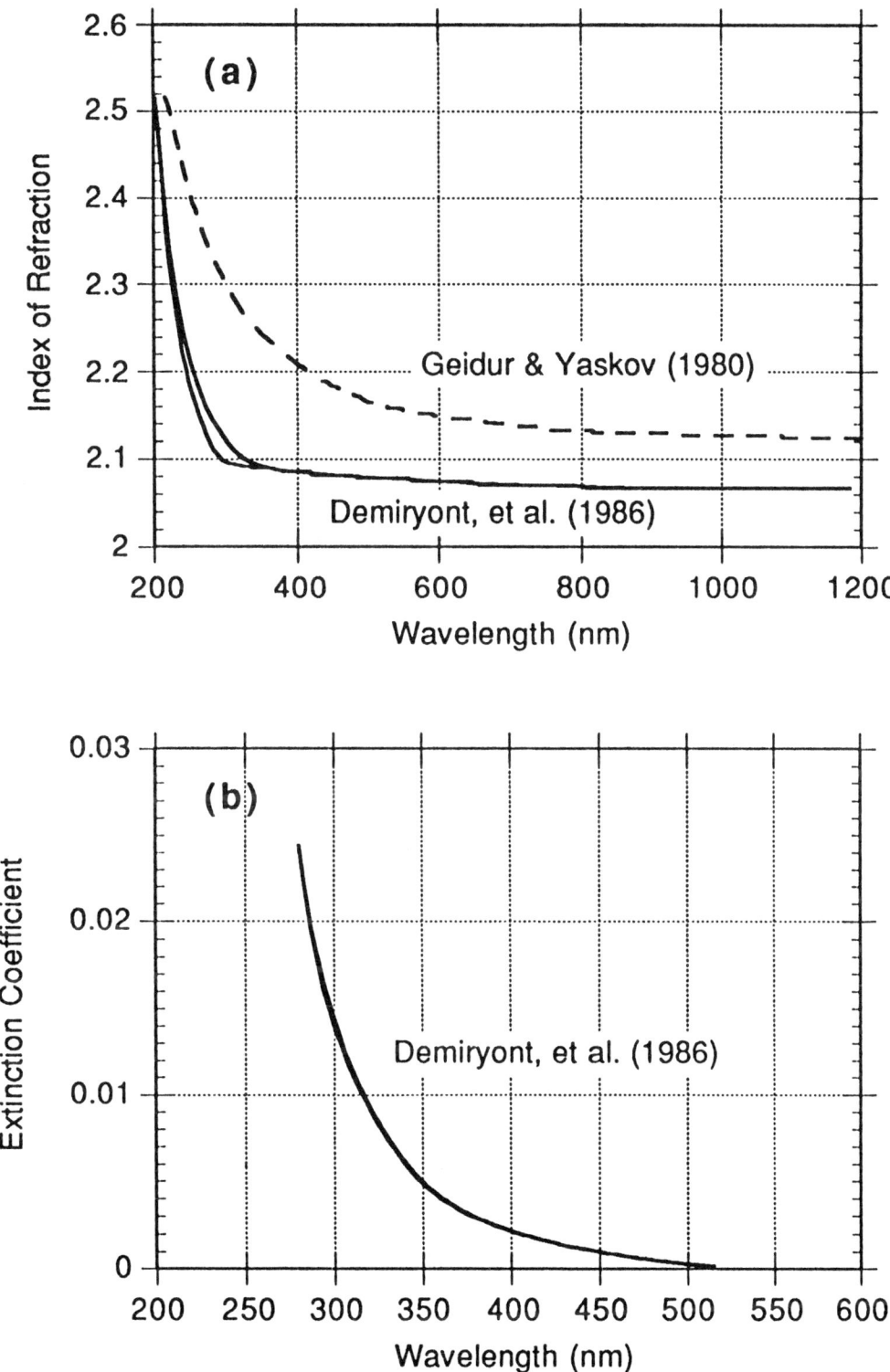

FIGURE 1 The results of Demiryont et al [10] are compared with those of Geidur and Yaskov [9] in (a) for the index of refraction, n, and shown alone in (b) for the extinction coefficient, k.

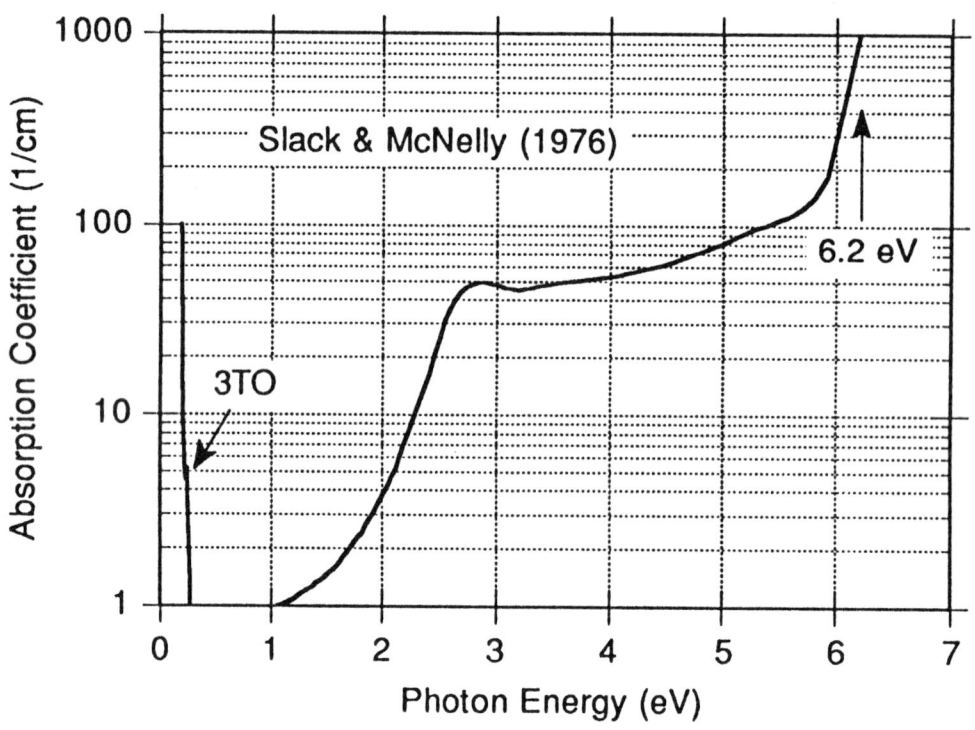

FIGURE 2 The optical absorption near the band edge after Slack and McNelly [2].

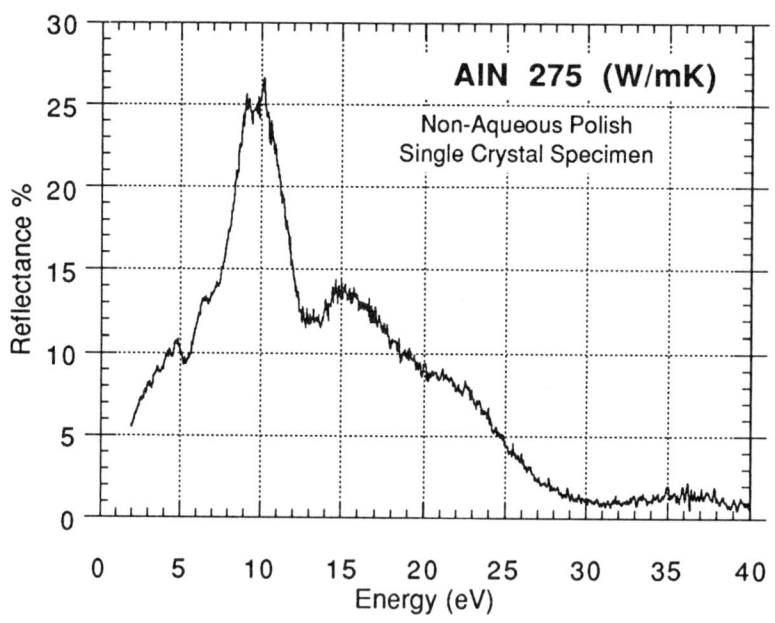

FIGURE 3 Vacuum ultraviolet reflectance of AlN single crystal material.

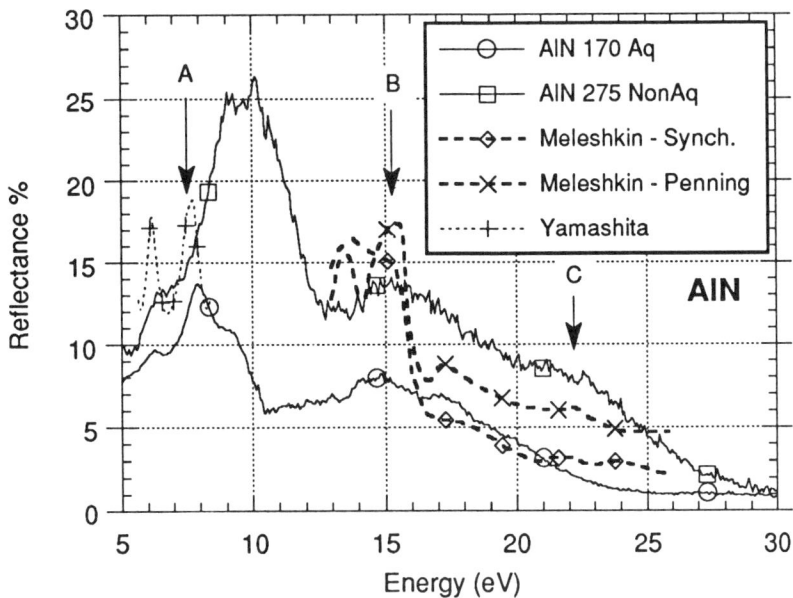

FIGURE 4 Comparison of AlN reflectivity with previous reports from the literature. Squares mark the non-aqueously polished single-crystal AlN which matches features reported by Meleshkin et al [21] at 'B' and 'C'. Circles mark the aqueously polished (hydrated) AlN which matches the second peak of Yamashita et al [18] at 'A'.

E   UV OPTICAL FUNCTIONS

A Kramers-Krönig analysis of the reflectance spectrum, R, (scaled to n = 2.1 at 633 nm), recovers the phase, ϕ, of the reflected light. From the pair {R,ϕ}, other representations of the optical functions have been calculated as described in Wooten [23]. FIGURE 5 shows the index of refraction, n + ik; FIGURE 6 shows the dielectric function, $\varepsilon_1 + i\varepsilon_2$; and FIGURE 7 shows the absorption coefficient, α, and loss function, Im[-1/ε]. FIGURE 8 shows J_{cv}, an optical function used in subsequent analysis, obtained from the dielectric function as $J_{cv} = (i E^2/8\pi^2) \times [\varepsilon_1 - i\varepsilon_2]$, where E is photon energy. Numerically, Re[J_{cv}] is related to the interband transition strength by a factor, $(m^2h^{-2}e^{-2})$, where m is electron mass, h is Planck's constant, and e is electron charge. For computation convenience, this factor is taken to have a value of 1.

Figure 9 compares the loss functions obtained by Olson et al [24] and by Gautier et al [25] with loss functions calculated from the reflectance measurements on the single crystal specimen (#302) and also on the aqueously-polished specimen (#300). Note that the y-axis for their curves was not specified so the y-axis is meaningful only for #300 and #302. The fact that the #300 curve shows a better qualitative agreement with the EELS measurements suggests that these previously studied samples were not entirely free of oxygen. The EELS results were also compared with the surface loss function calculated from VUV results, but that function peaks at a lower energy than either EELS result. Olson et al [23] reported a 0.1% bulk oxygen content for their thin films, although EELS is a surface measurement so the possibility of surface oxide is not precluded.

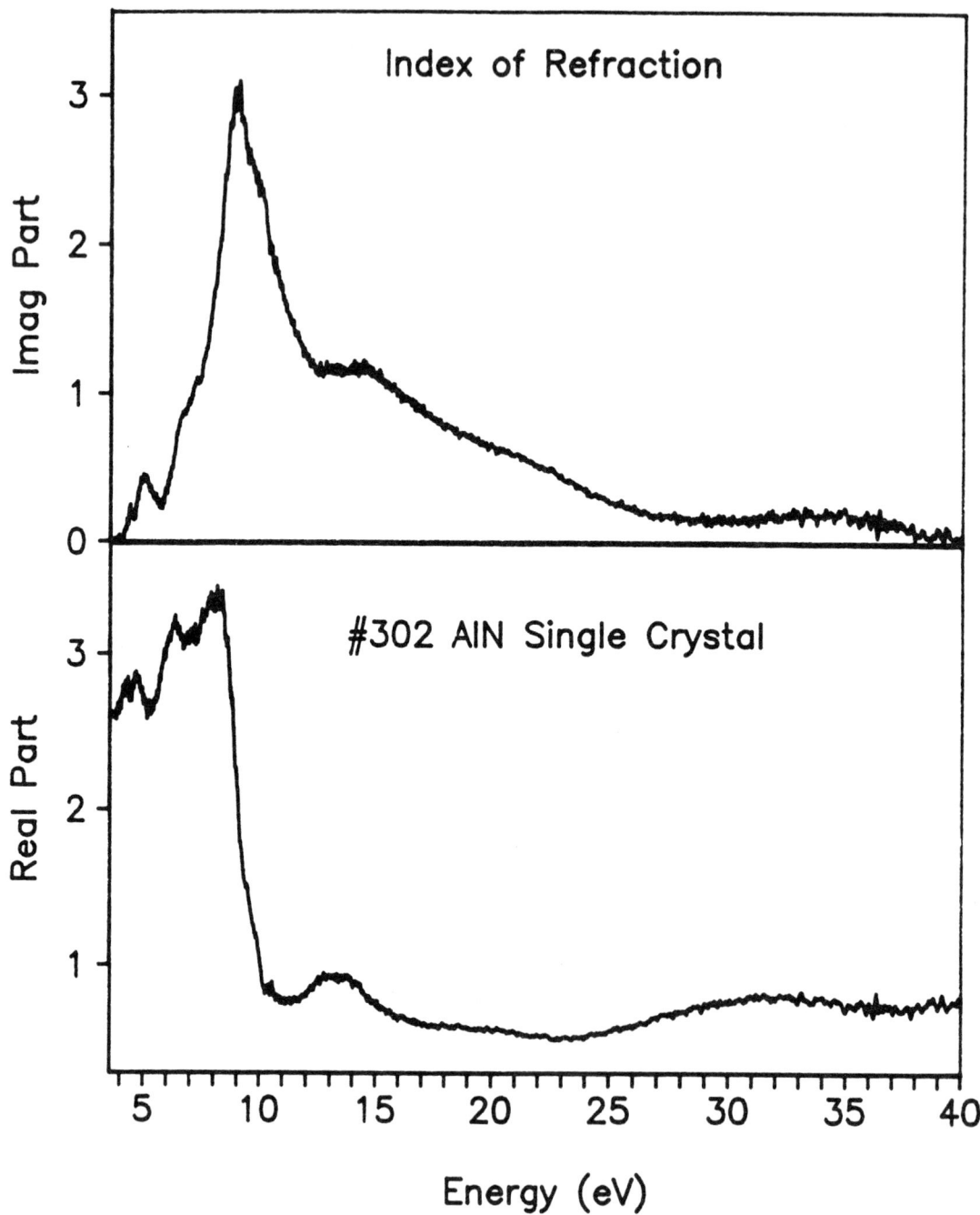

FIGURE 5 The index of refraction, n, and the extinction coefficient, k, are shown for high purity single crystal AlN.

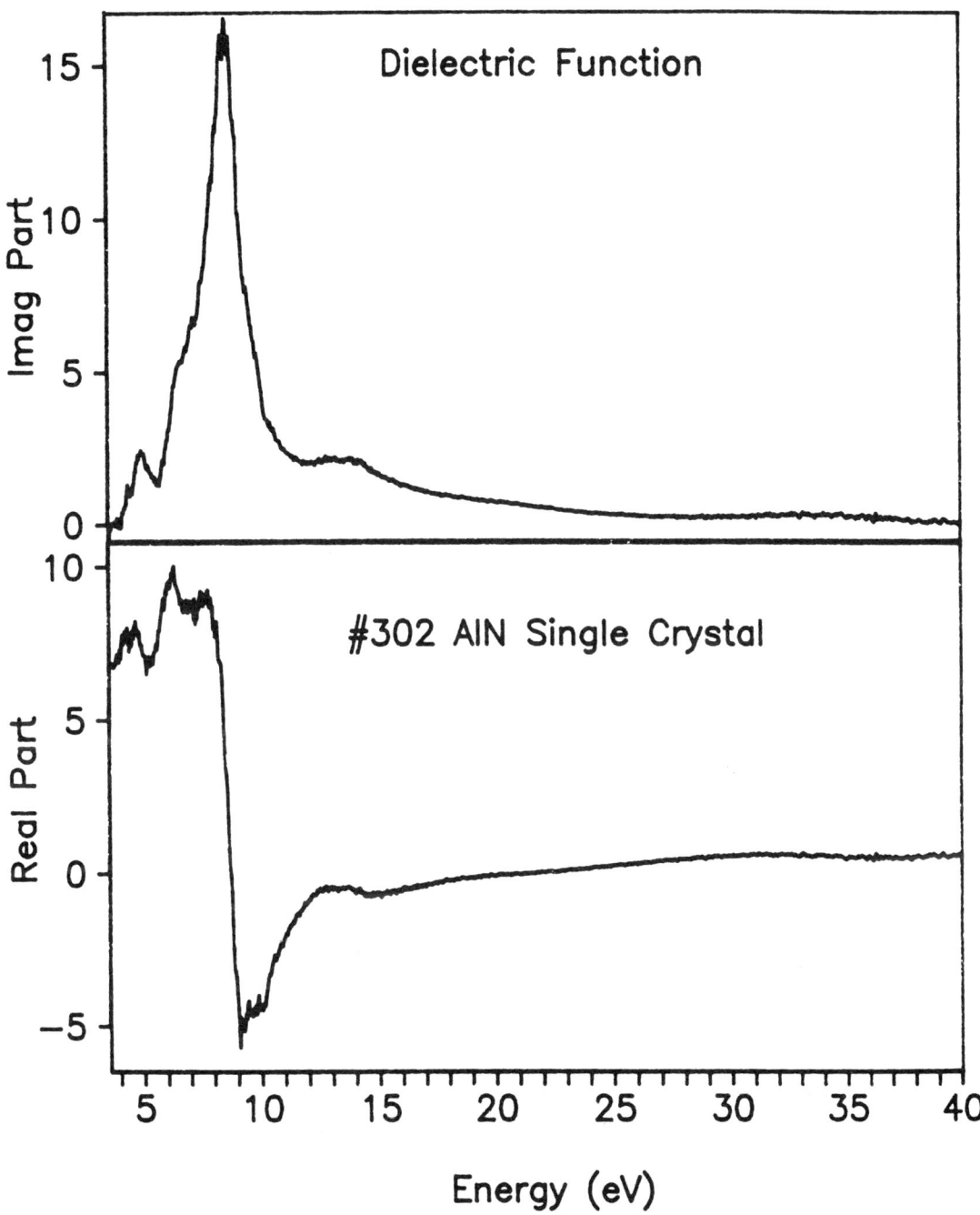

FIGURE 6 The real and imaginary parts of the dielectric function, ε_1 and ε_2, are shown for high purity single crystal AlN.

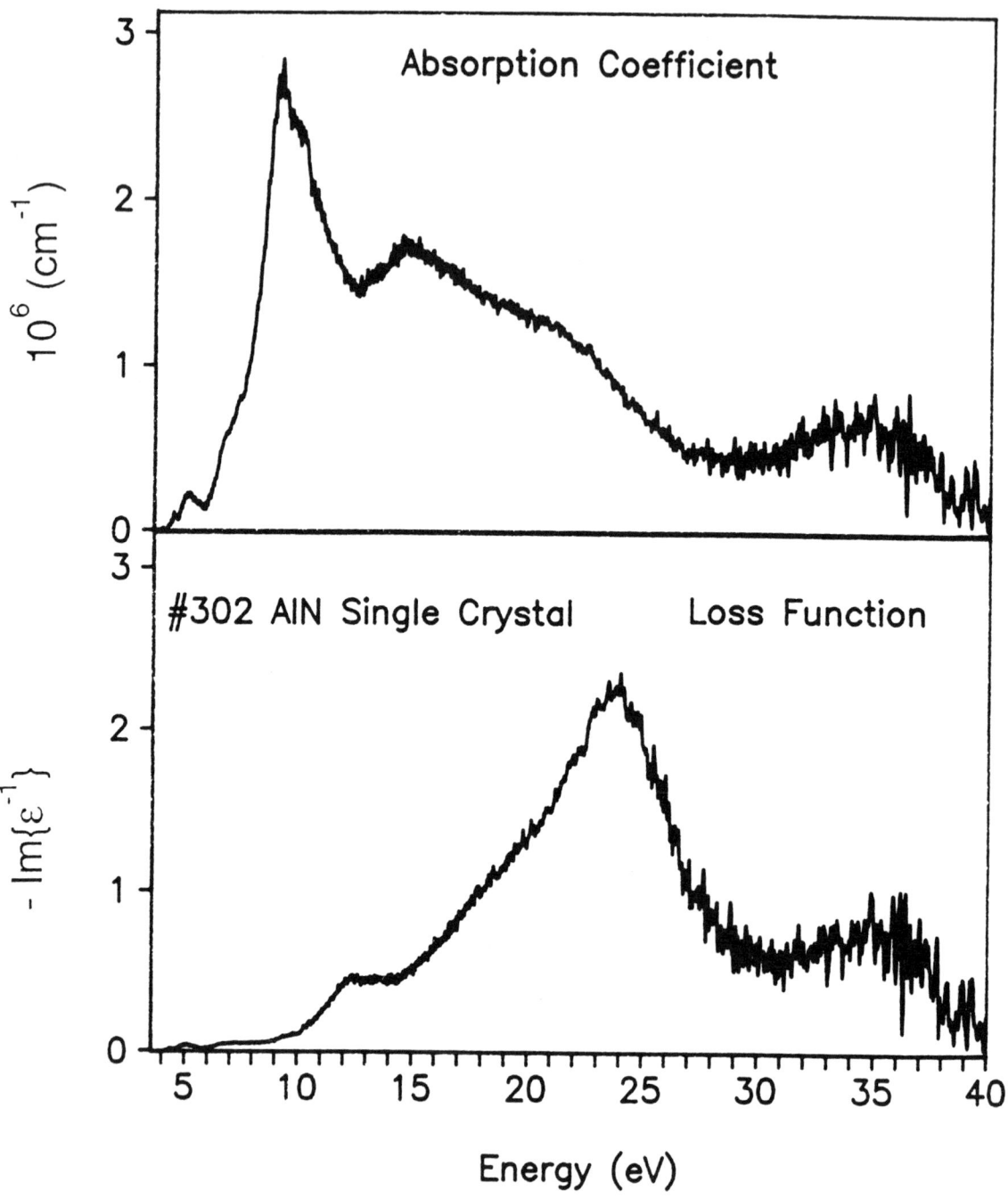

FIGURE 7 The absorption coefficient, α, and the loss function, Im[-1/ε], are shown for high purity single crystal AlN.

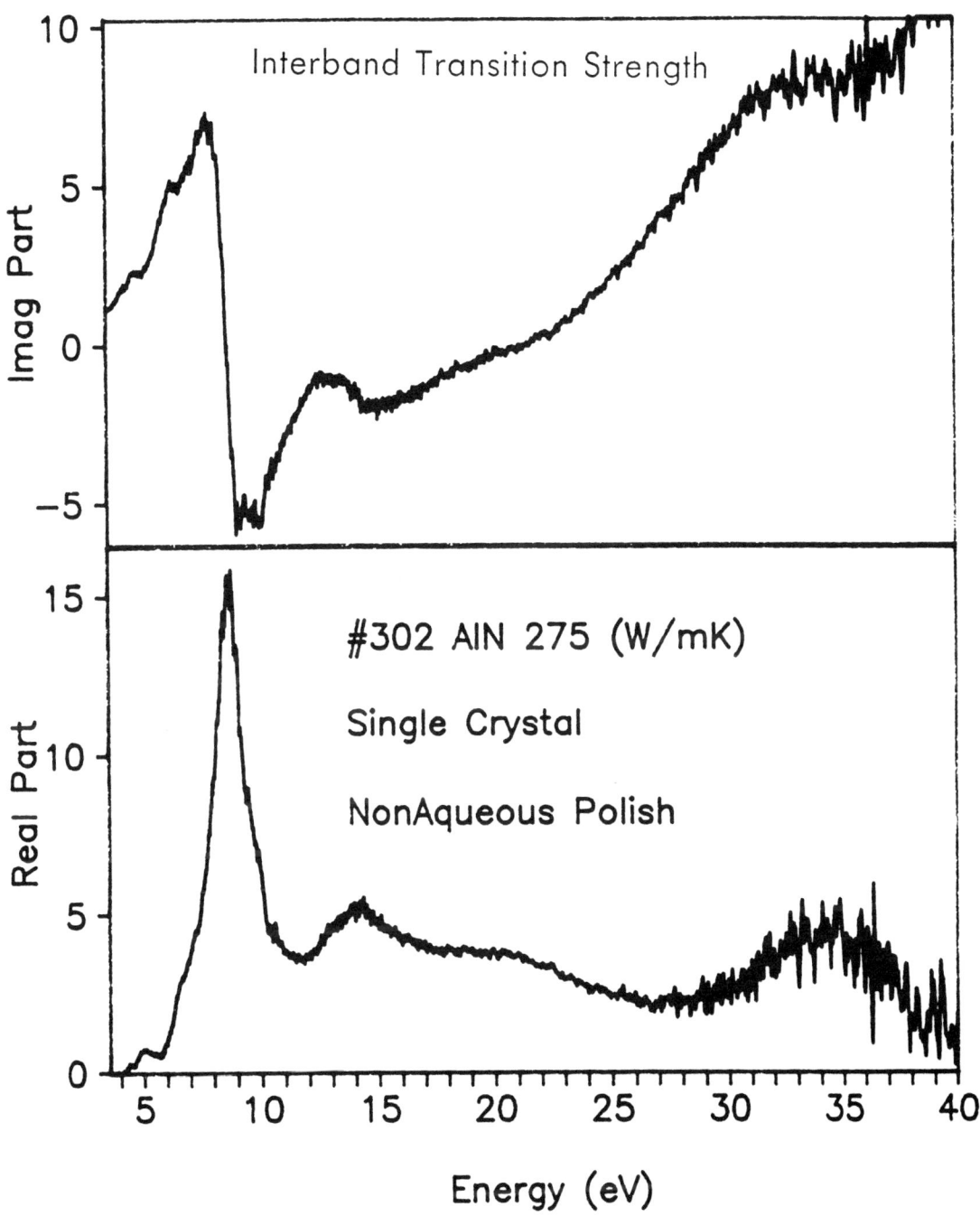

FIGURE 8 The real and imaginary parts of the interband transition strength are shown for high purity single crystal AlN.

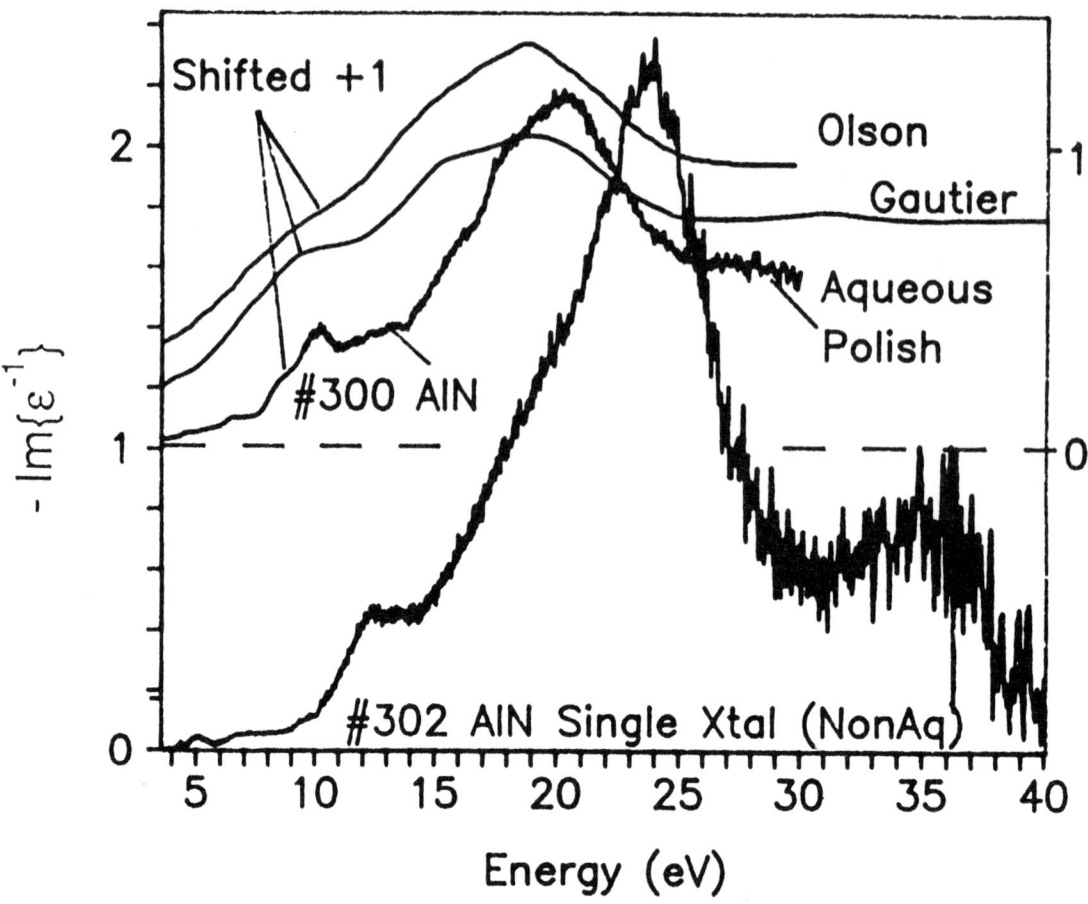

FIGURE 9 Comparison of AlN loss functions obtained for single crystal AlN (#302) and for aqueously polished polycrystalline AlN (#300) with those obtained by Olson et al [24] and by Gautier et al [25]. Note the better agreement between the EELS data and the VUV data on aqueously polished material.

F   ELECTRONIC STRUCTURE AND CRITICAL POINTS

Critical point analysis [21] fits the interband transition strength to the model

$$J_{cv} = i \left[C + (h\nu)^2 - \sum_k A_k e^{i\phi_k} (h\nu - E_k + i\Gamma_k)^{n_k} \right]^\dagger \tag{1}$$

where C is a constant, $h\nu$ is photon energy, k indexes the critical points (CPs) and † is the complex conjugate. The CP parameters are A_k, ϕ_k, E_k, Γ_k and n_k, respectively identified as the amplitude, phase, energy centre, width and dimensional exponent [16] for each CP. Critical points are grouped into sets corresponding to transitions between a valence and conduction band pair. Transitions between an isolated pair of valence and conduction bands have well defined energies for the onset and exhaustion of transition strength. FIGURE 10 shows the model constructed for the $E \perp c$ J_{cv} for single crystal AlN. The parameters for this model are given in TABLE 2.

Each of the CPs can be referenced to the calculated band structure of wurtzite AlN as calculated by Lambrecht and Segall as shown in Chapter 4. In addition, the reader is referred to the calculation by Ching and Xu [16] which agrees quite well with the Chapter 4 results, but also provides calculated optical properties and data on the effective mass asymmetry - specifically, for holes at the top of the valence band, the effective mass calculated in the c-axis direction ($m^* = m_h/m_e = 0.30 \, \Gamma{\rightarrow}A$) is an order of magnitude lower than in the basal plane directions ($m^* = 3.52 \, \Gamma{\rightarrow}M$ and $m^* = 3.40 \, \Gamma{\rightarrow}K$).

The calculated band structure reveals a direct bandgap occurring at the Γ-point. Due to band splitting at the valence and conduction band edges, a double, overlapping set is used to model the low energy feature with a single saddle point. A two dimensional (2D) CP has been assigned to the bandgap, reflecting the fact that the effective mass is considerably different in one of the three directions. TABLE 3 assigns the CP energies to specific transitions where the majuscule labels the symmetry as shown in the Brillouin zone and band diagram in Chapter 4, and the subscripts index the CP set membership and type of critical point. It is useful, if not formally correct, to infer orbital origins based on calculated partial density of states information [16] for the transitions in each set as shown in the rightmost column of TABLE 3.

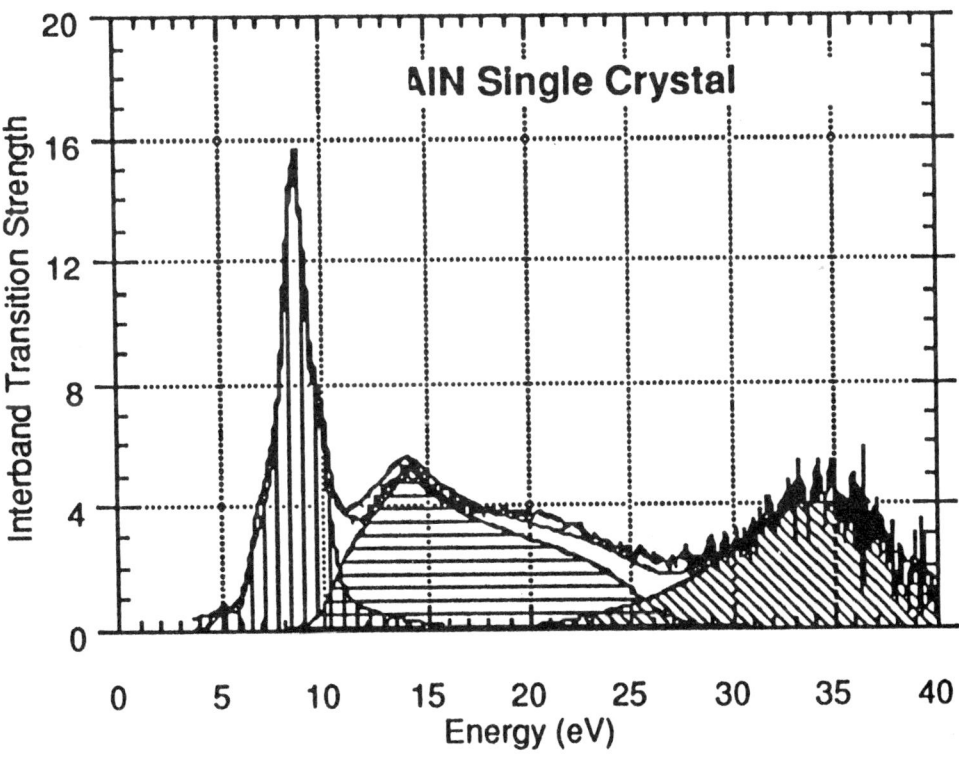

FIGURE 10 The individual sets of critical points for single crystal AlN (shaded regions) show how the experimental data can be fitted using balanced sets of critical points.

TABLE 2 Parameters for the CP model fit to single crystal AlN.

Critical point (k)	Dimension (n)	Energy E_k	Width Γ_k	Phase ϕ_k/π	Amplitude A_k
1	2D	6.29	0.38	1.85	1.32
2	2D	8.02	0.38	1.85	3.17
3	2D	8.68	0.38	0.50	4.45
4	2D	9.16	0.38	1.15	2.324
5	2D	10.39	0.38	1.12	2.18
6	2D	10.22	0.78	1.85	1.164
7	2D	14.00	0.78	0.50	1.08
8	2D	25.67	1.84	1.15	0.99
9	0D	33.85	5.64	0.00	26.13
10	3D	41.00	0.1	1.50	4.18
Constants:	Real:	20.08	Imaginary:	1.34	

TABLE 3 Assignment of features in the electronic structure of AlN.

Energy (eV)	Type	Assignment	Predominant orbital character
6.3	D_0	Γ_{01}	
8.0	D_0	Γ'_{01}	
8.7	D_1	A_{11}	$N\,2p \rightarrow Al\,3s$
9.2	D_2	H_{21}	
10.4	D_2	H'_{21}	
10.2	D_0	A_{02}	
14.0	D_1	Γ_{12}	$Al=N \rightarrow Al\,3p$
25.7	D_2	H_{22}	
33.9	S_0	Γ_{03}	$N\,2p \rightarrow Al\,3d$

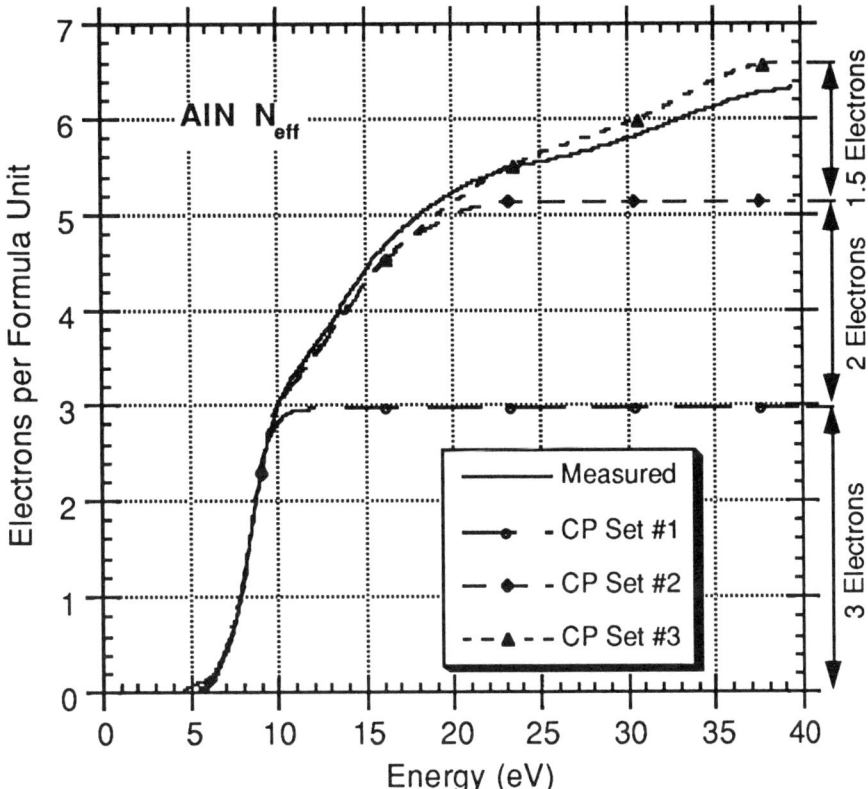

FIGURE 11 The effective number of electrons per AlN formula unit is estimated for each of the sets in the CP model by applying the sum rule to the optical conductivity contribution from each set (dashed lines), and also to the single crystal data (solid line).

The Thomas-Reiche-Kuhn f-sum rule relates the integrated optical conductivity to the plasma frequency which depends on the effective electronic density, n_{eff}. Smith and Shiles [26] have previously considered the application of sum rules over a finite energy range for valence electrons. Since the critical point model decomposes the interband transition strength into sets of transitions which arise from specific pairs of bands, one can apply the sum rule to the portion of the optical conductivity arising from each set in the model to estimate the number of electrons that are optically active between specific bands. Partial sum rules are shown in terms of number of electrons per formula unit for the single crystal in FIGURE 11, with successive sets offset to facilitate comparison with the total sum rule based on the original experimental data.

G CONCLUSION

Sections A-F provide an overview of the available data on the refractive index of AlN in the visible region, the absorption in the vicinity of the band edge at 6.2 eV and the reflectance, optical functions, and electronic structure arising from valence to conduction band transitions in the UV. These data provide a fairly complete picture of the overall electronic structure of AlN and reveal the fact that the optical effective mass is highly anisotropic with a quasi-two-dimensional character as evidenced by the logarithmic divergent peak in the interband transition strength, A_{11} at 9.9 eV. Comparison of data on high purity single crystal AlN with previous results emphasizes the fact that optical functions of AlN are strongly dependent on oxygen content.

REFERENCES

[1]   G.A. Slack [*J. Phys. Chem. Solids (UK)* vol.34 (1973) p.321-35]

[2]   G.A. Slack, T.F. McNelly [*J. Cryst. Growth (Netherlands)* vol.34 (1976) p.263-79]

[3]   G.A. Slack, T.F. McNelly [*J. Cryst. Growth (Netherlands)* vol.42 (1977) p.560-3]

[4]   K.H. Jack [*J. Mater. Sci. (UK)* vol.11 (1976) p.1135-58]

[5]   J.W. McCauley, N.D. Corbin [Report submitted to NATO Advanced Study Inst. on Nitrogen Ceramics (Aug 1981)]; see also N.D. Corbin [Master's Thesis, Mass. Inst. Tech. (June 1982)]

[6]   J.H. Harris, R.A. Youngman, R.G. Teller [*J. Mater. Res. (USA)* vol.5 (1990) p.1763-73]

[7]   T.J. Mroz Jr. [*Am. Ceram. Soc. Bull. (USA)* vol.70 (1991) p.849-50]

[8]   Y. Pauleau, J.J. Hantzpergue, J.C. Remy [*Bull. Soc. Chim. Fr. (France)* no.5-6 (1979) p.I-199-211]

[9]   S.A. Geidur, A.D. Yaskov [*Opt. Spectrosc. (USA)* vol.48 (1980) p.618-22]

[10]   H. Demiryont, L.R. Thompson, G.J. Collins [*Appl. Opt. (USA)* vol.25 (1986) p.1311-8]

[11]   J. Bauer, L. Biste, D. Bolze [*Phys. Status Solidi (Germany)* vol.39 (1977) p.173-81]

[12]   S.A. Kutolin, L.L. Lukina, R.N. Samoilova [*Inorg. Mater. (USA)* vol.9 (1973) p.862-4]

[13]   T.L. Chu, R.W. Kelm Jr. [*J. Electrochem. Soc. (USA)* vol.122 (1975) p.995-1000]

[14]   Z.-Q. Yin, G.L. Harding [*Thin Solid Films (Switzerland)* vol.120 (1984) p.81-108]

[15]   C.J.G. Kubiak, C.R. Aita, F.S. Hickernell, S.J. Joseph [*Mater. Res. Soc. Symp. Proc. (USA)* vol.47 (1985) p.75-84]

[16]   S. Loughin, R.H. French, W.Y. Ching, Y.N. Xu, G.A. Slack [*Appl. Phys. Lett. (USA)* vol.63 (1993) p.1182-4]

[17]   G.A. Slack, R.A. Tanzilli, R.O. Pohl, J.W. Vandersande [*J. Phys. Chem. Solids (UK)* vol.48 (1987) p.641-7] (Single crystal AlN sample #302 was designated W201 in their report)

[18]   H. Yamashita, K. Fukui, S. Misawa, S. Yoshida [*J. Appl. Phys. (USA)* vol.50 (1979) p.896-8]

[19]   V.E. Oranovski, J. Pasternák, S.I. Pacesová, M.V. Fock [*Phys. Status Solidi B (Germany)* vol.72 (1975) p.K39-K41]

[20]   S. Loughin, R.H. French, W.Y. Ching, Y.N. Xu, G.A. Slack [*Appl. Phys. Lett. (USA)* vol.63 (1993) p.1182-4]

[21]   V.N. Meleshkin et al [*Synchrotron Radiation* vol.80, Ed. N.G. Basov (Proc. (Trudy) of the P.N. Lebendev Physics Inst., Moscow, c.1975) p.169-74]

[22]   S. Loughin [PhD Diss., Univ. of Pennsylvania (UMI, Ann Arbor, MI, No. 9227713, 1992)]

[23]   F. Wooten [*Optical Properties of Solids* (Academic Press, San Diego, 1972) p.28, 81-2, 181-2]

[24]   C.G. Olson et al [*Solid State Commun. (USA)* vol.56 (1985) p.35-7]

[25]   M. Gautier, J.P. Duraud, C. LeGressus [*J. Appl. Phys. (USA)* vol.61 (1987) p.574-80]

[26]   D.Y. Smith, E. Shiles [*Phys. Rev. B (USA)* vol.17 (1978) p.4689-94]

6.3   Optical functions of GaN

J.A. Miragliotta

December 1993

A   INTRODUCTION

GaN, with a room temperature bandgap energy (E_g) of 3.4 eV in the wurtzite phase, has been the subject of optical investigations for nearly 25 years. In these studies, the measured optical response has provided structural and compositional information for thin film and bulk crystalline materials. In this Datareview, the optical properties associated with the refractive index (Section B), band edge absorption (Section C), and optical reflectance (Section D) of thin film and bulk crystalline GaN samples are presented. Most of the cited literature deals with undoped, intrinsic material. Modifications to the optical functions as a result of crystalline defects, alloy composition, or sample doping are referenced where appropriate. Concluding remarks are presented in Section E.

B   REFRACTIVE INDEX

Characterization of the dielectric properties of wurtzite GaN were first reported in the mid- to late-1960s. Typically, refractive index measurements ($\tilde{n} = n + ik$) of this birefringent phase have been performed in the visible and IR spectral range (370 to ~2000 nm), i.e. below the band edge of the undoped metal nitride [1-5]. In this wavelength region, the imaginary part of the dielectric function (k) is negligible and is typically ignored. A recent investigation by Amano et al [6] has evaluated the real part of the refractive index for photon energies above the band edge of GaN using spectroscopic ellipsometry (incident wavelength between ~250 and 600 nm). In addition, the absorption coefficient was also evaluated from the ellipsometric data so as to allow a determination of the imaginary part of the refractive index. Regarding other investigations of the real part of the refractive index, n, various techniques have been employed for the measurement of the ordinary (n_o) and/or extraordinary (n_e) parts of this optical function, most notably, linear transmission (UV/visible spectrophotometer) [1,7] and ellipsometry [8]. The most comprehensive treatment of the birefringence has been performed by Ejder [5] where the difference between n_e and n_o was found to be + 1.5 % at a wavelength of 500 nm, a result which is consistent with a positive uniaxial material. Analysis of IR and visible reflectance and transmission results have determined weak dispersion in GaN for wavelengths between 900 nm and 2000 nm; however, dispersion becomes increasingly important for wavelengths shorter than 800 nm as the photon energy approaches the band edge [8]. Dispersion in this optical region has also been shown to be very dependent on the oxygen content in the GaN sample, where increasing oxygen leads to a decrease in the observed dispersion [9]. An example of n in an undoped bulk sample is shown in FIGURE 1, where

$$ n = \left[1 + \frac{A}{E_0^2 - E^2} \right]^{\frac{1}{2}} \tag{1} $$

was used in modelling the experimental refractive index data in [7]. In this expression, A = 385, $E_o = 9$ eV, and E is the energy of the light source in eV.

Pressure dependence experimental studies by Perlin et al [7] revealed a linear pressure coefficient of the refractive index $\{(1/n)(\partial n/\partial P)\}$ of $-0.30 \pm 0.04 \times 10^{-2}$ GPa^{-1}. This value is consistent with their theoretical prediction of -0.24×10^{-2} GPa^{-1}, but much higher than the value predicted by Camphausen [10] of -0.05×10^{-2} GPa^{-1}. In regard to the temperature dependence, Ejder determined a value of 2.6×10^{-5} K^{-1} for the temperature coefficient of wurtzite GaN [5].

A comprehensive listing of a number of refractive measurements between the years of 1970 and 1993 is presented in TABLE 1. The sample type ranges from epitaxial films to bulk single-crystal materials. Typically, the values for high quality, single crystal films range from 2.1 at 1900 nm to 2.5 at 480 nm. It is noted that one cited reference [9] has quoted refractive index values for GaN samples where little or no experimental description of the measurements was given.

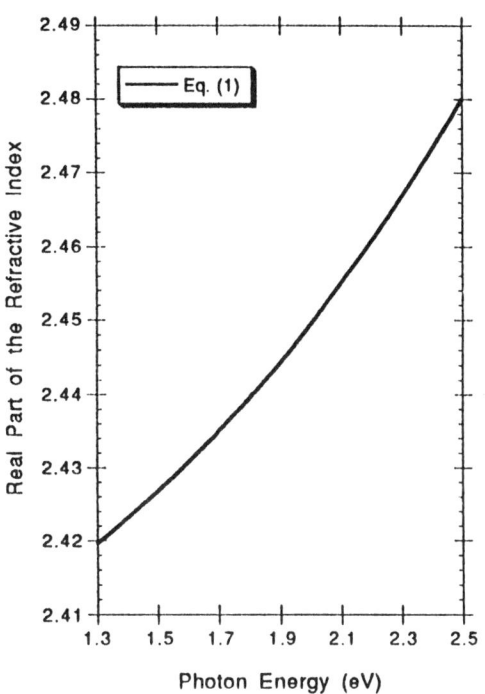

FIGURE 1 Calculated real part of the refractive index of an undoped GaN sample as determined from Eqn (1) [7].

TABLE 1 Refractive index values for GaN in the visible and IR. Where possible, the wavelength of the measurement and degree of dispersion are listed.

Index value	Wavelength of measurement	Ref
2.5	IR	[4]*
2.397	546.1 nm	[8]
2.1 - 2.5	no listed wavelength	[9]
2.53	IR	[3]*
2.42 to 2.48	497 to 956 nm	[7]
2.03	777 nm	[2]
2.1 to 2.4	800 to 2000 nm	[1]
~2.25 to 2.65	500 nm	[1]
2.41	532 nm	[5]
~2.79	~365 nm (E_g)	[6]

* No mention of dispersion in this reference.

C   BAND EDGE ABSORPTION

Over the past 25 years, a number of studies have been conducted concerning UV/visible absorption in both wurtzite and zinc blende GaN thin films. Most investigations of GaN have employed a standard UV/visible spectrophotometer for absorption measurements. From the experimental studies of intrinsically undoped material, the room temperature bandgap energy of the wurtzite form is observed to be ~3.4 eV [11-15], with a slightly lower value of ~3.2 eV for the zinc blende form [16,17]. In most of these studies, the square of the absorption coefficient for the wurtzite and zinc blende structures has been observed to be a linear function of the photon energy for energies above the bandgap ($E = E_g + 0.5\,eV$) which confirms the direct energy gap transition in GaN. An illustration of the square of the absorption coefficient dependence on photon energy is shown in FIGURE 2 [18]. In this plot, the following equation has been used to generate the curve:

$$\alpha^2 = (\alpha_0)^2 (E - E_g)$$

(2)

where α is the absorption coefficient, $\alpha_0 = 1.08 \times 10^5\,cm^{-1}$, and $E_g = 3.4\,eV$. The exponential tail for energies below the band edge is attributed to impurities and defects in the material.

Absorption measurements have also been employed in investigations of $Ga_{1-x}InN_x$ [19] and $Al_xGa_{1-x}N$ [20] ternary alloys which have determined the functional dependence of the band edge position on the x composition of these respective compounds. In addition to alloy effects, absorption measurements have been used for investigations of stoichiometry [11], oxygen content [14] and defect concentrations [21] in thin films deposited on sapphire substrates.

Temperature and pressure dependent absorption investigations of the band edge position in the wurtzite phase have also been performed. Camphausen [10] observed a temperature coefficient ($\partial E_g / \partial T$) of -0.67 meV K^{-1} and a pressure coefficient ($\partial E_g / \partial P$) of 3.7 ± 0.4 meV kbar^{-1}. Perlin et al [7] determined a pressure coefficient of 4.2 meV kbar^{-1}. These experimental values are consistent with the theoretical prediction of 4.2 meV kbar^{-1} [22]. Finally, absorption coefficients in excess of $10^4\,cm^{-1}$ for photon energies above the band edge can also be found in [1,2,9,23-26].

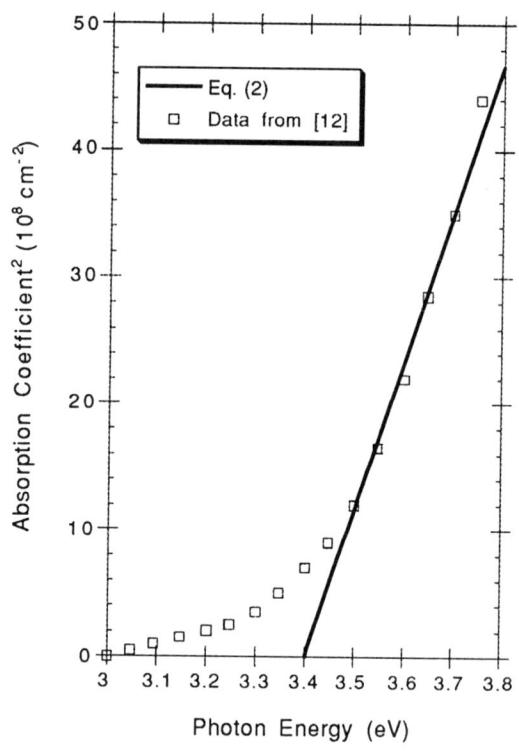

FIGURE 2 Square of the absorption coefficient for a polycrystalline GaN film [12]. The linear dependence above 3.4 eV is indicative of a direct bandgap in this material.

D   UV/VISIBLE REFLECTANCE

Although not as commonly employed as optical absorption or photoluminescence, the technique of UV/visible linear reflectance has been used in a number of experimental investigations on GaN [2,3,26-28]. The reflectance measurements have shown a sensitivity to excitonic and interband transitions in crystalline films and bulk samples for investigations that have spanned the photon energy range of 3 to 10 eV. For excitonic effects, Dingle and co-workers [26,28] provide a detailed experimental analysis of the reflectance measurements at liquid helium temperatures in the vicinity of the bandgap energy (~ 3.47 eV). The observation of three excitonic features in the spectra support the C_{6v} symmetry of the wurtzite GaN structure; in addition, the energy spacing between the excitonic features determined the spin-orbit and crystal-field splitting energies of 11 and 22 meV, respectively. In regard to interband transitions, Bloom et al [3] have proposed interband transitions for the optical resonances that were observed in UV/visible reflectivity data from a vapour deposited GaN thin film on a sapphire substrate. An example of the UV/visible reflectivity measurement from [3] is shown in FIGURE 3. The spectra were obtained from a wurtzite phase with the incident field polarized perpendicular to the c-axis of the film.

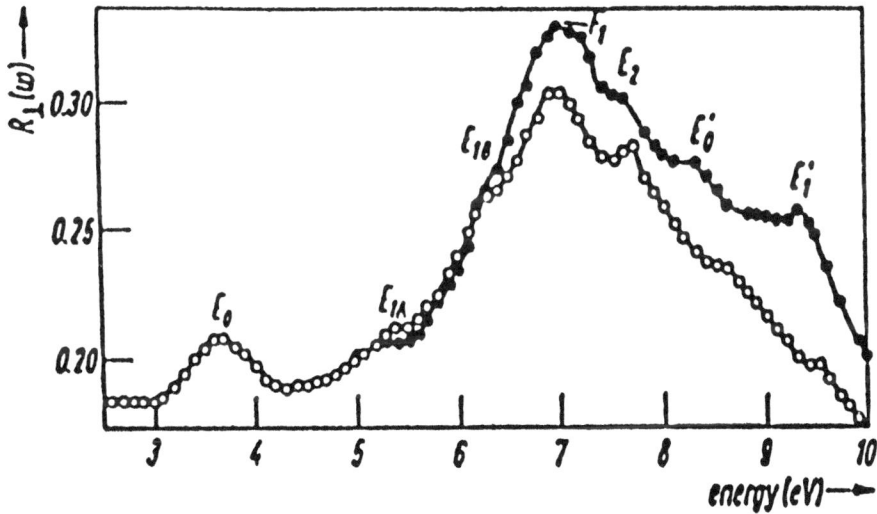

FIGURE 3   Experimental reflectivity spectra of wurtzite GaN for $E \perp$ c-axis. Open circles are for data at 295 K and filled circles are for data at 4.2 K [3].

E   CONCLUSION

Data and discussion related to the refractive index, UV absorption and UV/visible reflectance of GaN were presented in Sections A-D. The cited literature in the reference section demonstrates the need for a detailed characterization of the dielectric (real and imaginary refractive index) properties of GaN for photon energies above the band edge of this metal nitride. Optical absorption measurements have been the most frequently used optical probe of the GaN samples, providing valuable information regarding the bandgap energy position and profile for either intrinsic, alloyed or doped GaN samples. Reflectance measurements have demonstrated a sensitivity to the electronic structure (excitonic and interband) of the illuminated sample; however, there have been no recent (1983-1993) UV reflectivity studies of GaN epitaxial films for band structure analysis.

REFERENCES

[1]   H.J. Hovel, J.J. Cuomo [*Appl. Phys. Lett. (USA)* vol.20 (1972) p.71-3]

[2]   B.B. Kosicki, R.J. Powell, J.C. Burgiel [*Phys. Rev. Lett. (USA)* vol.24 (1970) p.1421-3]

[3]   S. Bloom, G. Harbeke, E. Meier, I.B. Ortenburger [*Phys. Status Solidi B (Germany)* vol.66 (1974) p.161-8]

[4]   D.D. Manchon Jr., A.S. Barker, P.J. Dean, R.B. Zetterstrom [*Solid State Commun. (USA)* vol.8 (1972) p.1227-31]

[5]   E. Ejder [*Phys. Status Solidi A (Germany)* vol.6 (1971) p.442]

[6]   H. Amano, N. Watanabe, N. Koide, I. Akasaki [*Jpn. J. Appl. Phys. (Japan)* vol.32 (1993) p.L1000-L1002]

[7]   P. Perlin, I. Gorczyca, N.E. Christensen [*Phys. Rev. B (USA)* vol.45 (1992) p.13307-13]

[8]   J.I. Pankove, J.E. Berkeyheiser, H.P. Maruska, J. Wittke [*Solid State Commun. (USA)* vol.8 (1970) p.1051-3]

[9]   K. Matsushita, Y. Matsuno, T. Hariu, Y. Shibata [*Thin Solid Films (Switzerland)* vol.80 (1981) p.243-7]

[10]   D.L. Camphausen, G.A.N. Connell [*J. Appl. Phys. (USA)* vol.42 (1971) p.4438-43]

[11]   E. Lakshmi, B. Mathur, A.B. Bhattacharya, V.P. Bhargava [*Thin Solid Films (Switzerland)* vol.74 (1980) p.77-82]

[12]   J.I. Pankove, H.P. Maruska, J.E. Berkeyheiser [*Appl. Phys. Lett. (USA)* vol.17 (1970) p.197-9]

[13]   K. Kubota, Y. Kobayashi, K. Fujimoto [*J. Appl. Phys. (USA)* vol.66 (1989) p.2984-8]

[14]   B-C. Chung, M. Gershenzon [*J. Appl. Phys. (USA)* vol.72 (1992) p.651-9]

[15]   H.P. Maruska, J.J. Tietjen [*Appl. Phys. Lett. (USA)* vol.15 (1969) p.327-9]

[16]   R.C. Powell, N.-E. Lee, Y.-W. Kim, J.E. Greene [*J. Appl. Phys. (USA)* vol.73 (1993) p.189-204]

[17]   T. Lei, T.D. Moustakas [*J. Appl. Phys. (USA)* vol.71 (1992) p.4933-43]

[18]   K. Osamura, K. Nakajima, Y. Murakami [*Solid State Commun. (USA)* vol.11 (1972) p.617-21]

[19]   T. Nagatomo, T. Kuboyama, H. Minamino [*Jpn. J. Appl. Phys. (Japan)* vol.28 (1989) p.L1334-L1336]

[20]   S. Yoshida, S. Misawa, S. Gonda [*J. Appl. Phys. (USA)* vol.53 (1982) p.6844-8]

[21]   T.L. Tansley, R.J. Egan [*Physica B (Netherlands)* vol.185 (1993) p.190-8]

[22]   L. Wenchang, Z. Kaiming, X. Xide [*J. Phys., Condens. Matter (UK)* vol.5 (1993) p.875-82]

[23]   J.C. Knights, R.A. Lujan [*J. Appl. Phys. (USA)* vol.49 (1978) p.1291-3]

[24]   S. Zembutsu, M. Kobayashi [*Thin Solid Films (Switzerland)* vol.129 (1985) p.289-97]

[25]   T.L. Tansley, R.J. Egan [*Thin Solid Films (Switzerland)* vol.164 (1988) p.441-8]

[26]   R. Dingle, D.D. Sell, S.E. Stokowski, M. Ilegems [*Phys. Rev. B (USA)* vol.4 (1971) p.1211-8]

[27]   R. Dai, S. Fu, J. Xie, G. Hu, H. Schrey, C. Klingshirn [*J. Phys. C, Solid State Phys. (UK)* vol.15 (1982) p.393-400]

[28]   R. Dingle, D.D. Sell, S.E. Stokowski, P.J. Dean, R.B. Zetterstrom [*Phys. Rev. B (USA)* vol.3 (1971) p.497-500]

6.4 Optical functions of InN

J.A. Miragliotta

December 1993

A INTRODUCTION

Optical investigations of wurtzite InN ($E_g \sim 2.0$ eV) thin films have not received the experimental attention afforded either GaN or AlN due to (1) difficulties associated with the growth of high quality, crystalline materials and (2) the existence of alternative, well-characterized semiconductors such as GaAs which have similar bandgap energies. There is, however, a growing database regarding the alloying of InN with either Ga or Al in attempts to tailor the bandgap energies of these composite materials in the photon energy range of $\sim$ 2.0 to 6.2 eV. However, this recent activity has provided only a handful of experimental and theoretical investigations concerning the optical properties of thin film InN during the past decade. This Datareview provides data and references concerning the dielectric properties, band edge absorption and optical reflectance which are presented in Sections B, C and D, respectively. Concluding remarks are presented in Section E.

B REFRACTIVE INDEX AND DIELECTRIC FUNCTION

Experimental and theoretical investigations of the dielectric properties of wurtzite InN have led to an initial characterization of the real (ε_1) and imaginary (ε_2) parts of the dielectric function ($\varepsilon = \varepsilon_1 + i\varepsilon_2$) in the photon energy range of ~ 1.5 eV to 16 eV [1-5]. In the two most detailed reports to date, Foley et al [2] and Wakahara et al [3] have assigned the peak structures in their experimentally obtained spectrum of ε_2 to interband transitions in InN between 2 and 11.4 eV. In the former study, the assignments were made in conjunction with a pseudopotential theoretical examination of ε_2 which found very good agreement between the position of the peaks in the experimental data and the calculated critical point energies in the band structure. The long-wavelength dependence for ε_1 and ε_2 has also been experimentally examined for incident wavelengths between 2.2 µm and the band edge [3,6,7]. In these reports, the authors determined that the optical response in this wavelength range is dominated by effects associated with the high carrier concentrations in the samples ($> 10^{19}$ cm^{-3}). In fact, Wakahara et al have calculated a plasma wavelength of 1.57 µm for an InN film with a carrier concentration of 6.5×10^{20} cm^{-3} assuming an effective electron mass of $0.18 m_0$ and a high frequency dielectric function ε of 7.8.

For photon energies between the band edge of InN and the free-carrier plasma resonance, the real part of the refractive index (n) has been characterized in a number of experimental investigations [7,8]. In these reports, there was little dispersion in dielectric properties from 600 to 800 nm; however, the dispersion became very large as the photon energy approached either interband transitions (shorter wavelengths) or free-carrier absorptions (longer wavelengths) in the InN films. TABLE 1 lists the results of Tyagai and co-workers [7], where n_r was measured from 620 nm to 1600 nm. In the experimental studies, ε_2 has been determined from an appropriate analysis of normal incidence reflectivity spectra [3,6] in the photon energy region of 1.5 to 6.5 eV.

TABLE 1 Real part of the refractive index as determined from reflectance measurements from an n-type InN film [7].

Wavelength (nm)	n
620	3.10
660	3.12
740	3.12
780	3.04
820	2.93
880	2.90
910	2.78
1000	2.56
1120	2.39
1480	2.12
1600	1.63

C BAND EDGE ABSORPTION

As was the case for GaN, optical absorption investigations have been more prevalent than either dielectric or reflectance measurements due to the ability of the former technique in accurately analyzing the band edge region in InN films [7-11]. As in the case of GaN absorption investigations, the investigations of InN films have employed standard UV/visible spectrophotometers for the optical measurements. In most of these studies, the absorption coefficients for the InN films were in excess of $10^4 \, cm^{-1}$ for photon energies near or above the band edge ($\sim 2\,eV$) of the metal nitride. In addition, the linear behaviour of the square of the absorption coefficient for photon energies $\sim 0.5\,eV$ above the band edge of InN suggests a direct energy gap in the band structure of the material. An illustration of the square of the absorption coefficient dependence on photon energy is shown in FIGURE 1 [12]. In this plot, the following equation has been used to generate the curve:

$$\alpha^2 = (\alpha_0)^2 (E - E_g) \tag{1}$$

where α is the absorption coefficient, $\alpha_0 = 1.23 \times 10^5 \, cm^{-1}$, and $E_g = 1.89\,eV$.

A listing of the more commonly referred publications concerning InN optical absorption studies is shown in TABLE 2. For all samples in this listing, the absorption coefficient was $> 10^4 \, cm^{-1}$ and displayed a linear behaviour for photon energies in the vicinity of the energy gap, $E_g \sim 2.0\,eV$. The effects of alloying [14] and defects [11,13] on the absorption profile of InN thin films are also cited.

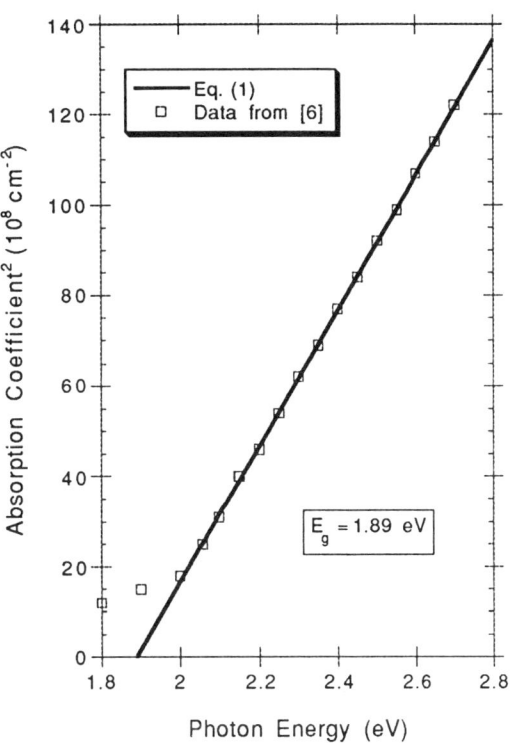

FIGURE 1 Square of the absorption coefficient for a polycrystalline InN film [12]. The linear dependence above 1.89 eV is indicative of a direct bandgap in this material.

TABLE 2 A selection of experimentally determined absorption coefficients near the band edge of InN. The photon energy refers to the approximate gap energy as determined from the absorption measurement.

α (cm^{-1})	Photon energy (eV)	Ref
$> 10^5$	2.09 to 2.21	[9]
$> 4 \times 10^4$	1.9 to 2.1	[3]
$> 3 \times 10^4$	1.89	[12]
$> 10^5$	2.0	[15]
$> 5.5 \times 10^4$	2.1	[16]
$> 10^4$	1.83	[17]
$> 2 \times 10^4$	2.05	[7]
10^4 to 10^5	$\sim$ 1.8 to 1.9	[8]
$> 3 \times 10^4$	1.95	[18]

D UV/VISIBLE REFLECTANCE

To date, only a handful of UV/visible reflectance investigations have been performed on InN thin films or alloys of this material [3,6,7,19]. The optical transitions that are present in the reflectivity spectrum from this metal nitride crystal permit the identification of the principle interband transitions in the illuminated sample. References [3,6,7] provide analysis of the ~ 0.5 to 6.5 eV photon energy region while [14] extends this range to ~ 12.5 eV. The combination of these reports has resulted in the identification of various interband transitions in InN which are listed in TABLE 3. In addition, the long-wavelength reflectance spectra ($\lambda > 1800$ nm) have been used to determine the parameters that are associated with the free-carrier excitations in the highly n-type InN films, e.g. electron plasma resonances (energy (0.6 eV), damping constant (0.18 eV), and effective mass (0.11 m_0) [7]).

TABLE 3 Reflectivity peak positions from an InN thin film on sapphire and the respective interband transitions associated with these features.

Interband transition	Reflectivity peak (eV)
E_0 Γ_6-Γ_1	1.89 [2], 1.8 [3]
E_{1a} Γ_5-Γ_3	5.0, 5.5 [19]
E_{1b} U_4-U_3	5.8 [19], 4.8 [3]
M_2-M_1	5.1 [3], ~7.3 [19]
F_1 M_4-M_3	7.3 [19]
E_2 K_3-K_2	7.2 [2], 7.3 [19]
E'_0 Γ_6-Γ_1	8.8 [19]
E'_1 Γ_5-Γ_6	8.8 [19]

E   CONCLUSION

Data and discussion related to the refractive index, UV absorption and UV/visible reflectance of InN were presented in Sections A-D. The cited literature in the reference section shows both experimental and theoretical characterization of the dielectric (real and imaginary refractive index) properties, for photon energies that scan the free-carrier resonances and interband transitions in InN. Optical absorption measurements have been used frequently in determining the bandgap energy position and profile for either intrinsic, alloyed or doped InN samples. UV/visible reflectance measurements have demonstrated a sensitivity to both the free-carrier concentration and interband properties of InN. Future investigations should continue in hope of characterizing the dielectric properties of InN samples that have relatively low carrier concentrations ($<10^{17}$ cm^{-3}). To date, all samples that have been investigated have suffered from carrier concentrations in excess of 10^{19} cm^{-3}.

REFERENCES

[1]   B. Sullivan, R.R. Parsons [*J. Appl. Phys. (USA)* vol.64 (1988) p.4144-9]
[2]   C.P. Foley, T.L. Tansley [*Phys. Rev. B (USA)* vol.33 (1986) p.1430-3]
[3]   A. Wakahara, T. Tsuchiya, A. Yoshida [*Vacuum (UK)* vol.41 (1990) p.1071-3]
[4]   O. Takai, J. Ebisawa, Y. Hisamatsu [*Proc. 7th ICVM* (Japan, 1982) p.137-44]
[5]   V.A. Tyabai, O.V. Snitko, A.M. Evstigneev, A.N. Krasiko [*Phys. Status Solidi B (Germany)* vol.103 (1981) p.589-94]
[6]   K.L. Westra, M.J. Brett [*Thin Solid Films (Switzerland)* vol.192 (1990) p.227-34]
[7]   V.A. Tyagai, A.M. Evstigneev, A.N. Krasiko, A.F. Andreeva, V.Ya. Malakhov [*Sov. Phys.-Semicond. (USA)* vol.11 (1977) p.1257-9]
[8]   H.J. Hovel, J.J. Cuomo [*Appl. Phys. Lett. (USA)* vol.20 (1972) p.71-3]
[9]   N. Puychevrier, M. Menoret [*Thin Solid Films (Switzerland)* vol.36 (1976) p.141-5]

[10] T.L. Tansley, R.J. Egan [*Thin Solid Films (Switzerland)* vol.164 (1988) p.441-8]

[11] T.L. Tansley, R.J. Egan [*Physica B (Netherlands)* vol.185 (1993) p.190-8]

[12] T.L. Tansley, C.P. Foley [*J. Appl. Phys. (USA)* vol.59 (1986) p.3241-4]

[13] T.L. Tansley, C.P. Foley [*J. Appl. Phys. (USA)* vol.60 (1986) p.2092-5]

[14] K. Kubota, Y. Kobayashi, K. Fujimoto [*J. Appl. Phys. (USA)* vol.66 (1989) p.2984-8]

[15] H. Takeda, T. Hada [*Toyama Kogyou Kute Semmon Gakko Kiyo (Japan)* vol.11 (1977) p.73-5]

[16] G.V. Samsonov, A.F. Andreeva [*Sci. Sinter. (Serbia)* vol.12 (1980) p.155-62]

[17] K.L. Westra, R.P.W. Lawson, M.J. Brett [*J. Vac. Sci. Technol. A (USA)* vol.6 (1988) p.1730-2]

[18] K. Osamura, K. Nakajima, Y. Murakami [*Solid State Commun. (USA)* vol.11 (1972) p.617-21]

[19] V.V. Sobolev, S.B. Kroitoru, A.F. Andreeva, V.Ya. Malakhov [*Sov. Phys.-Semicond. (USA)* vol.13 (1979) p.485-6]

CHAPTER 7

PHOTOLUMINESCENCE AND CATHODOLUMINESCENCE

7.1 Photoluminescence and cathodoluminescence of AlN

J.H. Harris and R.A. Youngman

January 1994

A   INTRODUCTION

In the following Datareview, the photoluminescence and cathodoluminescence characteristics of aluminium nitride (AlN) are reviewed. The forms of AlN covered include AlN powders (which contain a native oxide layer), sintered ceramics which typically contain an oxide second phase, and low oxygen content, single crystal samples. The powder and ceramic forms of this material are of particular interest, because sintered ceramic AlN is currently finding application as an important new high thermal conductivity packaging material for electronic devices. This review is organized into three main topics: photoluminescence of oxygen defects in AlN; photoluminescence of AlN doped with Mn; and cathodoluminescence of AlN. The large emphasis on oxygen-related luminescence results reflects both the technological importance of this defect in AlN applications and the large body of literature which exists investigating oxygen related properties in this material. Oxygen incorporation in the AlN lattice has been shown to degrade thermal conductivity, a key technological driver for many AlN applications, including advanced packaging. From a scientific viewpoint, both the oxygen related studies and the work on Mn-doped material yield key insights into the defect accommodation mechanisms which exist in this technologically important III-V material.

B   PHOTOLUMINESCENCE OF ALUMINIUM NITRIDE

B1   Oxygen-Related Spectra as a Function of Temperature

After irradiation with ultra violet (UV) light, aluminium nitride (AlN) samples doped with oxygen emit a series of broad luminescence bands in the near UV at 300 K. This is observed in AlN single crystal samples, sintered ceramics and powders prepared via a variety of synthesis routes, as will be discussed below. Since the AlN lattice has a very large affinity for oxygen dissolution, oxygen contamination is impossible to eliminate completely in AlN, and consequently the optical properties of all samples studied to date are influenced by oxygen-related defects. For example, currently available commercial AlN powders contain about 1.0 to 1.5 wt.% oxygen, some of which is dissolved in the AlN lattice with the remainder in the form of an oxide coating on each powder grain. Commercially produced sintered AlN ceramics contain between 0.5 and 1.0 wt.% dissolved oxygen.

In an early study of the light emission properties of AlN, Pacesova and Jastrabik [1] observed two broad luminescence lines centred near 3.0 and 4.2 eV and greater than 0.5 eV wide for samples containing between 1 and 6 at.% oxygen (no more specific concentrations were given) after steady state excitation at 5.5 eV. These workers also observed the intensity of this line to increase by approximately a factor of 2 upon cooling to 77 K [1].

Youngman et al [2] investigated photoluminescence of AlN powder and single crystal samples at 300 and 10 K after photoexcitation at 4.6 eV. The luminescence spectrum from the single crystal sample is shown in FIGURE 1. This large single crystal (>2 cm), which was grown by Slack et al [3] via a high temperature and high pressure vapour transport method under nitrogen deficient conditions, had an oxygen content of 380 ppm. Note that the single crystal spectrum is characterized by broad peaks centred at 2.7 and 3.8 eV [2]. This is to be contrasted with the luminescence spectrum shown in FIGURE 2 for AlN powder containing 1.53 wt.% dissolved oxygen. At 300 K, this consists of an extremely broad line centred at 3.3 eV which only sharpens slightly upon cooling to 10 K, in closer agreement with the observations of Pacesova and Jastrabik [1]. It will be shown below that the oxygen related luminescence spectra in AlN are very sensitive to sample preparation, particularly oxygen impurity content, thus explaining some of the differences in energy of the oxygen associated emission spectra discussed above. In each case, however, the emission spectrum from oxygen-doped AlN is dominated by a broad, intense line in the near UV.

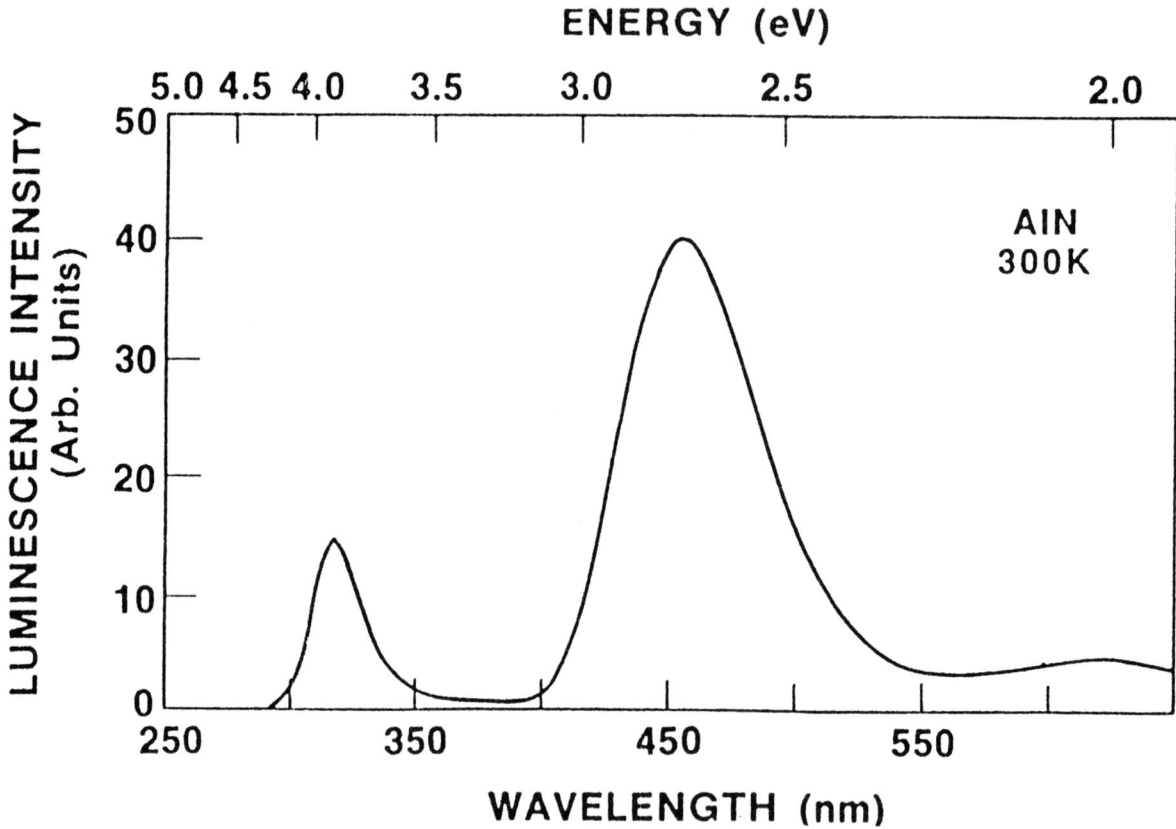

FIGURE 1 Photoluminescence spectrum at 300 K from a large single crystal of AlN grown under nitrogen deficient conditions.

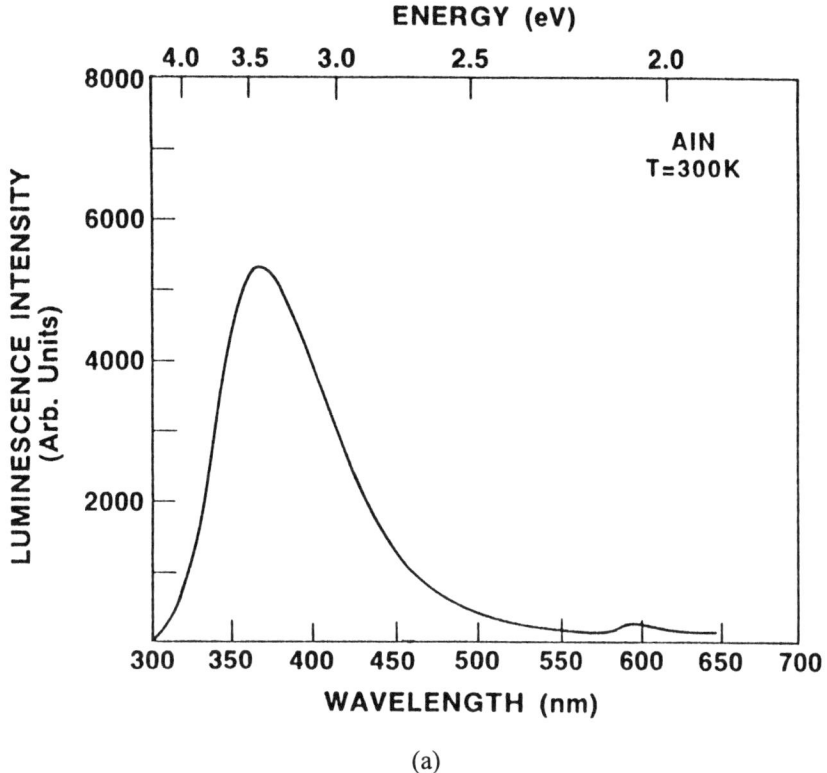

(a)

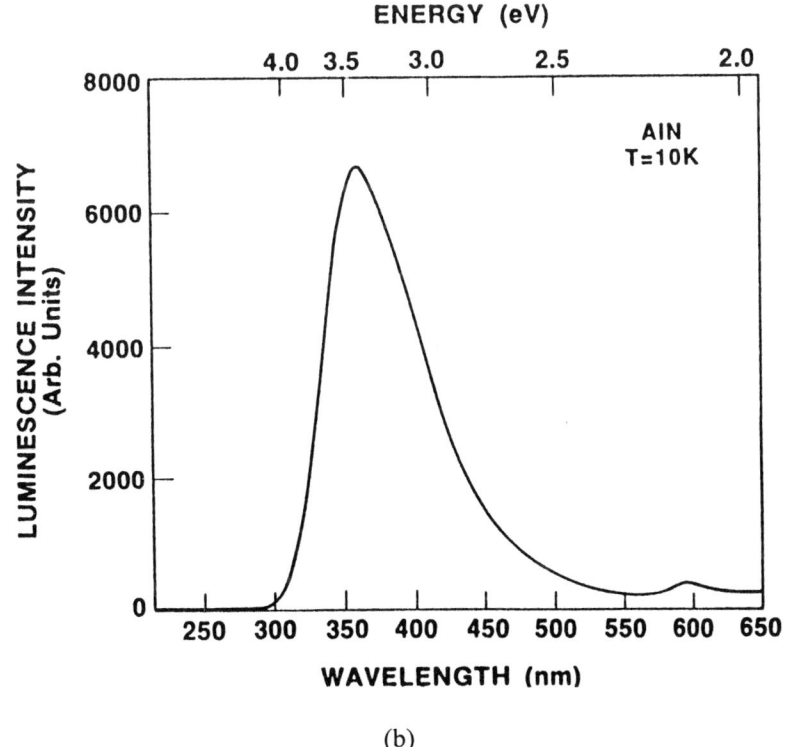

(b)

FIGURE 2 Photoluminescence spectrum at 300 K (a) and 10 K (b) from AlN powder with an oxygen content of 1.53 wt.%.

B2 Luminescence Spectra as a Function of Oxygen Content

In 1976, Oranovski et al [4] measured the excitation spectra for photoluminescence of various AlN powders with oxygen contents between 1.84 and 11.1 at.%. This measurement, which was made at 300 K, integrated luminescence within the range between 2.3 eV and 3.5 eV. These workers observed a large peak in the excitation spectra between 8.0 and 10 eV which increased significantly as powder oxygen content increased (the peak intensity for the 11.1 at.% samples was approximately twice as high as for the 1.84 at.% sample). Thus, these workers observed a scaling of the luminescence intensity with the sample oxygen content.

Pastrnak et al [5] measured the luminescence spectra for AlN samples containing $6.7 \times 10^{20} \, cm^{-3}$ and $5.4 \times 10^{21} \, cm^{-3}$ oxygen atoms after excitation at 254 nm and 313 nm. These workers observed very little spectral shift of the major luminescence line centred near 3 eV as oxygen concentration was varied. The centre of this peak did shift slightly with excitation energy [5].

In more recent work, Harris et al [6] and Youngman and Harris [7] investigated the luminescence spectra for AlN polycrystalline sintered samples over a wide range of oxygen concentrations. These samples were prepared with yttrium oxide as a sintering additive, and contained yttrium aluminate second phase particles. As shown in FIGURES 3 and 4, these workers observed a continuous shift in the peak position of the UV luminescence line as a function of oxygen content up to a critical concentration of about 0.75 at.%. For concentrations above this critical value, the luminescence line remained stationary. These workers also observed a dramatic increase in luminescence intensity at about this same critical value, as indicated by the dotted line in FIGURE 4. Harris et al discussed a microscopic model to understand these results which proposed that at concentrations below 0.75 %, oxygen goes into the AlN lattice substitutionally on nitrogen sites, with the subsequent formation of aluminium atom vacancies, and that at higher oxygen concentrations a new defect based on octahedrally coordinated aluminium forms. A key component of this model was the charged nature of the substitutional nitrogen defect and the aluminium vacancy which form at low oxygen concentrations.

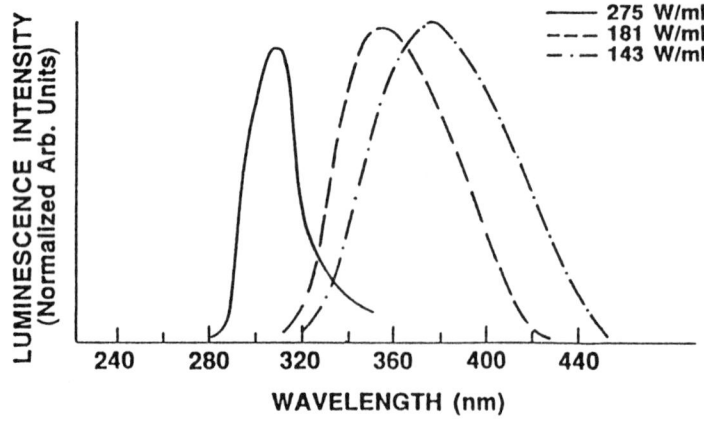

FIGURE 3 Luminescence intensity as a function of wavelength for AlN samples with different oxygen concentrations (and thus differing thermal conductivities).

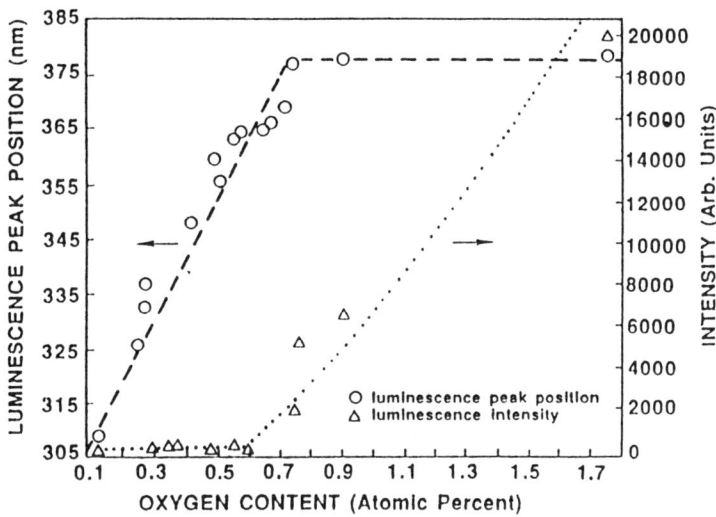

FIGURE 4 Steady-state luminescence peak position and intensity as a function of oxygen concentration showing a transition in both peak position and intensity.

B3   Time Evolution of the Luminescence Spectrum of Oxygen-Doped AlN

In 1974, Pastrnak et al [8] evaluated the time evolution of the 3.25 eV luminescence peak associated with the oxygen impurity, for AlN samples containing an oxygen concentration of 6.8×10^{20} cm^{-3}. This sample was excited at 254 nm and luminescence spectra were taken during excitation and 1.8 ms after excitation. During excitation, the luminescence spectra consisted of a broad, symmetrical line centred near 3.25 eV. The width of this line was approximately 0.5 eV. This spectrum was observed to change significantly after a delay of 1.8 ms, when the luminescence line was observed to be extremely broad (about 1.5 eV) and asymmetrical with a significant skew toward the low energy side and a peak shift toward lower energy.

These trends are consistent with recent experimental observations by Harris and Youngman [9] where time resolved luminescence spectra were collected over a range of delay times for a polycrystalline sintered AlN sample containing 2.5 at.% oxygen. As in [8] above, excitation was accomplished at 254 nm. In FIGURE 5 is shown the time evolution of the 375 nm luminescence line after delay times of 1, 10, 100 and 1000 ms. Note that, consistent with the observations of Pastrnak et al [8], the oxygen associated luminescence line is observed to become asymmetric with increased delay time and experience a peak shift toward lower energies.

In both of these studies, the experimental results are consistent with relaxation of photo-excited carriers to more localized states as a function of delay time. This implies the existence of a broad distribution of sites which may accommodate photo-excited carriers.

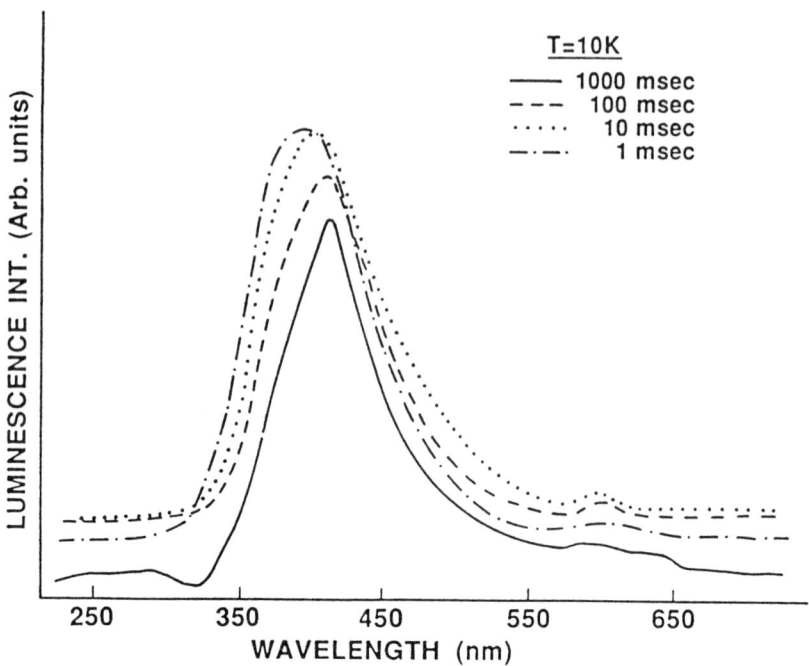

FIGURE 5   Time resolved luminescence spectra of the 375 nm luminescence peak in oxygen-doped AlN at 300 K, with delay times of 1, 10, 100 and 1000 ms after a 1 ms excitation pulse. A large shift of this peak toward lower energies is apparent.

B4   Relations between Optical Absorption and Luminescence for Oxygen-Doped AlN

Since UV absorption in AlN is covered as a separate topic, only a brief comment will be made here. In Section B2, trends in the emission spectra of UV excited AlN were discussed for AlN samples with various oxygen concentrations. Complementary trends have been observed by a number of workers in the absorption spectra for oxygen-doped AlN samples. Features in the UV absorption spectra that may be associated with the oxygen impurity have been observed by Pastrnak and Roskovcova [10], Kim et al [11], and Youngman and Harris [7].

B5   Energy Level Diagrams for the Oxygen-Related Defect in AlN

Energy level diagrams proposed to describe the results of luminescence experiments performed on oxygen-doped AlN samples have been described by a number of workers. Pastrnak et al [8] observed luminescence in oxygen-containing AlN from several overlapping bands between 2.37 and 3.51 eV. These workers interpreted these emissions as originating from various transitions between substitutional oxygen defects, which act as donors (D), and aluminium vacancies, which act as acceptors (A). An energy level diagram illustrating these transitions is shown in FIGURE 6.

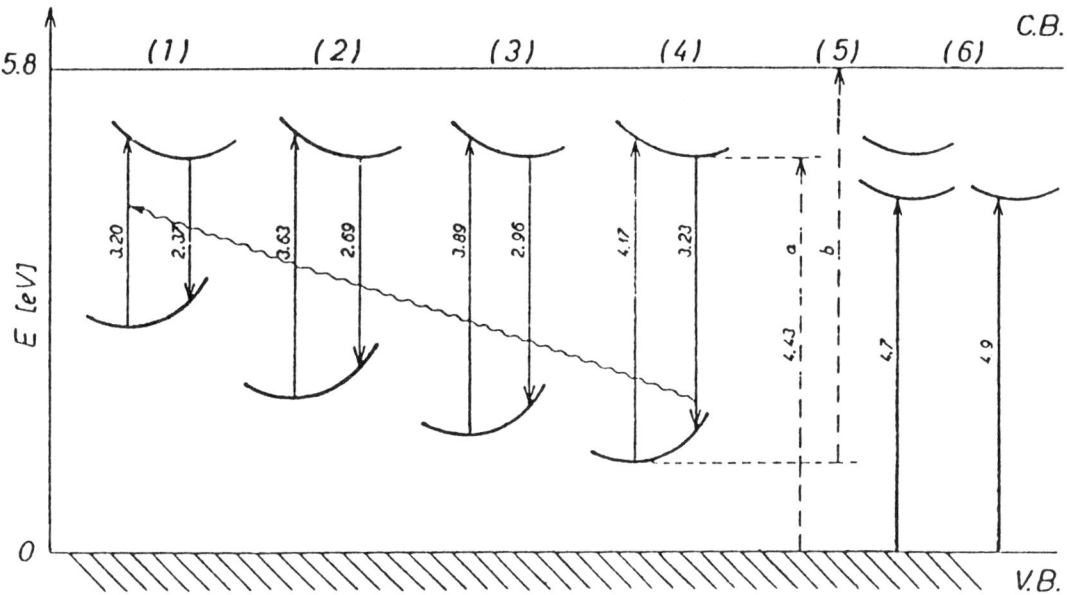

FIGURE 6 Optical transitions in oxygen luminescence centres of donor (D)-acceptor (A) pairs and D-A complexes. The wavy line denotes the possible energy transfer from V_{Al}-O_N pairs to the V_{Al}-$2O_N$ centres.

In 1979, Pacesova and Jastrabik [1] studied emission and absorption from AlN fabricated via nitridation of $AlF_6(BN)_3$ in ammonia at 1000-1100 °C, which had an oxygen concentration between 0.73 and 4.85 wt.%. These workers also interpreted luminescence spectra as originating from transitions between deep donor states (due to oxygen) and compensating aluminium vacancies. As in [8] above, this work also involved de-convoluting a number of overlapping luminescence bands. The energy level diagram proposed by these workers is shown in FIGURE 7.

Harris and Youngman utilized a combination of luminescence and photo-induced absorption measurements to construct a different energy level diagram for oxygen associated defects in AlN [12]. A representation of this energy scheme is shown in FIGURE 8. In this case, charged substitutional oxygen sites and charged aluminium vacancies act as traps for photo-generated carriers. These defects result in a broad band of energy levels deep within the AlN bandgap. In this picture, radiative recombination of these trapped carriers is responsible for the UV luminescence observed in oxygen-doped samples. Upon exposure to visible light, these photo-generated carriers are elevated from deep trap levels to the band edge, where efficient radiative recombination can occur. This later transition, which is indicated in FIGURE 8, is responsible for the observed UV-induced absorption of visible light (UV induced photo-darkening) seen in oxygen-doped AlN samples.

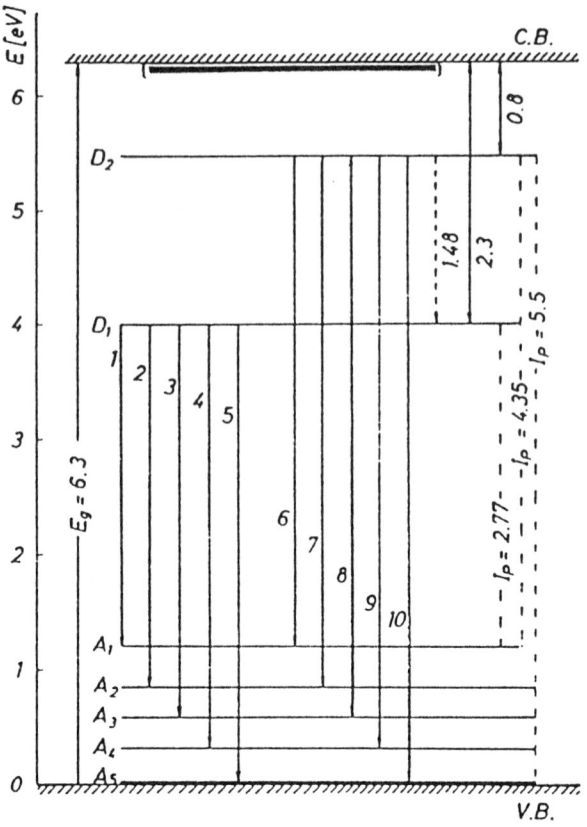

FIGURE 7 Energy diagram of impurity levels in the gap of AlN. Arrows (1)-(10) indicate optical transitions responsible for elementary emission and absorption bands. Dashed lines indicate transitions accompanied by photoconductivity.

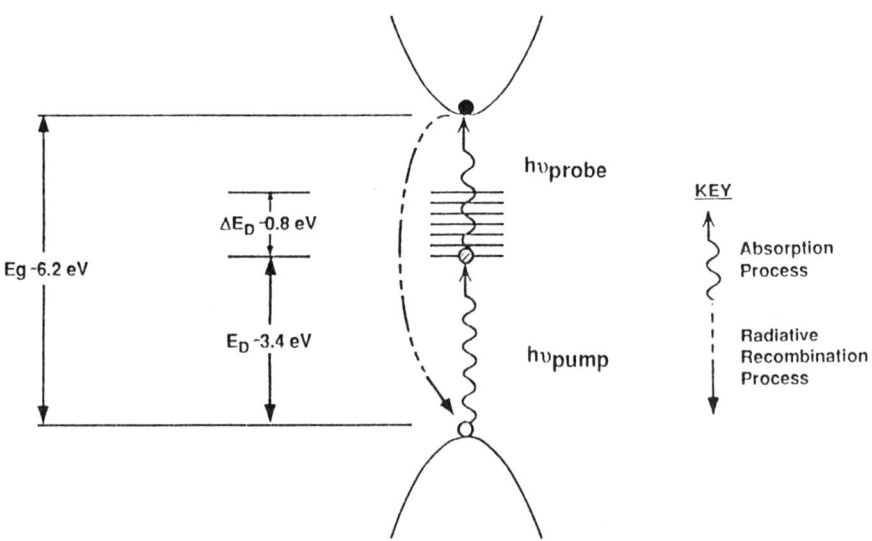

FIGURE 8 Schematic energy level diagram illustrating the location of the UV induced levels within the AlN bandgap. This diagram also describes the UV photoinduced absorption of visible light, with: UV pump induced excitation of a carrier from the valence band (open circle) to a localized trap state at the charged impurity (partially shaded circle); visible probe induced excitation of the localized, trapped carrier to the conduction band (dark circle); and finally, radiative recombination which returns the system to the unexcited state.

B6   Manganese Luminescence Centres in AlN as a Function of Temperature

As indicated in the previous five sections, oxygen is always present as an impurity in the AlN lattice, and thus a wide range of studies have been performed to interrogate the optical properties of oxygen-doped AlN. Another impurity that has been widely investigated is manganese. Manganese substitution in various wide bandgap materials has been routinely investigated for a number of years, particularly in situations where its luminescence spectra can be utilized as an effective probe of the host structure. As will be discussed below, Mn-doped AlN falls into this material category.

A detailed study of red luminescence from Mn-doped AlN is provided in an early study by Karel et al [13]. AlN samples were prepared for this study by reacting ammonium fluoaluminate and gaseous ammonia followed by an anneal in an atmosphere of $MnCl_2$ vapour at 1200 °C for 10 to 20 hr. This process produced AlN crystals with Mn concentrations between 0.01 and 1 wt.%. After these Mn-doped AlN samples were excited in the UV, a luminescence band consisting of a number of sharp emission peaks was observed in the visible. These results are shown in FIGURE 9 for a sample containing 0.01 % Mn and in FIGURE 10 for a sample with 1 % Mn, in both cases at 300 K, 77 K and 10 K. Note that the sharp spectral features evident in the 0.01 % Mn-doped sample all but disappear in the more heavily doped sample, except at the lowest temperature. Karel et al interpret the peaks in these spectra as arising from phonon emission, in all likelihood due to localized Mn ion vibration, associated with electronic transitions experienced by Mn^{4+} ions located on Al sites. They propose that these transitions are from the metastable energy levels ^2E_g to the basic level $^4A_{2g}$ for an octahedral site and $^4T_1$ for a tetrahedral site. This work was further expanded upon by Karel

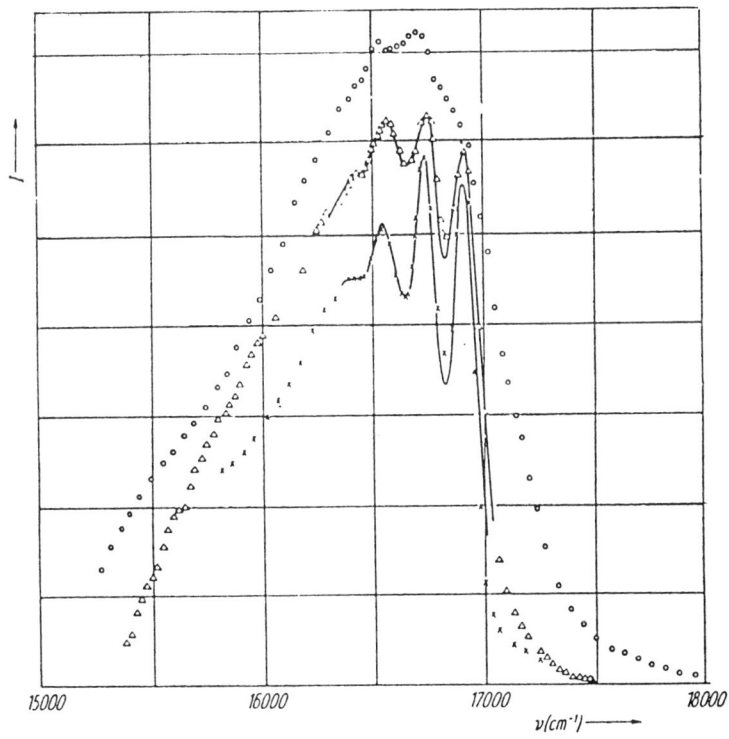

FIGURE 9 Emission spectra from AlN:Mn (0.01 % Mn) photoluminescence. Open circles are room temperature, triangles are at liquid nitrogen temperature, crosses are at liquid helium temperature.

and Mares in a 1972 paper [14], and Archangelskii et al, who associated an observed red band at 300 K (600 nm) with Mn^{4+} and a green band (515 nm) with Mn^{2+} [15]. These workers also noted that the position of the Mn-associated emission depends not only on the charge state of the Mn, but on the site symmetry and crystal field parameters.

In addition to Mn-doping, a number of other phosphors have been incorporated in AlN and have been investigated using luminescence. In FIGURE 11 the luminescence spectrum from Eu-doped AlN is shown at 77 K, as studied by Karel and Pastrnak [16]. Note that this results in a broad featureless emission band centred in the green near 2.36 eV. These workers propose that europium is incorporated in the AlN lattice as an Eu^{2+} ion, and that the energy of the emission depends strongly on the symmetry of the dopant site, thus explaining the featureless nature of the luminescence. These workers also investigated luminescence from samarium centres at 77 K, as shown in FIGURE 12.

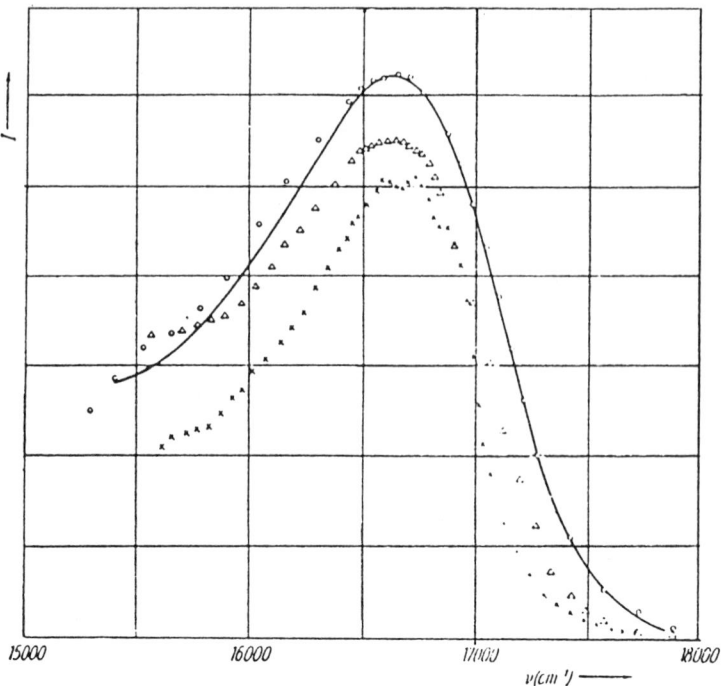

FIGURE 10   Emission spectra from AlN:Mn (1.0 % Mn) photoluminescence. Open circles are room temperature, triangles are at liquid nitrogen temperature, crosses are at liquid helium temperature.

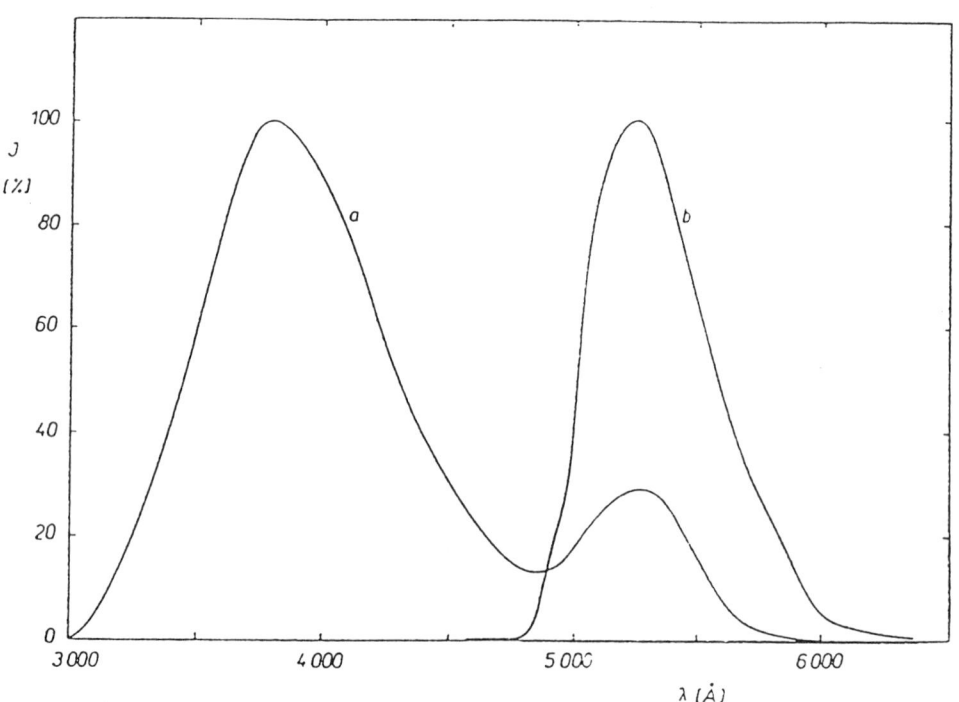

FIGURE 11 The luminescence spectrum of AlN doped with Eu at 77 K. Curve a is cathodoluminescence and b is photoluminescence under 3650 Å mercury line excitation.

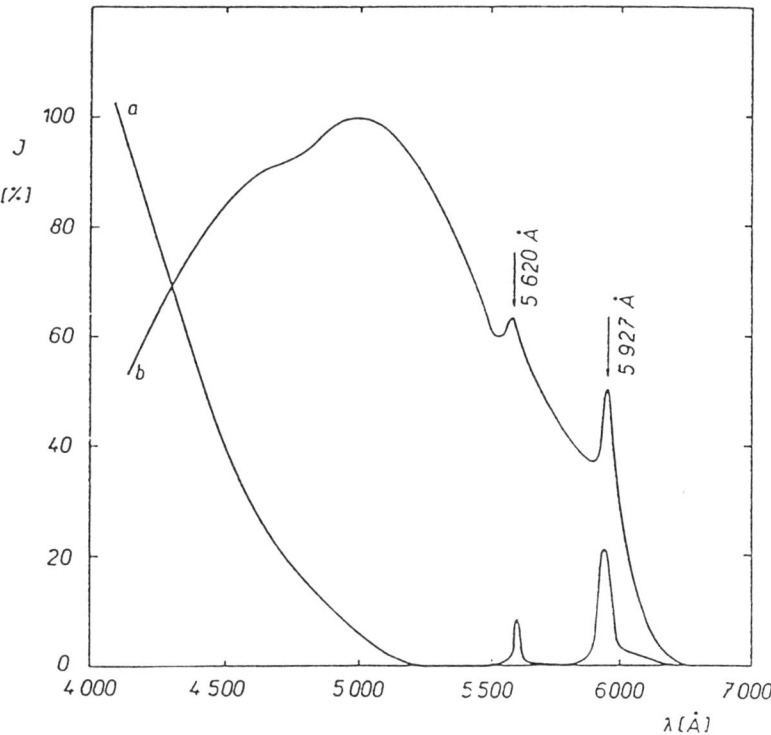

FIGURE 12 The luminescence spectrum of AlN doped with Sm at 77 K. Curve a is cathodoluminescence and b is photoluminescence under 3650 Å mercury line excitation.

B7 Time Evolution of Luminescence in Mn-Doped AlN

As discussed in Section B6, Mn-doped AlN produces luminescence in the red region of the visible spectra typically with significant structure, particularly at low temperature. Karel and Pastrnak studied the decay of this luminescence band at room temperature in a 1970 paper [16]. These workers observed luminescence lifetimes extending out to 30 ms. They interpreted the luminescence decay dynamics as being dominated by a complicated system of traps which are active over the entire temperature region investigated (- 190 to 500 °C).

The time evolution of Mn-associated luminescence was further pursued in a study by Pastrnak et al [17]. In this work, Mn enters the lattice as either Mn^{4+} or Mn^{2+}, depending on the activation conditions. In addition to Mn-doping, these samples also contained 0.3 to 3 wt.% oxygen. In this study, luminescence was either activated on the surface of the sample (using pulsed nitrogen discharge tubes) or within the sample bulk (by exciting the oxygen or Mn absorption bands with a frequency doubled die laser). In FIGURE 13 is shown the decay of luminescence associated with the Mn^{4+} centre for different excitation conditions. These workers observed that if the excitation occurred directly in the Mn^{4+} band (at an energy of 2.37 eV), the decay was exponential with a time constant of 1.3 ms; and if excitation occurs through the oxygen band (3.68, 4.75 eV), the decay can be resolved into two exponentials. Pastrnak et al explained these results by proposing the existence of an emission-free layer at the surface of the sample. This centre could be formed by a process of recombination stimulated diffusion of Mn toward the sample surface.

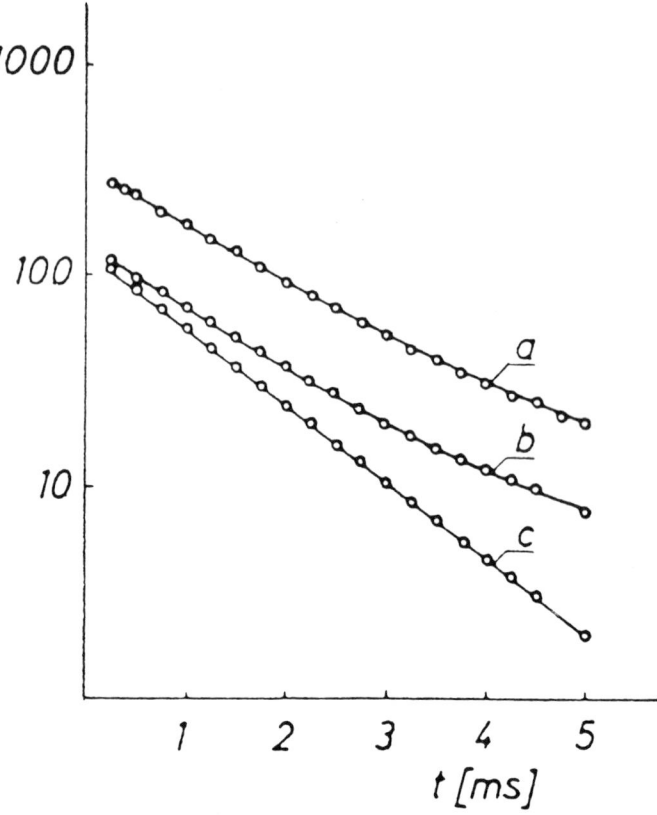

FIGURE 13 Decay curves for photoluminescence from Mn^{4+} centres in AlN:Mn at room temperature: (a) after excitation at 3.68 eV; (b) after excitation at 4.75 eV; and (c) after excitation at 2.37 eV.

The dynamics of Mn-associated luminescence was also investigated in a 1982 paper by Archangelskii et al [15]. In FIGURE 14 are shown time resolved luminescence spectra from this study showing the dynamics of both the Mn^{4+}-associated red band and the Mn^{2+}-associated green peak. These workers concluded that these decay curves have both an exponential and non-exponential component. They interpreted this as an indication that energy transfer processes and trap levels play a key role in controlling decay dynamics. These energy transfer processes may be between Mn^{2+} and Mn^{4+} ions, because of the overlap of the emission band of the former with the excitation band of the latter. These workers also noted that at high Mn concentrations, a new emission band in the red is seen, possibly due to interacting pairs of Mn^{4+} ions.

B8   Energy Level Diagrams for the Mn-Associated Centre in AlN

As discussed above, Karel and Mares studied Mn-associated luminescence centres in AlN using both excitation and emission spectra [14]. The scheme of energy levels which was surmised from this work is shown in FIGURE 15. This work was expanded upon in a more recent study by Pastrnak et al [17], whose energy level diagram is shown in FIGURE 16. Note that this diagram considers energy transfer between different Mn-associated levels and between oxygen and Mn-associated defects.

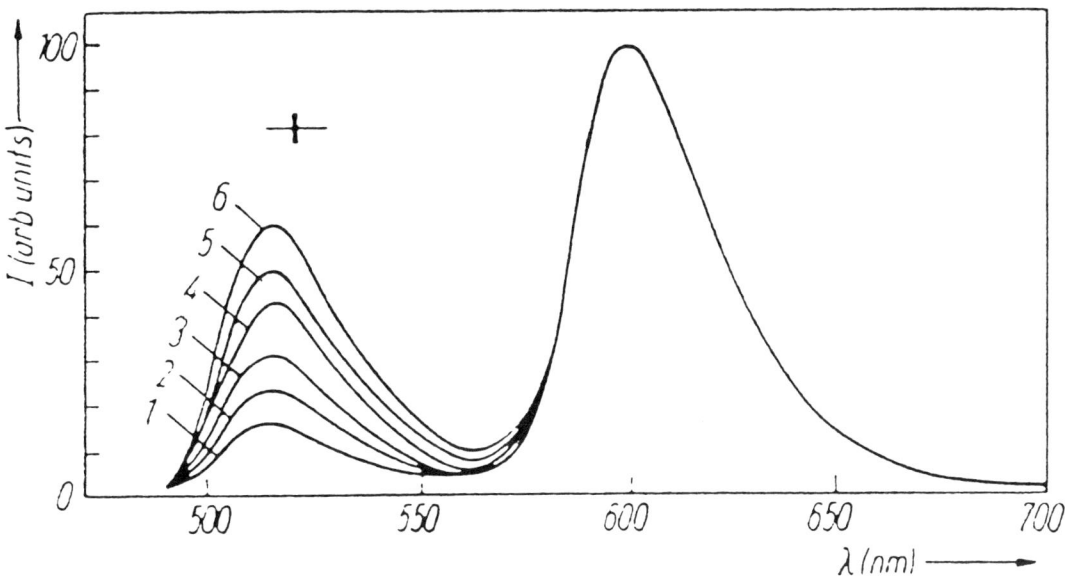

FIGURE 14 Time resolved normalized photoluminescence of AlN:Mn at 300 K taken with time delays 1, 2, 3, 4, 5 and 6 ms (curves 1 through 6, respectively) all excited by the 337.1 nm line of a pulsed nitrogen laser.

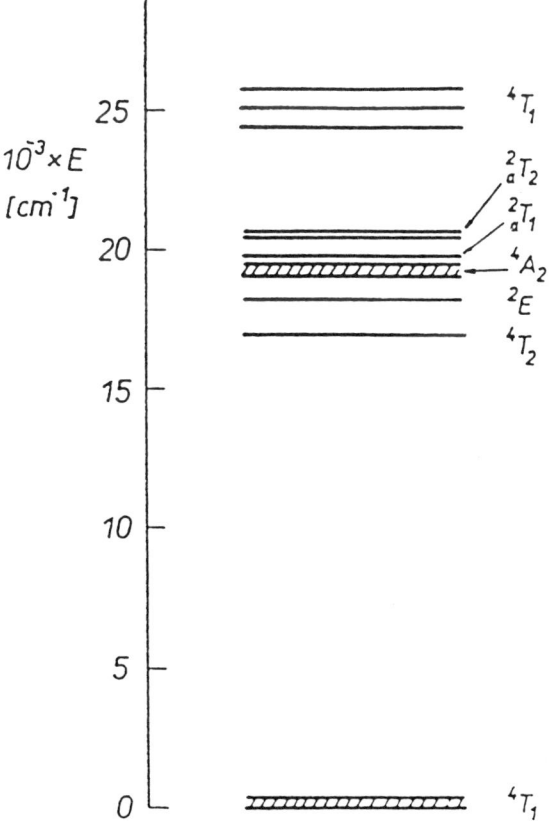

FIGURE 15 Scheme of energy levels for Mn centres in AlN (A - singlets, E - doublets, T - triplets).

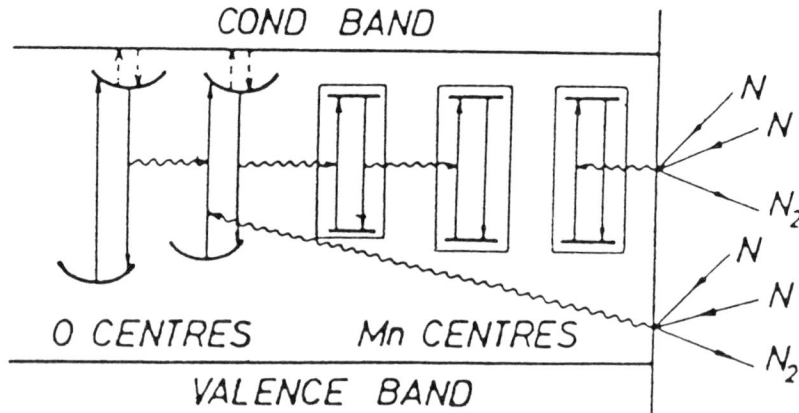

FIGURE 16 Model of energy transfer in AlN:Mn for photoluminescence.

C CATHODOLUMINESCENCE OF ALUMINIUM NITRIDE

C1 Oxygen-Related Spectra

Upon irradiation with an electron beam, AlN doped with oxygen emits a series of broad luminescence bands in the near-UV which are essentially the same as those discussed in Section B1. Such luminescence peaks were first reported by Fischer [18], Wolff et al [19], and Adams et al [20] utilizing electroluminescence. Later, work by Karel and Pastrnak [16], utilizing cathodoluminescence, correctly associated bands in this wavelength regime to oxygen impurities and they also demonstrated the similarity between the cathodoluminescence spectra and the photoluminescence spectra. This similarity is shown in FIGURE 17 where the photo- and cathodo-luminescence emission spectra are given for the same oxygen-doped AlN sample taken at room temperature.

The literature contains many references to the near-UV bands in AlN, without a realization of the correlation to oxygen impurities. Rutz [21] and later Kuznetsov et al [22] reported the existence of the broad near-UV band in electroluminescence and cathodoluminescence experiments on AlN thin films but they were apparently unaware of the work of Karel and Pastrnak and did not assign them to oxygen impurities. Yoshida et al [23] also observed these bands; however they did not believe that their data was exactly similar to that of Karel and Pastrnak. Sokolov et al [24] detected the bands but offered no explanation of their origin. Morita et al [25] prepared AlN samples with varying amounts of oxygen impurity (from <0.5 at.% to approximately 1.0 at.%) and showed shifts in the near-UV cathodoluminescence peaks which they postulated as being correlated with the presence of nitrogen and aluminium vacancies and not directly correlated with oxygen impurities.

Sections B1 through B5 report the work of Harris et al [6] which thoroughly delineates the nature of the photoluminescence peak in the UV-blue as being due to oxygen impurities in the AlN lattice. Youngman et al [2], Youngman et al [26], and Youngman and Harris [7] extended these photoluminescence studies to include cathodoluminescence and showed the similarity of the photoluminescence and cathodoluminescence spectra in this regime just as was found previously by Karel and Pastrnak. This similarity illustrated that the production of electron-hole pairs in the cathodoluminescence experiment does not change any essential

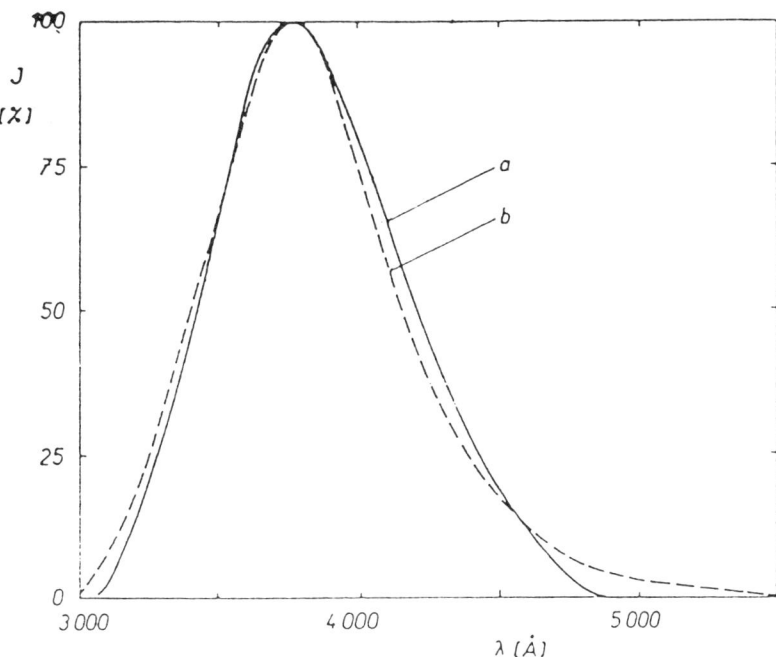

FIGURE 17 The luminescence spectra of oxygen-doped AlN at room temperature: a - cathodoluminescence, b - photoluminescence (from [16]).

feature of the luminescence process. One can therefore utilize the cathodoluminescence signal in an interchangeable manner with the photoluminescence signal. This is important because one of the most useful aspects of the cathodoluminescence experiment is that one can use the electron beam of a scanning electron microscope (SEM) or a transmission electron microscope (TEM) as the excitation source. This allows for spatial resolution of such luminescence phenomena on length scales from approximately 100 microns to approximately 100 nanometres. Thus, as pointed out by Youngman and Harris [7], investigations of distributions of impurities responsible for the luminescence can be conducted on a length scale that is meaningful for a typical, polycrystalline AlN material. Such microstructural studies were conducted by Youngman and Harris [7] and Youngman et al [26] for oxygen-related luminescence. An example is given in FIGURE 18 where a pair of micrographs obtained in the SEM from the same region of a polycrystalline AlN sample (sintered with yttria) are presented. FIGURE 18(a) presents the backscattered electron signal and FIGURE 18(b) presents the 375 nm (+ 40 nm/- 40 nm bandpass detection) cathodoluminescence signal combined with the backscattered signal. In FIGURE 18(a) the bright areas are the Y-Al-O compounds which form during sintering and the darker areas are the AlN grains. In FIGURE 18(b) the additional intensity is due to the luminescence of oxygen defect centres in the AlN. This intensity distribution in the cathodoluminescence micrograph is effectively 'mapping out' the oxygen distribution in the AlN grains, and represents a very powerful tool for microstructural and microchemical analysis in this system. At a higher level of detail, Youngman et al [26] and Youngman and Harris [7] have shown the use of such cathodoluminescence techniques in a TEM as useful in determining the distribution of oxygen within the AlN grains and the interaction of sub-grain features (e.g. grain boundaries, extended defects, etc.) with the incorporated oxygen. FIGURE 19 is a pair of TEM micrographs from the same area of the polycrystalline AlN specimen where the TEM image is presented in FIGURE 19(a) and the corresponding 375 nm (+ 40 nm/- 40 nm bandpass detection) cathodoluminescence image is given in FIGURE 19(b). The TEM image reveals

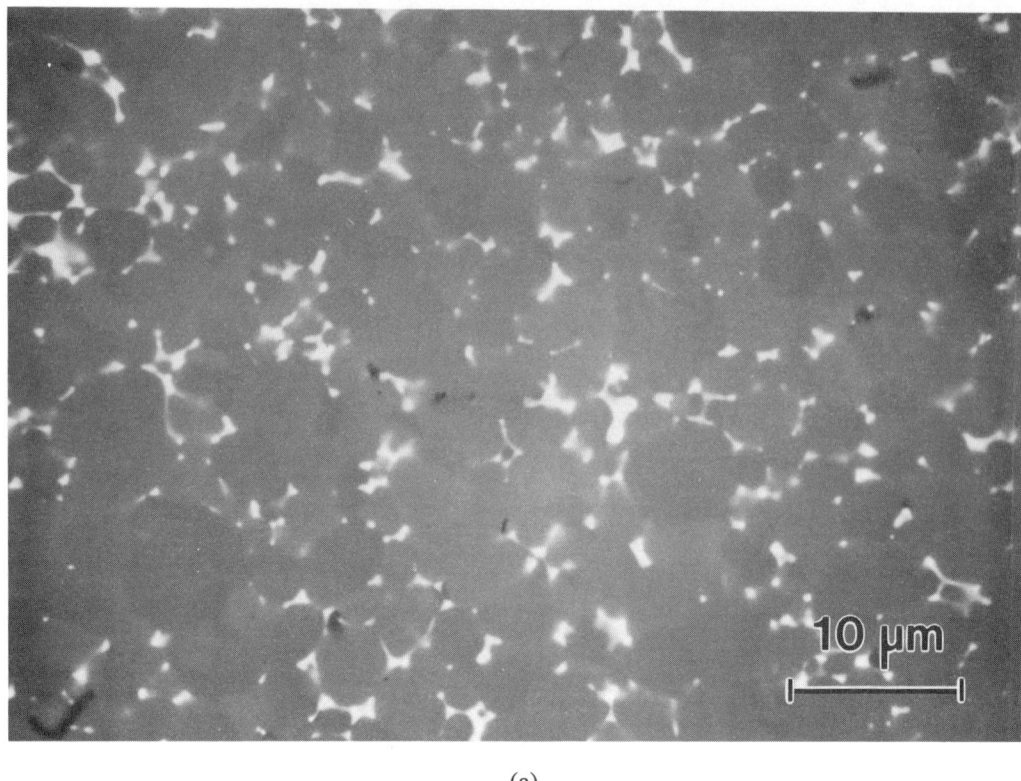

(a)

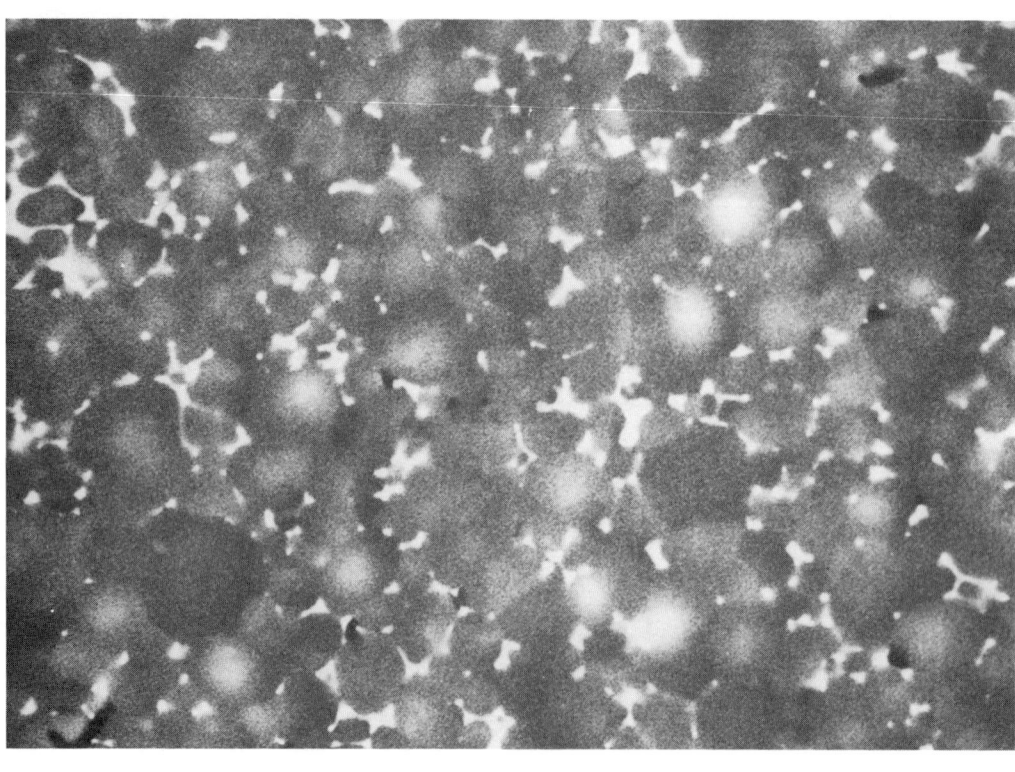

(b)

FIGURE 18 Backscattered electron (a) and cathodoluminescence plus backscattered (b) images of a sintered AlN ceramic sample (from [7]). Reprinted by permission of the American Ceramic Society.

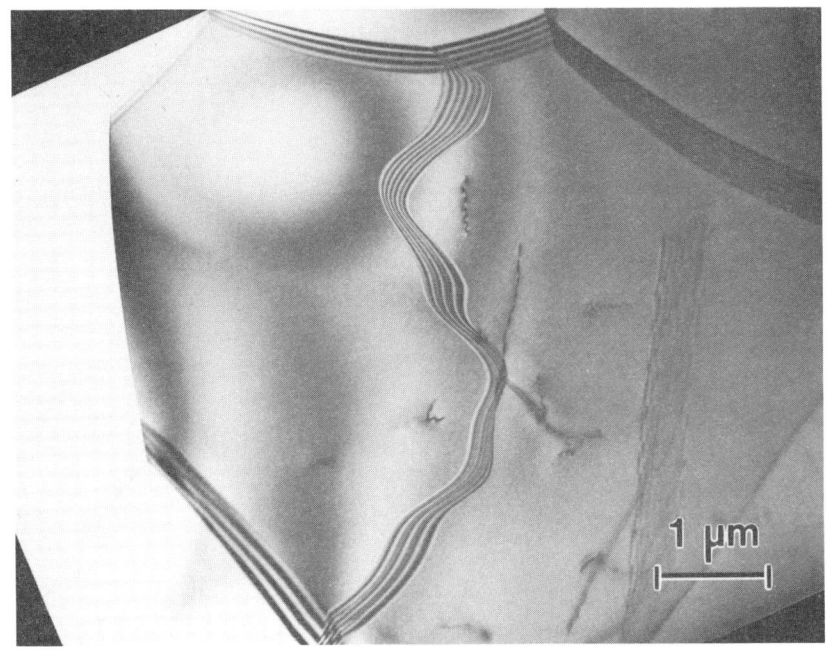

(a)

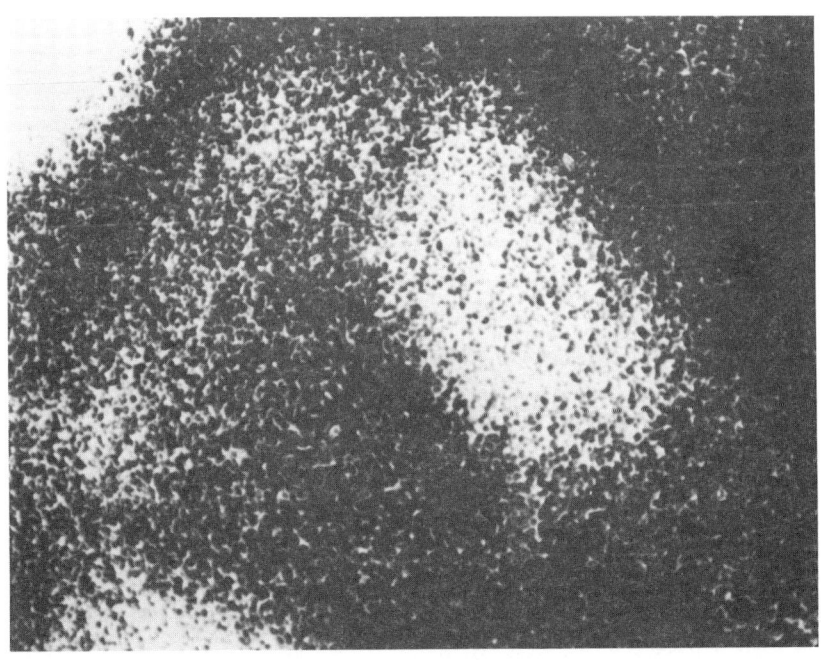

(b)

FIGURE 19 Transmission electron (a) and cathodoluminescence (b) micrograph of the same area in a sintered AlN ceramic sample containing an inversion domain boundary (from [7]).

the presence of an extended defect (in this case, an inversion domain boundary (IDB)) in the AlN grain. The cathodoluminescence image clearly shows a non-uniform intensity distribution which is spatially correlated with the position of the extended defect. This indicates a significant interaction between the extended defect and the dissolved oxygen in the AlN.

In summary, the oxygen-related cathodoluminescence spectra of AlN consist of a broad band centred at approximately 375 nm just as is found in photoluminescence of oxygen-doped AlN. Additionally it has been shown that the cathodoluminescence signal can be utilized in an SEM or a TEM to obtain information on the distribution and interaction of oxygen in the AlN microstructure.

C2   Manganese-Related, Rare Earth-Related, and Other Spectra

The first cathodoluminescence spectra of AlN with Mn-impurity were presented by Pastrnak and Karel [27] and shown to be essentially identical to those obtained utilizing photoluminescence and electroluminescence experiments [19]. These experiments and analysis are discussed in detail in Sections B6 through B8. The only differences which these authors noted were that the time decay of the luminescence in the case of cathodoluminescence is slightly more rapid than in photoluminescence (postulated to be due to possible heating of the sample during the cathodoluminescence experiment), and that an energy transfer process from the oxygen-related centres to the Mn-related centres takes place and may change the relative intensities of the emissions [16].

Other impurity-related cathodoluminescence spectra which have been documented include Eu with a broad peak at 525 nm [16] discussed earlier (section B6) and shown in FIGURE 11, Sm with two narrow peaks at 562 nm and 593 nm due to the $^4G_{5/2} \rightarrow {}^4H_{5/2}$ and $^4G_{5/2} \rightarrow {}^6M_{5/2}$ transitions respectively [16] and shown in FIGURE 12, Nb with a broad peak centred at 510 nm [23], and the curious report of very similar doublet peaks for Mg, Zn and Cu at 420 nm and 460 nm [23] (the authors question whether these particular impurities have any relationship to the emission spectrum). With the exception of the Eu and Sm impurities no thorough analysis of the origin of these luminescence peaks has been given.

D   CONCLUSION

This Datareview has provided a broad review of the photoluminescence and cathodoluminescence properties of aluminium nitride, an important technological material for electronic packaging applications. The photoluminescence results focus on the effects of two major impurities, oxygen and manganese. The oxygen studies in particular indicate how the details of the photoluminescence spectrum can be used to understand the nature of impurity accommodation in this material. For both impurities, proposed energy level diagrams illustrating impurity and associated defect energies within the bandgap have been reviewed. In the cathodoluminescence studies, the use of this technique to probe impurity related extended defects within the AlN microstructure was discussed. In this case, extended defects such as inversion domain boundaries and stacking faults associated with oxygen doping are revealed and connected to the photoluminescence results.

REFERENCES

[1]   S. Pacesova, L. Jastrabik [*Czech. J. Phys. B (Czechoslovakia)* vol.29 (1979) p.913-23]

[2]   R.A. Youngman, J.H. Harris, D.A. Chernoff [*Ceram. Trans. (USA)* vol.5 (1989) p.399-5]

[3]   G.A. Slack, T.F. McNelly [*J. Cryst. Growth (Netherlands)* vol.42 (1977) p.560]

[4]   V.E. Oranovskii, J. Pastrnak, S. Pacesova, M.V. Fok [*Sov. Phys.-Solid State (USA)* vol.18 (1976) p.444-6]

[5]   J. Pastrnak, S. Pacesova, J. Sanda, J. Rose [*Seriya Fizicheskaya (USSR)* vol.37 (1973) p.123-5]

[6]   J.H. Harris, R.A. Youngman, R.G. Teller [*J. Mater. Res. (USA)* vol.5 (1990) p.1763-73]

[7]   R.A. Youngman, J.H. Harris [*J. Am. Ceram. Soc. (USA)* vol.73 (1990) p.3238-46]

[8]   J. Pastrnak, S. Pacesova, L. Roskovcova [*Czech. J. Phys. B (Czechoslovakia)* vol.24 (1974) p.1149-61]

[9]   J.H. Harris, R.A. Youngman [*Mater. Res. Soc. Symp. Proc. (USA)* vol.167 (1990) p.253-8]

[10]   J. Pastrnak, L. Roskovcova [*Phys. Status Solidi (Germany)* vol.26 (1968) p.591-7]

[11]   W.M. Kim, E.J. Stofko, P.J. Zanzucchi, J.I. Pankove, M. Ettenberg, S.L. Gilbert [*J. Appl. Phys. (USA)* vol.44 (1973) p.292-6]

[12]   J.H. Harris, R.A. Youngman [*J. Mater. Res. (USA)* vol.8 (1993) p.154-62]

[13]   F. Karel, J. Pastrnak, J. Hejduk, V. Losik [*Phys. Status Solidi (Germany)* vol.15 (1966) p.693-9]

[14]   F. Karel, J. Mares [*Czech. J. Phys. B (Czechoslovakia)* vol.22 (1972) p.847-53]

[15]   G.E. Archangelskii, F. Karel, J. Mares, S. Pacesova, J. Pastrnak [*Phys. Status Solidi (Germany)* vol.69 (1982) p.173-83]

[16]   F. Karel, J. Pastrnak [*Czech. J. Phys. B (Czechoslovakia)* vol.20 (1970) p.46-55]

[17]   J. Pastrnak, F. Karel, J. Oswald [*J. Lumin. (Netherlands)* vol.18/19 (1979) p.805-8]

[18]   A. Fischer [*Physik. Verhandl. (Germany)* vol.7 (1957) p.204]

[19]   G.A. Wolff, I. Adams, J.W. Mellichamp [*Phys. Rev. (USA)* vol.114 (1959) p.1262-4]

[20]   I. Adams, T.R. Aucoin, G.A. Wolff [*J. Electrochem. Soc. (USA)* vol.109 (1962) p.1050-4]

[21]   R.F. Rutz [*Appl. Phys. Lett. (USA)* vol.28 (1976) p.379-81]

[22]   O.N. Kuznetsov, L.V. Lezheiko, E.V. Lyubopytora, L.S. Smirnov, Y.V. Shmartsev, F.L. Edelman [*Sov. Phys.-Semicond. (USA)* vol.10 (1976) p.1263-5]

[23]   S. Yoshida, S. Misawa, S. Gonda [*Thin Solid Films (Switzerland)* vol.58 (1979) p.55-9]

[24]   E.B. Sokolov, E.F. Uvarov, A.P. Khramtsov, V.P. Chegnov, M.V. Chukichev [*Sov. Phys.-Semicond. (USA)* vol.16 (1982) p.599-600]

[25]   M. Morita, K. Kazuo, N. Mikoshiba [*Jpn. J. Appl. Phys. (Japan)* vol.21 (1982) p.1102-3]

[26]   R.A. Youngman, J.H. Harris, P.A. Labun, R.J. Graham, J.K. Weiss [*Mater. Res. Soc. Symp. Proc. (USA)* vol.167 (1990) p.271-6]

[27]   J. Pastrnak, F. Karel [*Proc. Int. Conf. Lum.* (1966) p.1473-6]

7.2 Basic optical properties, photoluminescence and cathodoluminescence of GaN and AlGaN

I. Akasaki and H. Amano

April 1994

A INTRODUCTION

The successful fabrication of min-type GaN blue LEDs by Pankove et al [1] and Maruska et al [2] in the early 1970s attracted a great deal of interest in GaN luminescent devices. Important luminescence properties of both undoped and impurity doped GaN of the wurtzite polytype (WZ-GaN) were measured using photoluminescence by Pankove's group [3-5] and many others in the '70s. They used hydride vapour phase epitaxy (HVPE) for the growth of GaN. Difficulties in growing crack-free films, problems with thickness control, particularly at the sub-micron scale (important for the fabrication of low voltage driven min-type LEDs [6]), and difficulty in controlling conductivity, especially p-type conductivity, motivated the development of other epitaxial deposition techniques. Of these newer techniques, both metalorganic vapour phase epitaxy (MOVPE) [7] and molecular beam epitaxy (MBE) have been effective in improving the GaN epitaxial quality. Low-temperature deposited buffer layers enabled deposition of high-quality group III nitrides free from cracks on sapphire substrates, as first demonstrated by Amano et al [9] and later by Nakamura et al [10] and Moustakas et al [11]. p-type conductivity was also realized by Amano et al [12], later by Nakamura et al [13], and by many others. As a result of these improvements, high-efficiency p-n junction type blue and UV LEDs based on GaN and related compounds with efficiency exceeding 1 % at room temperature (RT) [14,15] are now commercially available.

In this Datareview, we will focus on the optical properties, especially luminescence properties, of GaN and $Al_xGa_{1-x}N$. Since GaN and $Al_xGa_{1-x}N$ are usually grown heteroepitaxially, we have to allow for the effect of heteroepitaxy in studying the intrinsic optical properties of these films. To date, nearly all measurements have been taken from WZ-GaN grown on sapphire substrates. Both microscopic [16,17] and macroscopic [18] characterization showed that the use of low temperature deposited buffer layers [9-11] is effective for achieving high-quality films on the dissimilar substrates. Nevertheless, the effect of strain on the luminescence properties due to the difference in the thermal expansion coefficients cannot be entirely ruled out [19,20]. The growth of high-quality bulk crystal free from a dissimilar substrate must be the logical approach for clarifying the intrinsic optical properties of GaN [21] and $Al_xGa_{1-x}N$.

B OPTICAL ABSORPTION EDGE

Maruska and Tietjen [22] and Pankove et al [23] determined that WZ-GaN has a direct energy bandgap of 3.39 eV at RT. The temperature dependence of the bandgap, dE_g/dT, of WZ-GaN was measured by Kauer and Rabenau [24] to be -3.9×10^{-4} eV K^{-1}, Camphausen and Connell [25] to be -6.7×10^{-4} eV K^{-1} and Pankove et al [5] to be -4.8×1 10^{-4} eV K^{-1} using transmission measurement. Later, Monemar et al experimentally formulated the temperature dependence of the bandgap using photoluminescence excitation measurement [26]. Eqn (1) was given as an empirical dependence of the bandgap of WZ-GaN on temperature.

$$E_g = 3.503 + (5.08 \times 10^{-4} \, T^2)/(T - 996) \text{ eV} \tag{1}$$

Hydrostatic pressure dependence of the bandgap dE_g/dP of WZ-GaN was measured by Camphausen and Connell [25] to be $4.2 \pm 0.4 \, \text{meV kbar}^{-1}$ on single crystal GaN the substrate of which was lapped away. Change of the bandgap induced by anisotropic stress, which is the combination of the hydrostatic and shear deformation potential, has been measured on heteroepitaxially grown WZ-GaN [19,20].

The reported optical absorption edge data on ZB-GaN was scattered due probably to problems in the heteroepitaxial layers such as containing some WZ-GaN. Powell et al measured the RT energy bandgap by transmission/reflection to be $3.21 \pm 0.02 \, \text{eV}$ [27] which is in close agreement with the bandgap calculation.

The compositional dependence of the bandgap of WZ-$Al_xGa_{1-x}N$ alloy at RT was determined by optical absorption measurement [28] from the MOVPE-grown AlGaN film with AlN molar fraction up to 0.4. The dependence of the bandgap on composition deviated downwards from linearity; b, the bowing parameter, was $1.0 \pm 0.3 \, \text{eV}$.

C   REFRACTIVE INDEX

The wavelength dispersion of the refractive index of WZ-GaN at RT was first measured in transmission and reflection by Ejder [29]. The refractive index of the $E \perp c$ component near the bandgap was found to be about 2.65. Later, Amano et al [30] measured the dispersion of the refractive index of the $E \perp c$ component using spectroscopic ellipsometry and they found a refractive index about 0.1 larger than Ejder's value possibly due to the difference in the effect of strain due to heteroepitaxy. The wavelength dispersion of the refractive index of ZB-GaN was measured by Lin et al [31] using reflection. The refractive index of ZB-GaN was about 0.1 to 0.15 higher than that of WZ-GaN in the wavelength range from $0.37 \, \mu\text{m}$ to $1 \, \mu\text{m}$. Amano et al measured that of WZ-$Al_xGa_{1-x}N$ [30]. Values of the $E \perp c$ component for GaN and $Al_{0.1}Ga_{0.9}N$ are presented in TABLE 1. The M_0 specific point shifts toward the higher energy side in proportion to the molar fraction of AlN.

TABLE 1 The refractive indices of the $E \perp c$ component for GaN and $Al_{0.1}Ga_{0.9}N$ [30].

	GaN	$Al_{0.1}Ga_{0.9}N$
370 nm	2.73	2.54
400 nm	2.58	2.50
500 nm	2.45	2.40
600 nm	2.39	2.35

D   PHOTOLUMINESCENCE AND CATHODOLUMINESCENCE

Extensive characterization of the luminescent properties of WZ-GaN has been performed by many researchers, especially WZ-GaN doped with Zn and Cd which form efficient blue luminescent centres. Excellent papers were given by Pankove et al [3-5]. Doping was usually accomplished by adding impurities during film growth, but ion implantation was also employed [32]. Characterization of luminescence properties of both undoped and doped WZ-AlGaN has also been performed. In contrast, to date, there are few reports concerning the effect of impurities on the luminescence properties of ZB-GaN and ZB-AlGaN. In this section, we focus on the luminescent properties of WZ-GaN and WZ-AlGaN.

At first, we discuss the luminescence properties of WZ-GaN. Roughly speaking, the luminescence of WZ-GaN is divided into three regions: the excitonic edge emission region, the D-A pair emission region, and the deep level related emission region. Characteristic properties in each region are described.

D1   Excitonic Edge Emission

Peak energies of excitonic near band edge emissions, free excitons, excitons bound to neutral donors (the so-called I_2-line) and excitons bound to neutral acceptors (the so-called I_1-line), show similar temperature dependence to dE_g/dT. They are also affected by the strain in the film. To determine the exciton binding energy of the impurity in the heteroepitaxially grown WZ-GaN from low temperature photoluminescence (PL) and cathodoluminescence (CL) data, we must consider the effect of strain in the film.

D1.1   Free exciton

Exciton binding energy, G_{ex}, of WZ-GaN is estimated to be about 22.7 meV assuming the static dielectric constants $\varepsilon_s(E \perp c)$ and $\varepsilon_s(E \parallel c)$ to be 9.5 and 10.4, respectively, and the effective mass for conduction band electrons and the effective mass for the valence band holes to be $0.2m_0$ and $0.8m_0$, respectively. Up to now, there is only one report in which free exciton emission can be clearly observed by low temperature PL [33]. The PL peak energy of the free exciton emission at 4.2 K is listed in TABLE 2.

D1.2   Exciton bound to neutral donor

Nominally undoped WZ-GaN usually contains residual donors. At low temperature, strong edge emission due to the annihilation of excitons bound to neutral donors was clearly observed from reasonably high quality undoped WZ-GaN. TABLE 2 summarizes the reported I_2-line energy and the corresponding energy difference from that of free excitons [33-36].

TABLE 2 Near-edge photoluminescence from GaN.

Origin	Peak energy (eV)	Binding energy (meV)	Ref
Free exciton (A-exciton)	3.483 (T = 4.2 K)		[33]
Exciton bound to neutral donor donor:residual	3.467 3.469 (T = 1.6 K) 3.477 (T = 4.2 K)	6 ~ 8 6.4 ± 0.4 6.0	[34] [35] [33,36]
Exciton bound to neutral acceptor	(Mg) 3.455 (Zn) 3.450 (T = 4.2 K) 3.471 (T = 4.2 K) (Cd) 3.4553 (T = 4.2 K) 3.454 (T = 1.6 K)	 18.0 19 ± 2	 [37] [38] [19] [38] [35]
Donor-acceptor pair donor:residual	(Be) 3.264 (T = 4.2 K) (with two LO-phonon replicas) (Mg) 3.264 (T = 4.2 K) (with four LO-phonon replicas) 3.27 (T = 4.2 K) (with two LO-phonon replicas) (Zn) 3.268 (T = 4.2 K) (with one LO-phonon replica) (Cd) 3.263 ± 0.001 (with one and two LO-phonon replicas) 3.268 (with one LO-phonon replica)		 [37] [37] [42] [38] [35] [38]

D1.3 Exciton bound to neutral acceptor

WZ-GaN lightly doped with group II impurities sometimes showed emission due to annihilation of excitons bound to neutral acceptors at temperatures below 77 K. Up to now, I_1-lines due to three different kinds of impurity, that is Cd, Zn and Mg, have been reported [19,35,37,38]. Peak positions and the energy differences from that of free excitons are listed in TABLE 2.

D1.4 Near band edge emission at room temperature

WZ-GaN having high residual donors or that doped with donor impurity shows near band edge (NBE) emission at RT. For example, Murakami et al showed that doping with Si enhances the CL-intensity of NBE emission from WZ-GaN [39].

D2   D-A Pair Emission

WZ-GaN lightly doped with group II impurities showed D-A pair emission with strong one, two or more LO-phonon replicas at low temperature. D-A pair emission has been identified using time resolved spectroscopy [40]. The donor may be a residual donor, the acceptor being the doping acceptor. The characteristic feature of the D-A pair emission from WZ-GaN is that the temperature dependence of the peak energy of D-A pair emission does not follow that of the bandgap [35]. The temperature dependence of the integrated intensity of D-A pair emission shows the activation energy of impurities. For example, the activation energy of Mg acceptor was estimated to be about $0.155 - 0.165\,eV$ [41]. Peak energies of D-A pair emission with LO-phonon replicas at low temperature from WZ-GaN doped with different group II impurities, that is, Be, Mg, Zn and Cd, are summarized in TABLE 2 [35,37,38,42].

D3   Emissions From Deep Levels

Emissions from deep levels are also important from the practical point of view. Group II atoms, Mg, Zn and Cd, form luminescence centres and act as shallow acceptors in WZ-GaN. Especially, Zn-related luminescence centres have been investigated by many researchers. A Zn-related luminescence centre was also employed for the fabrication of GaN based blue-LEDs [1]. As shown by Jacob et al [43], Zn forms four distinctive emission bands, the violet-blue band (VB-band), the blue-green band (BG-band), the yellow-green band (YG-band) and the red band (R-band). Pankove et al [4] revealed that the peak energy of the Zn-related luminescence centre depends on the concentration of Zn in GaN, growth conditions, etc. A model for the structures of four luminescence centres was presented by Monemar et al [44]. They suggested that the origin of the VB-band is Zn_{Ga}, a substitutional acceptor, while that of another band is Zn_N, Zn occupying N sites. TABLE 3 summarizes the Zn-related deep-level PL peak wavelengths in the visible region. TABLE 3 also summarizes the reported emission peaks in the visible region from WZ-GaN doped with several kinds of impurity in the temperature range from $1.6\,K$ to $300\,K$. Since deep level related emission is broad, there is some ambiguity about the emission peak energy of deep levels. As shown in TABLE 3, the 'doping' forms four distinctive emission bands irrespective of the impurity species [1,35,37,43-54]. This result indicates that the origin of these deep levels is intrinsic in nature.

WZ-GaN doped with transition metal impurities shows different luminescent behaviour from that of WZ-GaN doped with the above mentioned impurities. Fe-doped GaN shows a very sharp IR PL with peak energy of $1.299\,eV$ at $2\,K$, which may originate from a charge transfer process through d-shell electrons [55].

D4   Emissions From WZ-$Al_xGa_{1-x}N$

Compared to WZ-GaN, reports concerning the optical properties of WZ-$Al_xGa_{1-x}N$ are sparse, probably due to the difficulty in growing high-quality crystal.

TABLE 3 Photoluminescence and cathodoluminescence peak energies from deep levels in impurity doped WZ-GaN. In this table, unless otherwise noted, peak energy means the PL peak energy.

Dopant	Peak energy (eV)	Temperature (K)	Ref
Li	2.23	4.2, 300	[45]
Be	2.16	4.2	[37]
	2.2	4.2, 300	[45]
C	2.1 - 2.2	4.2, 77	[46]
Mg	2.96	4.2	[37]
	3.03 (CL)	RT	[47]
	2.95	RT	[48]
P	2.8	300	[49]
	2.85 (CL)	77, 294	[50]
	2.2 (CL)	77, 294	
Zn	2.55	RT	[1]
	2.85 (CL)	RT	[43,53]
	2.5 (CL)	RT	
	2.2 (CL)	RT	
	1.9 (CL)	RT	
	2.87	4.2 - 293	[44]
	2.6	4.2 - 293	
	2.2	4.2 - 293	
	1.8	4.2 - 293	
	2.87	78	[51]
	2.6	10, 300	[52]
	2.87 (CL)	100, 295	[54]
As	2.58 (CL)	73, 294	[50]
	2.2 (CL)	73, 294	
Cd	2.72	1.6 - 300	[35]
	2.85	100 - 300	[38]

$Al_xGa_{1-x}N$ with AlN molar fraction $x < 0.4$ shows strong I_2-line emission at low temperature similar to that of WZ-GaN. Khan et al [56] and Baranov et al [57] reported the compositional dependence of the peak energy of I_2-line as shown in the Eqns (2) (from [56]) and (3) (from [57]).

$$Epeak_{I_2} = 3.47 + 1.75x + 0.98x^2 \ (PL, 4.2\,K) \qquad (2)$$

$$Epeak_{I_2} = 3.465 + 1.069x + 1.667x^2 \ (PL, 77\,K, \text{ best fitted to data}) \qquad (3)$$

The bowing parameter $b = 0.98\,eV$ of Eqn (2) coincides quite well with that measured by optical absorption at RT [28].

As in the case of WZ-GaN, doping with Si does not bring about a new emission band in the UV or visible region, but it enhances the intensity of NBE emission from $WZ-Al_xGa_{1-x}N$ [39].

The compositional dependence of the peak energy of the VB-band and the BG-band at RT has been studied on MOVPE-grown $WZ-Al_xGa_{1-x}N$ using CL. The peak energies are 2.95 eV (x = 0) to 3.13 eV (x = 0.23) for the VB-band and 2.50 eV (x = 0) to 2.7 eV (x = 0.15) for the BG-band [58].

Another group reported many PL peaks from undoped and Zn-doped $Al_xGa_{1-x}N$ [59].

D5 Emission From $Al_xGa_{1-x}N$/GaN Multi-Layered Structure

Two groups have reported PL properties from an $Al_xGa_{1-x}N$/GaN multi-layered structure with GaN well layer thickness down to 2.5 nm [60], or 15 nm [61] on sapphire substrates using a low temperature deposited AlN buffer layer. The lattice constant of $Al_xGa_{1-x}N$ is smaller than that of GaN. Therefore, for the evaluation of the PL-peak shift, the effect of strain should be accounted for in addition to the quantum size effect. In the case of a multi-layered structure fabricated on a thick $Al_xGa_{1-x}N$ layer [61], the GaN well layer should be strained. Therefore, the strain induced PL-peak shift is dominant. On the other hand, in the case of a multi-layered structure fabricated on a thick GaN layer, the $Al_xGa_{1-x}N$ barrier layer may be strained by the tensile stress, and the GaN well layer may be relaxed. In the latter case, the shift of the PL-peak energy is mainly caused by the quantum size effect [60].

E CONCLUSION

A survey of the optical properties of WZ-GaN and WZ-$Al_xGa_{1-x}N$ has been given in this Datareview. Luminescence properties of both undoped and impurity doped WZ-GaN were mostly clarified in the '70s, although there are still some outstanding problems, for example, the structure of deep level luminescence centres in impurity doped GaN. Furthermore, the effect of strain on the luminescence properties requires further work. The preparation of high-quality free-standing bulk crystals should answer this question. Compared to WZ-GaN, the luminescence properties of WZ-AlGaN, ZB-GaN and ZB-AlGaN are generally not established. Progressive fundamental research work, both experimental and theoretical, together with improvements in growing much higher quality crystals, will surely lead to the clear understanding of the intrinsic luminescence properties of GaN and $Al_xGa_{1-x}N$.

REFERENCES

[1]   J.I. Pankove, E.A. Miller, D. Richman, J.E. Berkeyheiser [*J. Lumin. (Netherlands)* vol.4 (1971) p.63-6]

[2]   H.P. Maruska, D.A. Stevenson, J.I. Pankove [*Appl. Phys. Lett. (USA)* vol.22 (1973) p.303-5]

[3]   J.I. Pankove, J.E. Berkeyheiser, H.P. Maruska, J. Wittke [*Solid State Commun. (USA)* vol.8 (1970) p.1051-3]

[4]   J.I. Pankove [*J. Lumin. (Netherlands)* vol.7 (1973) p.114-26]

[5]   J.I. Pankove, S. Bloom, G. Harbeke [*RCA Rev. (USA)* vol.36 (1975) p.163-76]

[6]   Y. Ohki, Y. Toyoda, H. Kobayashi, I. Akasaki [*Inst. Phys. Conf. Ser. (UK)* no.63 (1981) p.479-84]

[7]   H.M. Manasevit, F.M. Erdmann, W.I. Simpson [*J. Electrochem. Soc. (USA)* vol.118 (1971) p.1864-8]

[8]   S. Yoshida, S. Misawa, S. Gonda [*Appl. Phys. Lett. (USA)* vol.42 (1983) p.427-9]

[9]   H. Amano, N. Sawaki, I. Akasaki, Y. Toyoda [*Appl. Phys. Lett. (USA)* vol.48 (1986) p.353-5]

[10]   S. Nakamura [*Jpn. J. Appl. Phys. (Japan)* vol.30 (1991) p.L1705-L1707]

[11]   T.D. Moustakas, R.J. Molnar, T. Lei, G. Menon, C.R. Eddy Jr. [*Mater. Res. Soc. Symp. Proc. (USA)* vol.242 (1992) p.427-32]

[12]   H. Amano, M. Kito, K. Hiramatsu, I. Akasaki [*Jpn. J. Appl. Phys. (Japan)* vol.28 (1989) p.L2112-L2114]

[13]   S. Nakamura, N. Iwasa, M. Senoh, T. Mukai [*Jpn. J. Appl. Phys. (Japan)* vol.31 (1992) p.1258-66]

[14]   I. Akasaki, H. Amano, K. Itoh, N. Koide, K. Manabe [*Inst. Phys. Conf. Ser. (UK)* no.129 (1992) p.851-6]

[15]   S. Nakamura, T. Mukai, M. Senoh [*Appl. Phys. Lett. (USA)* vol.64 (1994) p.1687-9]

[16]   N. Kuwano et al [*J. Cryst. Growth (Netherlands)* vol.115 (1991) p.381-7]

[17]   K. Hiramatsu et al [*J. Cryst. Growth (Netherlands)* vol.115 (1991) p.628-33]

[18]   I. Akasaki, H. Amano, Y. Koide, K. Hiramatsu, N. Sawaki [*J. Cryst. Growth (Netherlands)* vol.98 (1989) p.209-19]

[19]   H. Amano, K. Hiramatsu, I. Akasaki [*Jpn. J. Appl. Phys. (Japan)* vol.27 (1988) p.L1384-L1386]

[20]   T. Detchprohm, K. Hiramatsu, K. Itoh, I. Akasaki [*Jpn. J. Appl. Phys. (Japan)* vol.31 (1992) p.L1454-L1456]

[21]   P. Perlin, I. Gorczyca, N.E. Christensen, I. Grzegory, H. Teisseyre, T. Suski [*Phys. Rev. B (USA)* vol.45 (1992) p.13307-13]

[22]   H.P. Maruska, J.J. Tietjen [*Appl. Phys. Lett. (USA)* vol.15 (1969) p.327-9]

[23]   J.I. Pankove, H.P. Maruska, J.E. Berkeyheiser [*Appl. Phys. Lett. (USA)* vol.17 (1970) p.197-9]

[24]   E. Kauer, A. Rabenau [*Z. Nat.forsch. A (Germany)* vol.12 (1959) p.942]

[25]   D.L. Camphausen, G.A.N. Connell [*J. Appl. Phys. (USA)* vol.42 (1971) p.4438-43]

[26]   B. Monemar [*Phys. Rev. (USA)* vol.10 (1974) p.676-81]

[27]   R.C. Powell, N.E. Lee, Y.W. Kim, J.E. Greene [*J. Appl. Phys. (USA)* vol.73 (1993) p.189-204]

[28]   Y. Koide, H. Itoh, M.R.H. Khan, K. Hiramatsu, N. Sawaki, I. Akasaki [*J. Appl. Phys. (USA)* vol.61 (1987) p.4540-3]

[29]   E. Ejder [*Phys. Status Solidi A (Germany)* vol.6 (1975) p.445-8]

[30]   H. Amano, N. Watanabe, N. Koide, I. Akasaki [*Jpn. J. Appl. Phys. (Japan)* vol.32 (1993) p.L1000-L1002]

[31]   M.E. Lin, B.N. Sverdlov, S. Strite, H. Morkoc, A.E. Drakin [*Electron. Lett. (UK)* vol.29 (1993) p.1759-61]

[32]   J.I. Pankove, J.A. Hutchby [*J. Appl. Phys. (USA)* vol.47 (1976) p.5387-90]

[33]   K. Naniwae, S. Itoh, H. Amano, K. Itoh, K. Hiramatsu, I. Akasaki [*J. Cryst. Growth (Netherlands)* vol.99 (1990) p.381-4]

[34]   R. Dingle, D.D. Sell, S.E. Stokowski, M. Ilegems [*Phys. Rev. B (USA)* vol.4 (1971) p.1211-8]

[35]   O. Lagerstedt, B. Monemar [*J. Appl. Phys. (USA)* vol.45 (1974) p.2266-72]

[36]   T. Detchprohm, H. Amano, K. Hiramatsu, I. Akasaki [*J. Cryst. Growth (Netherlands)* vol.128 (1993) p.384-90]

[37]   M. Ilegems, R. Dingle [*J. Appl. Phys. (USA)* vol.44 (1973) p.4234-5]

[38]   M. Ilegems, R. Dingle, R.A. Logan [*J. Appl. Phys. (USA)* vol.43 (1972) p.3797-800]

[39]   H. Murakami, T. Asahi, H. Amano, K. Hiramatsu, N. Sawaki, I. Akasaki [*J. Cryst. Growth (Netherlands)* vol.115 (1991) p.648-51]

[40]   R. Dingle, M. Ilegems [*Solid State Commun. (USA)* vol.9 (1971) p.175-80]

[41]   I. Akasaki, H. Amano, M. Kito, K. Hiramatsu [*J. Lumin. (Netherlands)* vol.48&49 (1991) p.666-70]

[42]   H. Amano, M. Kito, K. Hiramatsu, I. Akasaki [*J. Electrochem. Soc. (USA)* vol.137 (1990) p.1639-41]

[43]   G. Jacob, M. Boulou, M. Furtado [*J. Cryst. Growth (Netherlands)* vol.42 (1977) p.136-43]

[44]   B. Monemar, O. Lagerstedt, H.P. Gislason [*J. Appl. Phys. (USA)* vol.51 (1980) p.625-39]

[45]   J.I. Pankove, M.T. Duffy, E.A. Miller, J.E. Berkeyheiser [*J. Lumin. (Netherlands)* vol.8 (1973) p.89-93]

[46]   T. Ogino, M. Aoki [*Jpn. J. Appl. Phys. (Japan)* vol.19 (1980) p.2395-405]

[47]   S.S. Liu, T.R. Cass, D.A. Stevenson [*J. Electron. Mater. (USA)* vol.6 (1977) p.237-51]

[48]   H. Amano, M. Kitoh, K. Hiramatsu, I. Akasaki [*Inst. Phys. Conf. Ser. (UK)* no.106 (1989) p.725-30]

[49]   T. Ogino, M. Aoki [*Jpn. J. Appl. Phys. (Japan)* vol.18 (1979) p.1049-52]

[50]   R.D. Metcalfe, D. Wickenden, W.C. Clark [*J. Lumin. (Netherlands)* vol.16 (1978) p.405-15]

[51]   J.I. Pankove, J.E. Berkeyheiser, E.A. Miller [*J. Appl. Phys. (USA)* vol.45 (1974) p.1280-6]

[52]   V.V. Rossin, V.G. Sidorov, A.D. Shagalov, Yu.K. Shalabutov [*Sov. Phys.-Semicond. (USA)* vol.15 (1981) p.589-90]

[53]   M. Boulou, M. Furtado, G. Jacob, D. Bois [*J. Lumin. (Netherlands)* vol.18/19 (1979) p.767-70]

[54]   L.A. Marasina, A.N. Pikhtin, I.G. Pichugin, A.V. Solomonov [*Phys. Status Solidi A (Germany)* vol.38 (1977) p.753-60]

[55]   K. Meier et al [*Mater. Sci. Forum (Switzerland)* vol.143-147 (1994) p.93-8]

[56]   M.R.H. Khan, Y. Koide, H. Itoh, N. Sawaki, I. Akasaki [*Solid State Commun. (USA)* vol.60 (1986) p.509-12]

[57]   B.V. Baranov, V.B. Gutan, U. Zhumakulev [*Sov. Phys.-Semicond. (USA)* vol.16 (1982) p.819-21]

[58]   K. Itoh, H. Amano, K. Hiramatsu, I. Akasaki [*Jpn. J. Appl. Phys. (Japan)* vol.30 (1991) p.1604-8]

[59]   H.G. Lee, M. Gershenzon, B.L. Goldenberg [*J. Electron. Mater. (USA)* vol.20 (1991) p.621-5]

[60]   K. Itoh, T. Kawamoto, H. Amano, K. Hiramatsu, I. Akasaki [*Jpn. J. Appl. Phys. (Japan)* vol.30 (1991) p.1924-7]

[61]   S. Krishnankutty, M. Kolbas, M.A. Khan, J.N. Kuznia, J.M. Van Hove, D.T. Olson [*J. Electron. Mater. (USA)* vol.21 (1992) p.437-40]

7.3 Photoluminescence of InN and InGaN

T. Matsuoka

January 1994

A INTRODUCTION

The quaternary InGaAlN system with a direct bandgap tunable from 2 to 6.2 eV is expected to be used for double-heterostructure laser diodes, transistors operated at high temperature and so on. InN growth in this system is most difficult because of the high equilibrium vapour pressure of the group V elements at typical growth temperatures. A key factor is that the components are supplied at pressures higher than the equilibrium vapour pressure. The equilibrium vapour pressures of N_2 over AlN, GaN and InN [1] are shown in FIGURE 1 [2], compared with those of the group V elements over GaAs and InP [3,4]. These pressures correspond to partial pressures of group V elements dissolved in group III melts which are in equilibrium with the solid phases. FIGURE 1 indicates that nitrogen equilibrium vapour pressures over InN are several orders of magnitude larger than over the other materials. Many attempts had been made to grow InN single crystals and had not been successful [5-7]. Recently, InN growth has been accomplished by using metalorganic vapour phase epitaxy (MOVPE) [2].

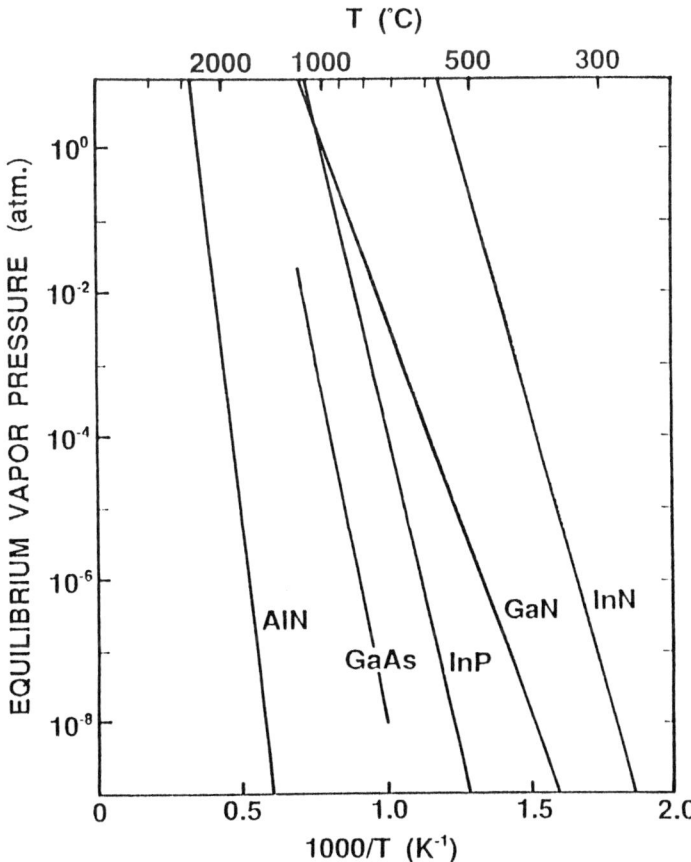

FIGURE 1 Equilibrium vapour pressures of N_2 over AlN, GaN and InN, the sum of As_2 and As_4 over GaAs, and the sum of P_2 and P_4 over InP (from [2]).

In this Datareview, photoluminescence (PL) characteristics of Ga-rich InGaN alloys are reviewed since PL is not observed for InN and In-rich alloys. Unfortunately, there are still no reports on cathodoluminescence from either InN or InGaN.

B   PHOTOLUMINESCENCE OF UNDOPED InGaN

FIGURE 2 shows the dependence of the photoluminescence spectra on the indium mole fraction for films grown at 800 °C [8]. The GaN films had a sharp, strong peak at 365 nm corresponding to the band-to-band emission. As the indium mole fraction was increased, this peak became weaker and shifted toward a longer wavelength.

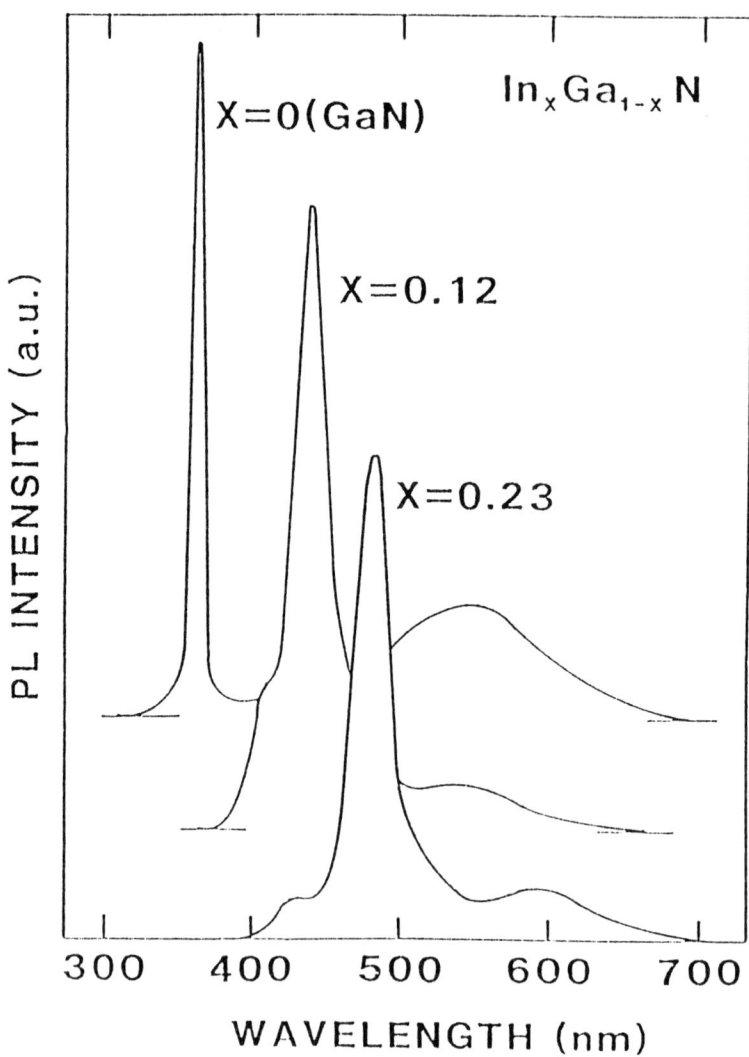

FIGURE 2 Dependence of photoluminescence spectra on indium mole fraction x of InGaN (a) x = 0, (b) x = 0.12, (c) x = 0.23 (from [8]).

Recently, InGaN crystalline quality has been improved by growing a thick GaN layer using a GaN buffer layer grown at low temperature. FIGURE 3 shows PL spectra of the InGaN films based on this growth method [9], measured at room temperature. In FIGURE 3, (a) and (b) correspond to indium mole fractions of 0.14 and 0.24, respectively. The growth temperatures of (a) and (b) were 830 °C and 780 °C. The other growth conditions were constant. In comparison with FIGURE 2, both spectra show sharp band edge emission and emission from deep levels cannot be observed.

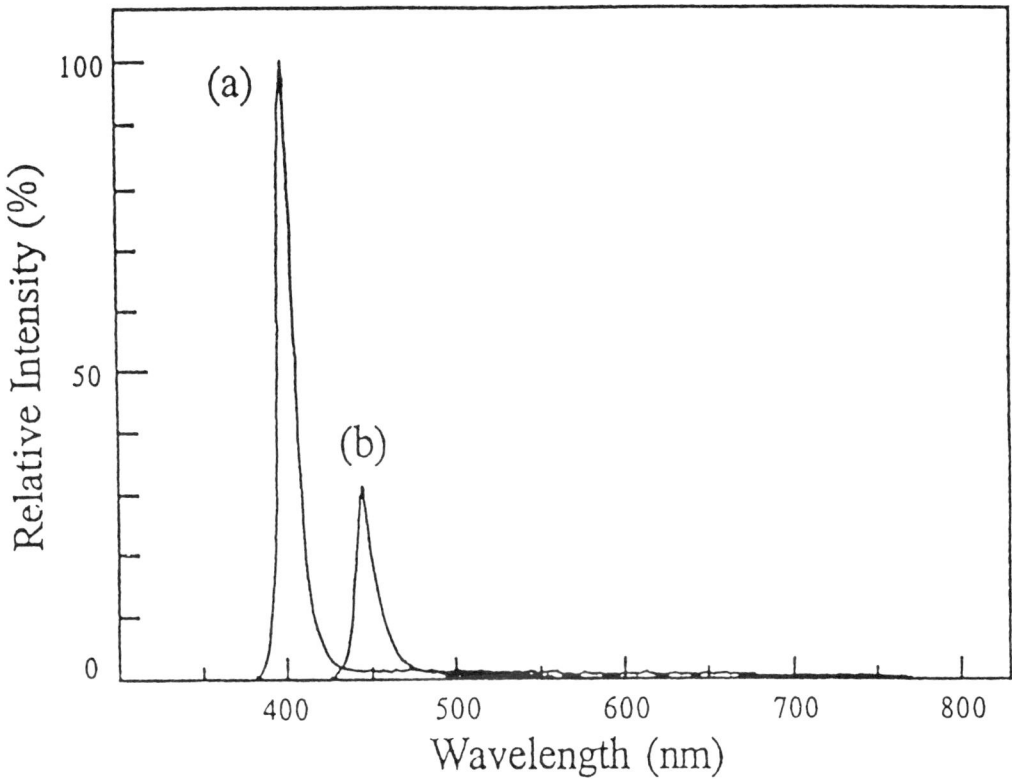

FIGURE 3   PL spectra of undoped InGaN at room temperature, grown on a thick GaN layer. The indium mole fractions of (a) and (b) are 0.14 and 0.24, respectively (from [9]).

C   PHOTOLUMINESCENCE OF DOPED InGaN

Doping into InGaN has been attempted by Nakamura et al [10,11]. Dopants are silicon and cadmium.

FIGURE 4 shows room-temperature PL spectra of Cd-doped InGaN films grown using a thick GaN layer and a GaN buffer layer [10]. In FIGURE 4, (a) and (b) correspond to indium mole fractions of 0.07 and 0.17, respectively. Both spectra (a) and (b) show a sharp peak and a broad one in the shorter and longer wavelength regions, respectively. The sharp peak corresponds to the band edge. On the other hand, a broad peak is the Cd-related emission. In the mole fraction region of indium from 0.01 to 0.19, the difference in peak emission energy between the band edge emissions and Cd-related ones is about 0.5 eV.

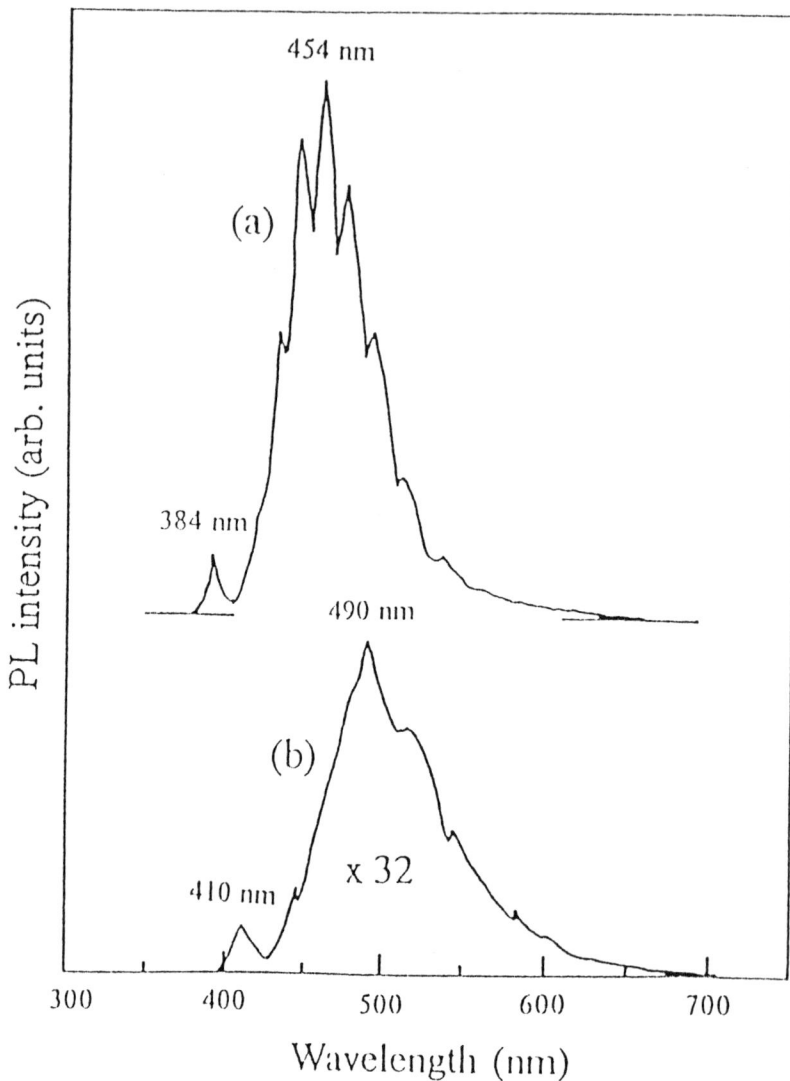

FIGURE 4 PL spectra of Cd-doped InGaN at room temperature, grown on a thick GaN layer. The indium mole fractions of (a) and (b) are 0.07 and 0.17, respectively (from [10]).

In Si-doped InGaN, the sharp peak from a band edge emission and the broad peak from a Si-related one were observed as shown in FIGURE 5 [11]. Silane gas was used for Si-doping. When the doping level was low, this broad peak was not observed. In FIGURE 6, the relationship between the relative intensity of the band edge emission and the silane gas flow rate is shown. The silane gas flow rate corresponds to the doping level of silicon. The PL intensity becomes strong due to the doping concentration.

D   CONCLUSION

Recently, it has become possible to grow single crystal InN using MOVPE although photoluminescence has not been observed. On the other hand, InGaN can be grown and its composition controlled. Band edge emission is readily observed for the Ga-rich alloy. Recently, introducing a thick GaN layer grown on a GaN buffer layer, the crystalline quality

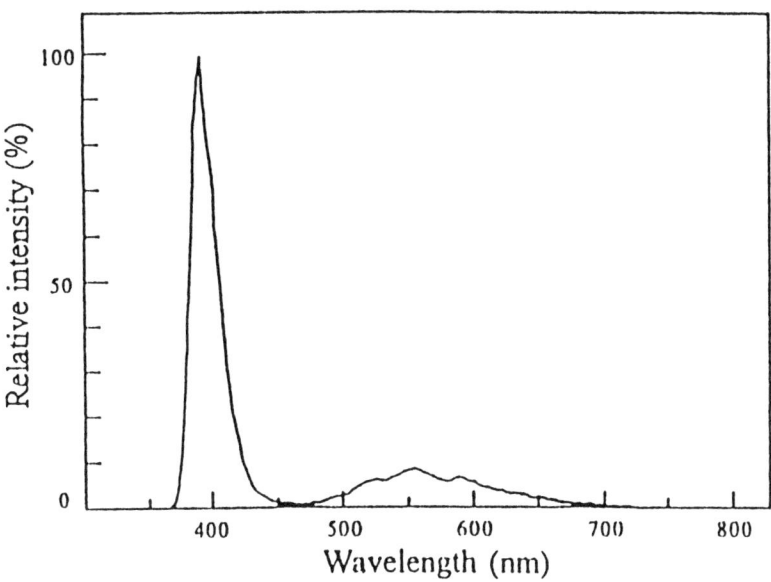

FIGURE 5 PL spectra of Si-doped InGaN at room temperature, grown on a thick GaN layer (from [11]).

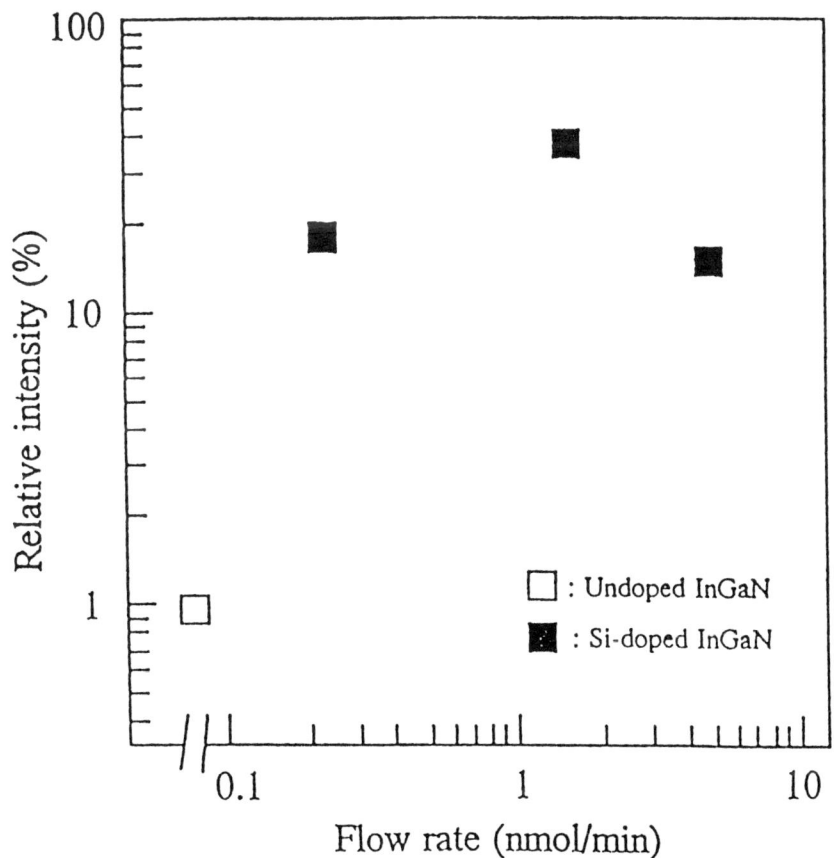

FIGURE 6 The relative PL intensity of the band edge emission of Si-doped InGaN films as a function of the silane gas flow rate (from [11]).

of InGaN has been further improved. Also n-type doping with dopants such as silicon and cadmium has recently been achieved. The effect of doping on the photoluminescence has been reported. In this Datareview, the photoluminescence of undoped and doped InGaN was reviewed.

REFERENCES

[1]   J.B. MacChesney, P.M. Bridenbaugh, P.B. O'Connor [*Mater. Res. Bull. (USA)* vol.5 (1970) p.783-91]

[2]   T. Matsuoka, H. Tanaka, T. Sasaki, A. Katsui [*Inst. Phys. Conf. Ser. (UK)* no.106 (1990) p.141-6]

[3]   J.R. Arthur [*J. Phys. Chem. Solids (UK)* vol.28 (1967) p.2257-67]

[4]   M.B. Panish, J.R. Arthur [*J. Chem. Thermodyn. (UK)* vol.2 (1970) p.299-318]

[5]   J.W. Trainor, K. Rose [*J. Electron. Mater. (USA)* vol.3 (1974) p.821-8]

[6]   K. Osamura, S. Naka, Y. Murakami [*J. Appl. Phys. (USA)* vol.46 (1975) p.3432-7]

[7]   C.P. Foley, T.L. Tansley [*Appl. Surf. Sci. (Netherlands)* vol.22/23 (1985) p.663-9]

[8]   N. Yoshimoto, T. Matsuoka, T. Sasaki, A. Katsui [*Appl. Phys. Lett. (USA)* vol.59 (1991) p.2251-3]

[9]   S. Nakamura, T. Mukai [*Jpn. J. Appl. Phys. (Japan)* vol.31 (1992) p.L1457-L1459]

[10]   S. Nakamura, N. Iwasa, S. Nagahama [*Jpn. J. Appl. Phys. (Japan)* vol.32 (1993) p.L338-L341]

[11]   S. Nakamura, T. Mukai, M. Senoh [*Jpn. J. Appl. Phys. (Japan)* vol.32 (1993) p.L16-L19]

CHAPTER 8

RAMAN AND IR REFLECTION SPECTROSCOPY

8.1 General remarks on Raman and IR spectroscopy of group III nitrides

L.E. McNeil

August 1993

The Raman scattering and infrared (IR) spectra of a crystal can be used to determine the frequencies of the normal modes of vibration (phonon modes) of the crystal. The number, symmetry type, and selection rules for the phonon modes of a crystal are determined by its space group. Where the orientation of the crystal axes relative to the polarization of the incident and scattered or absorbed photon is known (as with a single crystal sample), these selection rules can be used to identify the symmetry of an observed vibrational mode. Detailed discussions of the symmetry of vibrational modes and their relation to optical spectra are available in the standard literature [1,2].

The three crystal structures in which the group III nitrides crystallize are of different space groups and therefore have different symmetries for their phonon spectra. The hexagonal phase of BN (space group D_{4h}^6, $P6_3/mmc$) has the irreducible representation

$$\Gamma = 2E_{2g} + A_{2u} + E_{1u} + 2B_{1g} \tag{1}$$

The E_{2g} modes are Raman active, the A_{2u} and E_{1u} modes are infrared active, and the B_{1g} modes are silent (optically inactive). The E_{1u} and A_{2u} modes are polar modes, i.e. the interaction of the vibrations with the long-range Coulomb field leads to a higher energy for vibration polarized parallel to the direction of propagation of the phonon (longitudinal-optical or LO mode) than for that polarized perpendicular to that direction (transverse-optical or TO mode).

The cubic phase also observed for BN is the zinc blende type and has space group $T_d^2(F\bar{4}3m)$. Its phonon modes are of T_2 symmetry, and here also the LO vibration has higher energy than the TO polarization.

The remaining group III nitrides crystallize in the wurtzite structure, with space group C_{6v}^4, $P6_3/mc$. The irreducible representation is

$$\Gamma = A_1 + E_1 + 2E_2 + 2B_1 \tag{2}$$

The A_1, E_1 and E_2 modes are Raman active, the A_1 and E_1 modes are infrared active, and the B_1 modes are silent. Here, too, the A_1 and E_1 modes are polar, with different energies for the LO and TO polarizations.

The Raman spectrum of a crystal gives the frequencies of the vibrational modes excited in the particular geometry directly, as the frequencies of the resonances in the spectrum of the inelastically scattered light. The infrared reflection spectrum for light polarized in a given direction relative to the axes of the crystal will exhibit reststrahlen bands corresponding to the absorption allowed by symmetry for the particular geometry. The frequencies of the normal modes of vibration can be extracted by fitting the dielectric function obtained from the infrared spectrum to the classical dispersion formula

$$\varepsilon(\omega) = \varepsilon(\infty) + \sum_j \frac{s_j^2}{\omega_j^2 - \omega^2 + i\gamma_j\omega} \tag{3}$$

where $\varepsilon(\infty)$ is the high-frequency dielectric constant, ω is the frequency of the incident light, and s_j^2, ω_j and γ_j are the oscillator strength, frequency and damping constant of mode j.

REFERENCES

[1]   E.B. Wilson Jr., J.C. Decius, P.C. Cross [*Molecular vibrations: The theory of infrared and Raman vibrational spectra* (McGraw-Hill, 1955)]
[2]   W. Hayes, R. Loudon [*Scattering of light by crystals* (John Wiley & Sons, 1978)]

8.2 Raman and IR reflection spectra of BN

G.L. Doll

February 1994

A   INTRODUCTION

Raman scattering and infrared reflection studies performed on boron nitride are discussed in this Datareview. Ideally, optical lattice mode identification should be performed on high purity, single crystal samples. With one execption [1], these do not exist for the boron nitride phases. However, symmetry assignments from optical lattice modes have been determined for the h-BN and c-BN phases. This is possible since c-BN is cubic and therefore optically isotropic, and because c-axis oriented, polycrystalline h-BN samples are available. Neither of these conditions applies to w-BN or r-BN, so no mode assignments currently exist for these materials. In this Datareview, data will be presented for the h-BN, c-BN and w-BN phases. Since few data have been published on r-BN, no results will be presented for that phase. The temperature and pressure dependence of the h-BN and c-BN optical phonon frequencies and the dynamic effective charges will be discussed. Finally, as a practical example, infrared absorption is shown to be a sensitive characterization of the chemical bonding in BN films.

B   HEXAGONAL BORON NITRIDE

Raman spectra of a c-axis oriented, polycrystalline h-BN sample shown in FIGURES 1(a) and 1(b) from [2] have two modes at 1366 ± 2 and 52 ± 2 cm^{-1}. Since the sample in this study was oriented with its c-axis parallel to the incident and scattered light directions, the Raman-active modes are excited by in-plane (or xy) polarized light and therefore correspond to the E_{2g} symmetry vibrations. The low and high frequency modes are sometimes referred to as the E_{2g}^1 and E_{2g}^2 modes, respectively. Both modes are due to in-plane atomic displaccments, but the E_{2g}^1 mode is characterized by whole planes sliding against each other (termed rigid layer shear mode), and the E_{2g}^2 mode is due to B and N atoms moving against each other in-plane.

As is the case with graphite, finite crystallite size effects can be observed in the Raman spectrum of h-BN. Nemanich et al [3] performed an analysis of the Raman spectral dependence on the crystalline grain size of h-BN powders. The frequency and line width of the E_{2g}^2 mode were observed to correlate with the planar domain size (L_a) by the relationships,

$$\Delta \approx C/L_a \tag{1}$$

$$\Gamma_{\frac{1}{2}} \approx 9\,cm^{-1} + 1420\,cm^{-1}\,\text{Å}/L_a \tag{2}$$

where $\Delta = (\omega - 1366\,cm^{-1})$, L_a is in Å, C is equal to $380\,cm^{-1}\,\text{Å}$ and $\Gamma_{\frac{1}{2}}$ is defined to be the full width at half maximum of the E_{2g}^2 Lorentzian lineshape. Unlike graphite, second order Raman lines or density of state features were not observed in the Raman spectra of micro-crystalline BN [3].

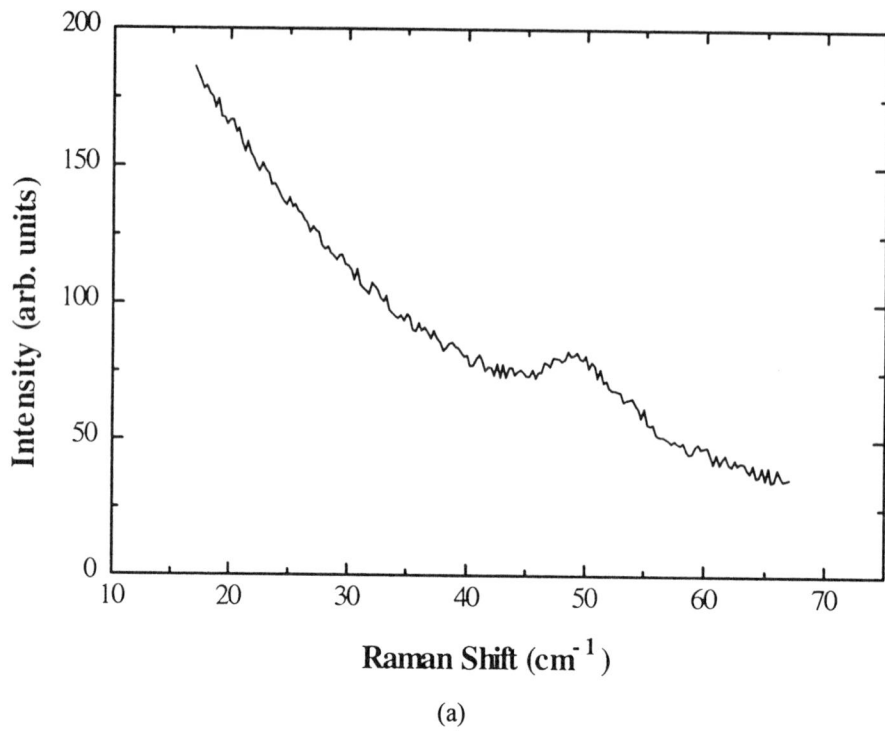

(a)

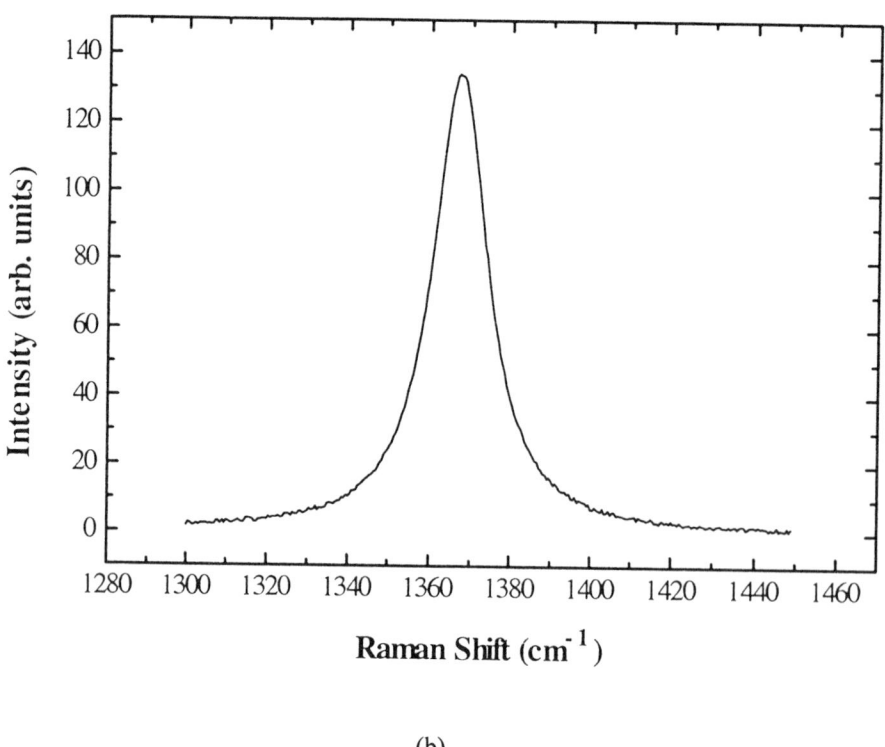

(b)

FIGURE 1   Raman spectra of (a) low-frequency E_{2g}^1 mode and (b) high-frequency E_{2g}^2 mode of h-BN from [2].

Kuzuba et al [4] have examined the effects of pressure on the Raman spectra of h-BN powders. They found that as the pressure was increased from 1 bar to 110 kbar, the frequency shifts in the E_{2g} modes corresponded to a 6% increase in the intralayer force constant, and a 400% increase in the interlayer shear force constant.

The infrared-active E_{1u} mode of h-BN is characterized by the in-plane B-N-B stretching vibration and is optically excited with $E \perp c$ polarization. The A_{2u} mode is an out-of-plane B-N-B bending vibration excited by $E \parallel c$ polarized light. Infrared reflectance spectra taken on a c-axis oriented, polycrystalline h-BN sample from Geick et al [5] are shown in FIGURES 2(a) and 2(b) for $E \parallel c$ and $E \perp c$ polarized light. The spectra exhibit two restrahlen bands for each polarization direction, with the weaker band in each polarization arising from a small c-axis misorientation in the polycrystalline h-BN sample. The ω_{LO} and ω_{TO} frequencies were determined for each restrahlen band by fitting a dielectric response to the reflectance spectrum in each polarization. The ω_{LO} and ω_{TO} values for the A_{2u} mode are 828 and 783 cm^{-1}, while ω_{LO} and ω_{TO} for the E_{1u} mode are 1610 and 1367 cm^{-1}. There is an accidental degeneracy of the E_{1u} TO infrared-active mode and the E_{2g}^2 Raman-active mode at 1367 cm^{-1}. Two-phonon absorption in the infrared spectra is sometimes observed, and has been interpreted [5] in terms of averaged phonon frequencies for modes residing at the edge of the Brillouin zone. Two-phonon absorptions have been observed to occur at 2600, 1540, 920, 820, 500 and 210 cm^{-1} [2,5]. These multi-phonon absorptions are not usually seen in well-crystallized samples [2].

The splitting of LO and TO modes belonging to the same irreducible representation is a measure of the dynamic effective charge (e^*) associated with each ion. Ramani et al [6] have used the E_{1u} and A_{2u} LO-TO mode splitting of h-BN to deduce that the in-plane effective charge per B-N bond (about 1.02e) is three times higher than the effective charge associated with out-of-plane vibrations (0.34e). The optical mode frequencies and the dynamic effective charges of h-BN are summarized in TABLE 1.

TABLE 1 Raman- and infrared-active frequencies and dynamic effective charges for three crystalline phases of BN.

	h-BN	c-BN	w-BN
Raman-active modes	E_{2g} 52 cm^{-1} E_{2g} 1366 cm^{-1}	$T_2(\omega_{LO})$ 1304 cm^{-1} $T_2(\omega_{TO})$ 1056 cm^{-1}	A_1 E_1 E_2
Infrared-active modes	$A_{2u}(\omega_{LO})$ 828 cm^{-1} $A_{2u}(\omega_{TO})$ 783 cm^{-1} $E_{1u}(\omega LO)$ 1610 cm^{-1} $E_{1u}(\omega_{TO})$ 1367 cm^{-1}	$T_2(\omega_{LO})$ 1340 cm^{-1} $T_2(\omega_{TO})$ 1065 cm^{-1}	A_1 E_1
e^*	in-plane: 0.6e out-of-plane: 0.2e	0.5e	0.2e

C   CUBIC BORON NITRIDE

A Raman spectrum taken by the author on a polycrystalline c-BN sample is shown in FIGURE 3.   The frequencies of the ω_{LO} and ω_{TO} modes are 1305 ± 5 and 1055 ± 5 cm^{-1}, respectively, and are in good agreement with values reported by Brafman et al [7]. Sanjurju et al [8] have used Raman scattering to study the LO and TO mode frequencies and the dynamic effective charge of c-BN as a function of pressure. They found that the ionicity of c-BN decreases with pressure, as is the case with most III-V compounds. Alvarenga at al [9] have studied the c-BN Raman-active phonons as a function of temperature, and observed that at 1600 K the c-BN interatomic force constant is only about 10 % weaker than at room temperature. This suggests that c-BN is a good candidate for applications requiring high-temperature mechanical strength.

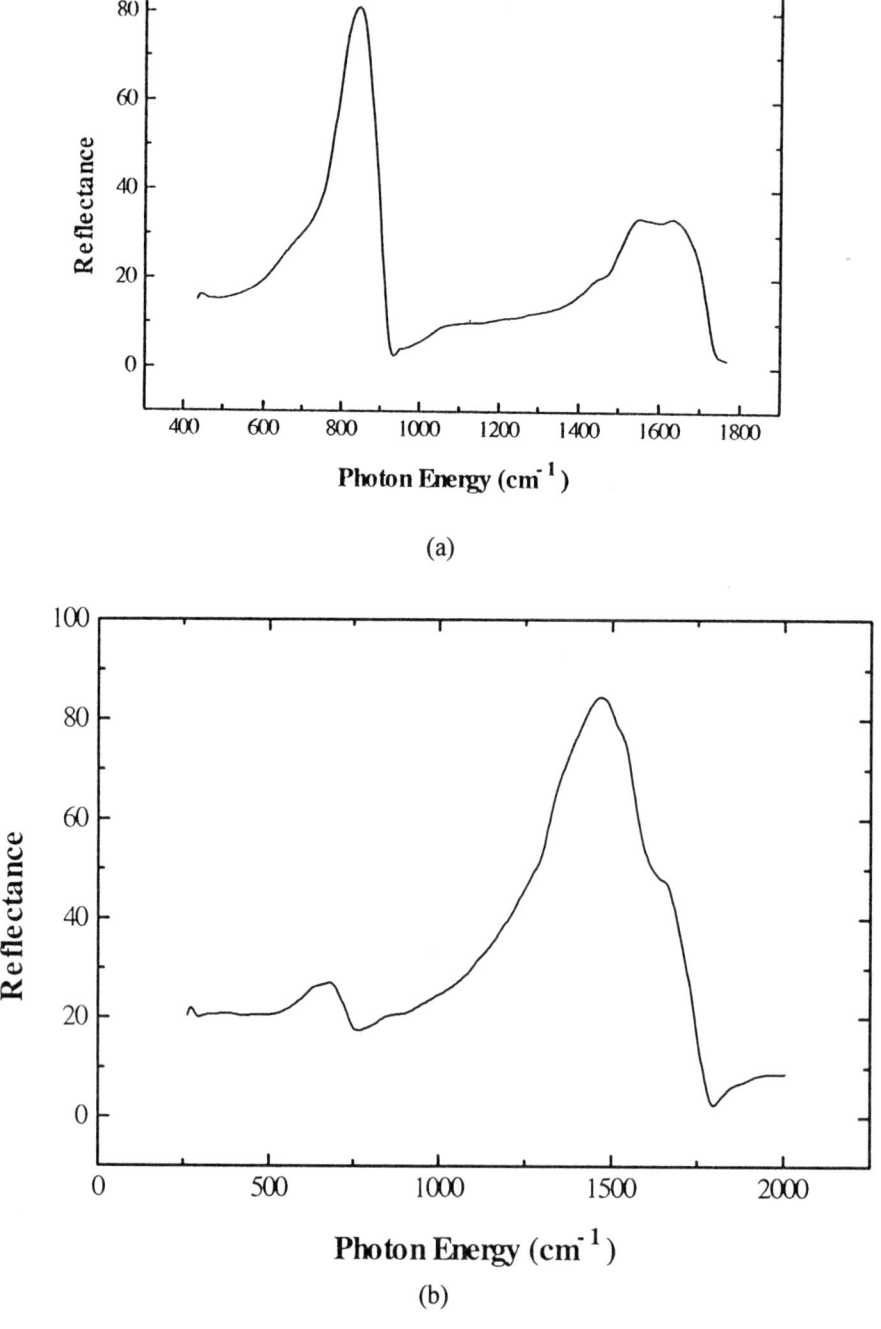

FIGURE 2 Infrared reflectance spectra of h-BN for (a) E ∥ c and (b) E ⊥ c polarizations from [5].

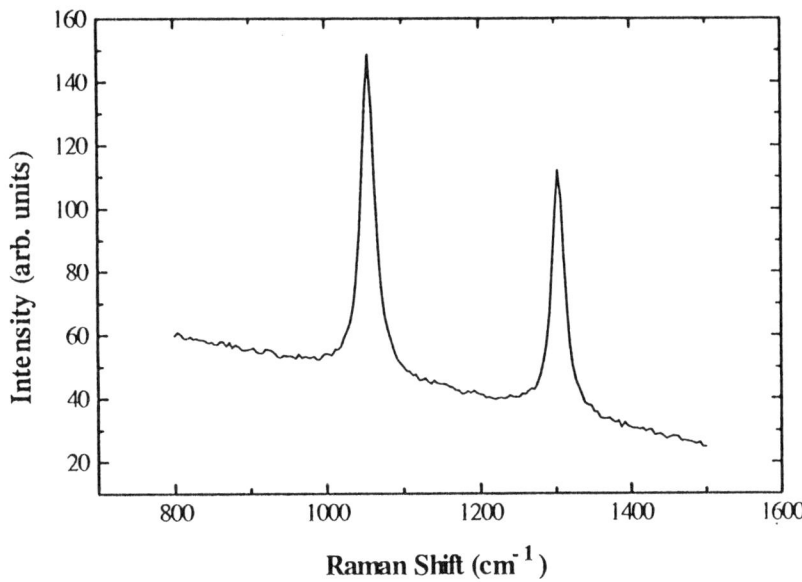

FIGURE 3 Raman spectrum of c-BN, plotted as intensity versus cm^{-1}. The TO and LO mode frequencies are 1056 and 1304 cm^{-1}, respectively.

An infrared reflectance spectrum taken by the author on a <111> face of a 10 μm size c-BN crystal [10] is shown in FIGURE 4. The frequencies of ω_{LO} and ω_{TO} were determined to be 1340 and 1065 cm^{-1} by fitting the reflectance spectrum with a dielectric response and by using the Lyddane-Sachs-Teller relationship. These values are in good agreement with results published by Gielisse et al [11]. Multiphonon absorption is sometimes observed in the region between 600 and 3000 cm^{-1}, and the frequencies of these absorptions were attributed to combinations of four zone boundary modes: LO 1232, TO 1000, LA 685, and TA 348 cm^{-1} [11]. Gielisse et al [11] also observed a very strong absorption at 1830 cm^{-1} which they claimed could be an impurity band. Chrenko [12] has argued that most of the high frequency absorptions observed by Gielisse et al [11] are not intrinsic to c-BN.

The optical mode frequencies and the dynamic effective charges of c-BN are summarized in TABLE 1.

D WURTZITE BORON NITRIDE

A Raman spectrum taken by the author on randomly oriented, sub-micrometer size w-BN powder [10] is shown in FIGURE 5. Three distinct peaks with Raman shifts of 950 ± 5, 1015 ± 5, and 1295 ± 5 cm^{-1} are evident in the spectrum. There is also a shoulder on the high energy side of the most intense peak near 1050 cm^{-1}. The w-BN Raman-active modes have A_1, E_1 and E_2 symmetry but at present no mode assignments are made to the Raman data in FIGURE 5.

An infrared transmission spectrum taken by the author on w-BN powder [10] pressed in a KBr pellet is shown in FIGURE 6. Three distinct peaks with energies of 1090, 1120 and 1230 cm^{-1} comprise the large absorption band near 1100 cm^{-1}. Weaker features reside near 1450, 1530 and 3500 cm^{-1}. The 3500 cm^{-1} feature is not due to a single-phonon absorption since its energy is so large. The w-BN infrared-active modes have A_1 and E_1 symmetry. The

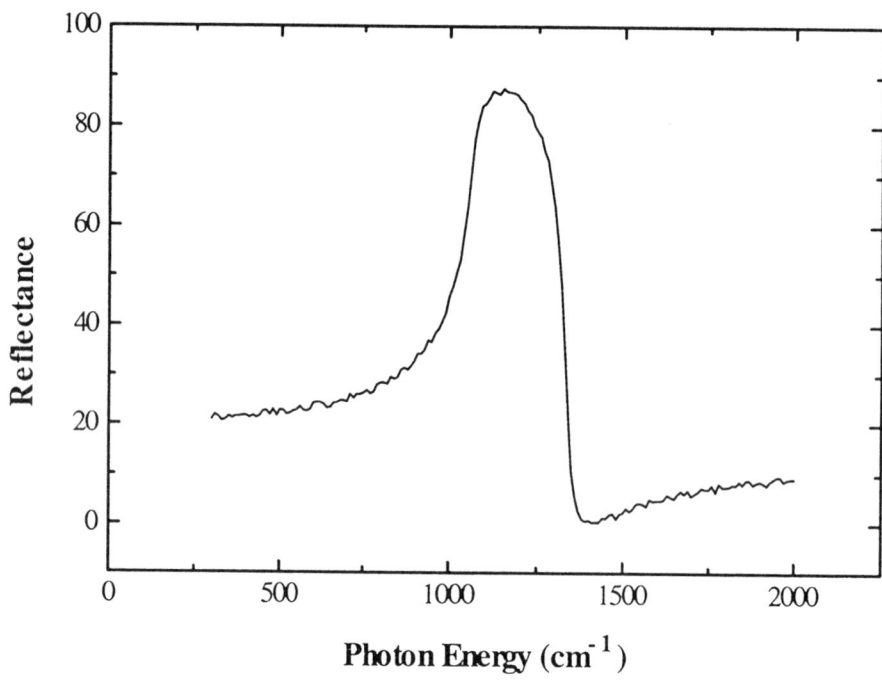

FIGURE 4 Infrared reflectance spectrum of c-BN plotted as intensity versus cm⁻¹.

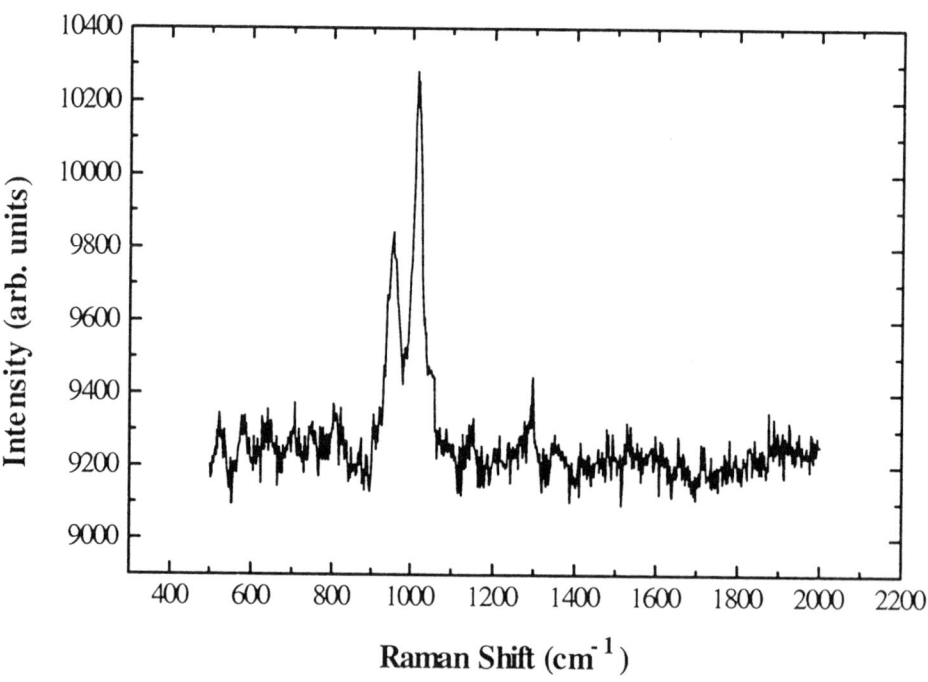

FIGURE 5 Raman spectrum of randomly oriented, polycrystalline w-BN powder.

material should be heteropolar so there will be LO-TO mode splitting and an effective charge associated with the B-N bond. Xu and Ching [13] have used the Mulliken scheme (which usually underestimates the charge) to calculate the fractional ionic charge of w-BN as 0.2e. The pressure and temperature dependence of the w-BN optically-active phonons have not been examined.

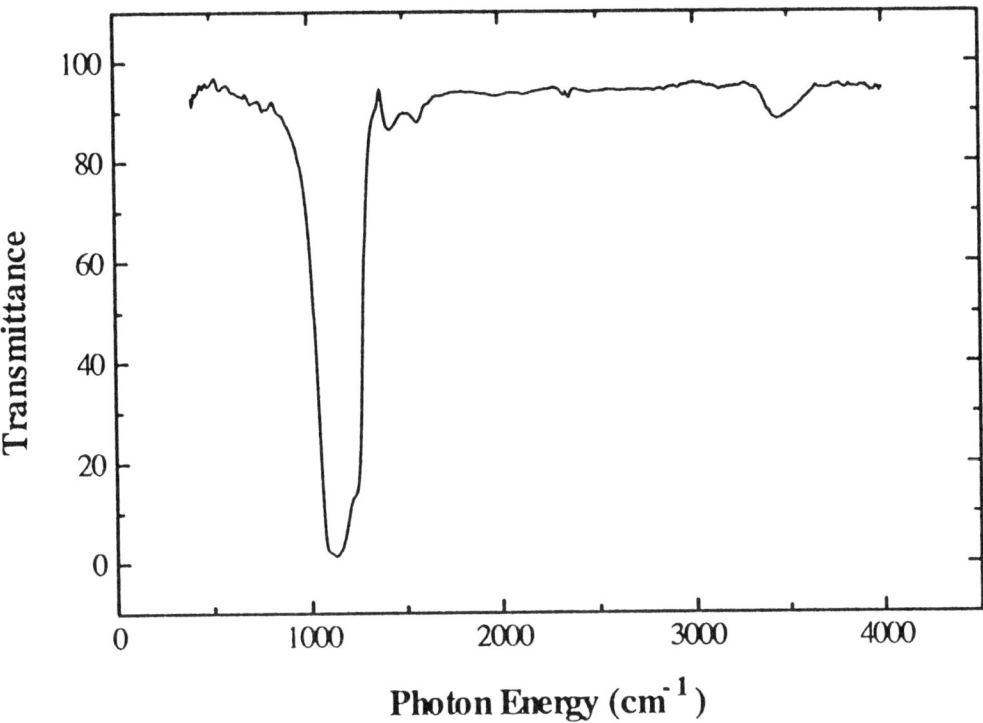

FIGURE 6 Infrared transmittance spectrum of w-BN powder.

E   BORON NITRIDE FILMS

Because of its desirable physical properties, there is considerable interest in growing c-BN films and coatings. Infrared absorption has been used extensively to characterize the chemical bonding (sp^2 vs. sp^3) in BN films [14]. An example is shown in FIGURE 7 where infrared transmittance spectra from two BN films are plotted. The top spectrum is taken from a film of microcrystalline h-BN (sp^2 bonded), and the bottom spectrum from a preferentially oriented, polycrystalline c-BN film (sp^3 bonded) [15]. The two features in the top spectrum with maxima near 780 and $1400\,cm^{-1}$ occur because of the h-BN restrahlen bands excited by $E \parallel c$ and $E \perp c$ polarized light. The positions of the maxima will vary with film properties such as compressive stress, stoichiometry, orientation and grain size. The bottom spectrum has one very strong maximum near $1080\,cm^{-1}$, which is characteristic of sp^3-bonded c-BN, and two weak maxima near 780 and $1400\,cm^{-1}$ that arise from sp^2-bonded BN that resides near the surface of the film. Although no trace of h-BN could be found in the c-BN film by diffraction methods, the presence of the weak sp^2 absorption features in the spectrum illustrates the sensitivity of infrared absorption in characterizing the chemical bonding of the entire BN film.

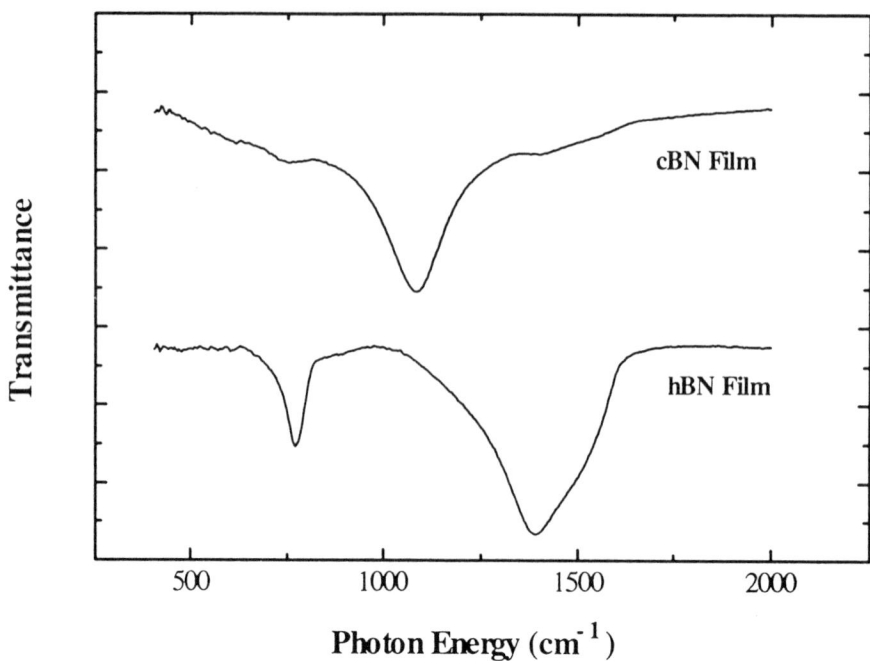

REFERENCES

[1]   N. Miyata, K. Moriki, O. Mishima, M. Fujisawa, T. Hattori [*Phys. Rev. B (USA)* vol.40 (1989) p.12028]

[2]   D.M. Hoffman, G.L. Doll, P.C. Eklund [*Phys. Rev. B (USA)* vol.30 (1984) p.6051]

[3]   R.J. Nemanich, S.A. Solin, R.M. Martin [*Phys. Rev. B (USA)* vol.23 (1981) p.6348]

[4]   T. Kuzuba, Y. Sato, S. Yamaoka, K. Era [*Phys. Rev. B (USA)* vol.18 (1978) p.4440]

[5]   R. Geick, C.H. Perry, G. Rupprecht [*Phys. Rev. (USA)* vol.146 (1966) p.543]

[6]   R. Ramani, K.K. Mani, R.P. Singh [*Phys. Status Solidi B (Germany)* vol.86 (1978) p.759]

[7]   O. Brafman, G. Lengyel, S.S. Mitra, P.J. Gielisse, J.N. Pendl, L.C. Mansur [*Solid State Commun. (USA)* vol.6 (1968) p.532]

[8]   J.A. Sanjurju, E. Lopez-Cruz, P. Vogl, M. Cardona [*Phys. Rev. (USA)* vol.28 (1983) p.4579]

[9]   A.D. Alvarenga, M. Grimsditch, A. Polian [*J. Appl. Phys. (USA)* vol.72 (1992) p.1955]

[10]   The c-BN and w-BN samples were provided by Dr. B. Sweeting of G.E.Superabrasives.

[11]   P.J. Gielisse et al [*Phys. Rev. (USA)* vol.155 (1967) p.1039]

[12]   R.M. Chrenko [*Solid State Commun. (USA)* vol.14 (1974) p.511]

[13]   Y.-N. Xu, W.Y. Ching [*Phys. Rev. B (USA)* submitted 18/1/93]

[14]   J.J. Pouch, S.A. Alterovitz [*Synthesis and Properties of Boron Nitride* (Trans. Tech. Publications, Brookfield, Vermont, 1990)]

[15]   A.K. Ballal, L. Salamanca-Riba, G.L. Doll, C. A. Taylor II, R. Clarke [*J. Mater. Res. (USA)* vol.7 (1992) p.1618]

8.3 Raman and IR reflection spectra of AlN

L.E. McNeil

August 1993

A RAMAN SPECTRA

Raman spectroscopic studies of AlN have been hampered by the lack of reasonably-sized single-crystal specimens from which complete sets of data could be obtained. Most of the measurements to date have been made on small whisker crystals or on thin films. Neither type of specimen offers a full complement of propagation and polarization directions for the phonon, making the symmetry identification of the peaks difficult. This has led to some confusion in the literature [1-4]. Recently, McNeil et al [5] obtained a set of Raman measurements made on a large single crystal grown by a sublimation-recondensation technique [6]. The values obtained for the Raman-active phonons at room temperature are given in TABLE 1.

TABLE 1 Raman-active phonons in AlN at room temperature.

Symmetry	Frequency (cm^{-1})
E_2	252
E_2	660
A_1 (TO)	614
A_1 (LO)	893
E_1 (TO)	673
E_1 (LO)	916

The dependence of three of these modes on pressure has been obtained by Perlin et al [7]. The frequencies were found to obey a quadratic law:

$$\nu = \nu_0 + \nu_1 P + \nu_2 P^2 \qquad (1)$$

where ν is the frequency in cm^{-1} and P is the pressure in GPa. The values of the coefficients are given in TABLE 2.

TABLE 2 Values of coefficients in Eqn (1).

Mode	ν_o (cm^{-1})	ν_1 ($cm^{-1} GPa^{-1}$)	ν_2 ($cm^{-1} GPa^{-2}$)
E_2	660	3.99	0.035
A_1 (TO)	607	4.63	-0.01
E_1 (LO)	924	1.67	0.207

An earlier study by Sanjurjo et al [3] fitted the pressure dependence of four of the modes to a linear expression:

$$\nu = \nu_0{}^* + \nu_1{}^*P \tag{2}$$

The mode observed near 660 cm^{-1} was identified by Sanjurjo et al as A$_1$ (TO) rather than E$_2$, but the pressure dependences of the Raman lines observed in both experiments are in reasonable agreement: for the two modes observed in both experiments, the pressure shift $\Delta\nu = \nu(P) - \nu(0)$ disagrees by ~ 3 - 7 cm^{-1} at P = 10 GPa. The agreement at P = 0 is poorer for the E$_1$ (LO) mode, but the two LO modes were resolved poorly or not at all in the two experiments. The coefficients obtained by Sanjurjo et al are shown in TABLE 3.

TABLE 3 Values of coefficients in Eqn (2).

Symmetry	$\nu_0{}^*$ (cm^{-1})	$\nu_1{}^*$ (cm^{-1} GPa^{-1})
E$_2{}^\dagger$	659.3 ± 0.6	4.97 ± 0.06
A$_1$ (LO)	888 ± 2	3.8 ± 0.2
E$_1$(TO)	671.6 ± 0.8	4.84 ± 0.09
E$_1$ (LO)	895 ± 2	4.0 ± 0.2

† Identified by Sanjurjo et al as A$_1$ (TO).

McNeil et al [5] also noted a correlation between the oxygen content of the AlN sample and the width of the Raman peaks, which parallels changes in the unit-cell volume, mid-gap luminescence and thermal conductivity noted by Harris et al [8, 9]. The widths of the A$_1$ and E$_1$ TO modes and the higher-frequency E$_2$ mode increase abruptly from 2 - 3 cm^{-1} at low oxygen concentrations to 4 - 5.5 cm^{-1} for samples with oxygen concentrations above approximately 1 atomic %. Harris et al [8, 9] proposed a model which postulates that as the oxygen concentration is increased, a shift occurs in the dominant defect from an isolated oxygen atom on a nitrogen site (with associated aluminium vacancy) to an aluminium atom octahedrally bound to a number of oxygen atoms.

B   IR REFLECTION SPECTRA

The availability of samples of suitable size and orientation has also limited the measurements of IR reflectivity in AlN. In particular, there appear to have been no reports of measurements made with the incident light polarized parallel to the c-axis, which would allow the observation of the A$_1$ modes. Reports of the E$_1$ frequencies agree reasonably well with the Raman data: Collins et al [10] obtained TO and LO frequencies of TO = 666.7 cm^{-1} and LO = 916.3 cm^{-1}. The close coincidence of this measured TO frequency and the frequency of the E$_2$ mode is merely a reflection of the uncertainty in the IR measurement, as the latter mode is not IR active. Carlone et al [4] used IR absorption to obtain frequencies of TO ~ 610 cm^{-1} and LO ~ 800 cm^{-1}, but the range of crystal orientations across their thin-film sample suggests that they observed a mixture of A$_1$ and E$_1$ modes.

REFERENCES

[1]   O. Brafman, G. Lengyel, S.S. Mitra, P.J. Gielisse, J.N. Plendl, L.C. Mansur [*Solid State Commun. (USA)* vol.6 (1968) p.523-6]

[2]   R. Tsu, R.F. Rutz [*3rd International Conference on Light Scattering in Solids*, Campinas, Brazil, 1975, Eds M. Balkanski, R.C.C. Leite, S.P.S. Porto (J. Wiley & Sons) p.393-5]

[3]   J.A. Sanjurjo, E. López-Cruz, P. Vogl, M. Cardona [*Phys. Rev. B (USA)* vol.28 (1983) p.4579-84]

[4]   C. Carlone, K.M. Lakin, H.R. Shanks [*J. Appl. Phys. (USA)* vol.55 (1984) p.4010-4]

[5]   L.E. McNeil, M. Grimsditch, R.H. French [*J. Am. Ceram. Soc. (USA)* vol.76 (1993) p.1132-6]

[6]   G.A. Slack, T.F. McNelly [*J. Cryst. Growth (Netherlands)* vol.42 (1977) p.560-3]

[7]   P. Perlin, A. Polian, T. Suski [*Phys. Rev. B (USA)* vol.47 (1993) p.2874-7]

[8]   J.H. Harris, R.A. Youngman, R.G. Teller [*J. Mater. Res. (USA)* vol.5 (1990) p.1763-73]

[9]   R.A. Youngman, J.H. Harris [*J. Am. Ceram. Soc. (USA)* vol.73 (1990) p.3238-46]

[10]   A.T. Collins, E.C. Lightowlers, P.J. Dean [*Phys. Rev. (USA)* vol.158 (1967) p.833-8]

8.4 Raman and IR reflection spectra of GaN

L.E. McNeil

August 1993

A RAMAN SPECTRA

The measurements of the Raman spectrum of GaN in its wurtzite form [1-7] have displayed remarkable unanimity, with disagreements of no more than $1 - 2\,cm^{-1}$ for the E_2, A_1 (TO) and E_1 (TO) modes. The LO modes have proved more difficult to observe (as is the case for AlN), and have been reported less frequently [3,4,6] and with some disagreement. Measurements made at $77\,K$ [3] and $20\,K$ [1] show temperature shifts of $1 - 3\,cm^{-1}$ for all modes. The room temperature values of the Raman lines for GaN are given in TABLE 1.

TABLE 1 Raman-active phonons for GaN at room temperature.

Symmetry	Frequency (cm^{-1})
E_2	144
E_2	568
A_1 (TO)	532
A_1 (LO)	710[†]
E_1 (TO)	560
E_1 (LO)	741 749[††]

[†] [4]
[††] [3]

Perlin et al [7] have measured the pressure dependence of the four lowest-frequency modes and find a quadratic dependence for the frequency:

$$\nu = \nu_0 + \nu_1 P + \nu_2 P^2 \tag{1}$$

where ν is the frequency in cm^{-1} and P is the pressure in GPa. The values of the coefficients are given in TABLE 2.

TABLE 2 Values of coefficients in Eqn (1).

Symmetry	ν_0 (cm^{-1})	ν_1 (cm^{-1}/GPa)	ν_2 (cm^{-1}/GPa^2)
E_2	144	-0.25	-0.0017
E_2	568	4.17	-0.0136
A_1 (TO)	531	4.06	-0.0127
E_1 (TO)	560	3.68	-0.0078

These values are in good agreement with the earlier measurement of Eremets et al [5] who found ν_0 = 567.9, ν_1 = 4.42 and ν_2 = -0.051 for the higher-frequency E_2 mode.

B   INFRARED REFLECTION SPECTRA

The available IR reflection spectra [1,8,9] show good agreement for the TO modes, with the values reported differing by no more than 8 cm^{-1} among different measurements and with the Raman values. For the E_1 (LO) mode there is similar agreement (particularly when the large uncertainties of Manchon et al [1] are considered). For the A_1 (LO) mode the values of 744 cm^{-1} obtained by Barker and Ilegems [8] and 738 cm^{-1} obtained by Sobotta et al [9] (which agree within uncertainty with the 770 ± 70 cm^{-1} value of Manchon et al) disagree with the Raman value of Cingolani et al [4] and Hayashi et al [6] of 710 cm^{-1}. It seems likely that the Raman scattering efficiency of the LO modes is low (which would explain the failure of most workers to observe them), but the identity of the peak at 710 cm^{-1} observed by the latter two groups remains a mystery.

REFERENCES

[1]   D.D. Manchon Jr., A.S. Barker Jr., P.J. Dean, R.B. Zetterstrom [*Solid State Commun. (USA)* vol.8 (1970) p.1227-31]

[2]   V. Lemos, C.A. Argüello, R.C.C. Leite [*Solid State Commun. (USA)* vol.11 (1972) p.1351-3]

[3]   G. Burns, F. Dacol, J.C. Marinace, B.A. Scott, E. Burstein [*Appl. Phys. Lett. (USA)* vol.22 (1973) p.356-7]

[4]   A. Cingolani, M. Ferrara, M. Lugarà, G. Scamarcio [*Solid State Commun. (USA)* vol.58 (1986) p.823-4]

[5]   M.I. Eremets, V.V. Struzhkin, A.M. Shirkov, J. Jun, I. Grzegory, P. Perlin [*Acta Phys. Pol. A (Poland)* vol.75 (1989) p.875-7]

[6]   K. Hayashi, K. Itoh, N. Sawaki, I. Akasaki [*Solid State Commun. (USA)* vol.77 (1991) p.115-8]

[7]   P. Perlin, C. Jauberthie-Carillon, J.P. Itie, A. San Miguel, I. Grzegory, A. Polian [*Phys. Rev. B (USA)* vol.45 (1992) p.83-9]

[8]   A.S. Barker Jr., M. Ilegems [*Phys. Rev. B (USA)* vol.7 (1973) p.743-50]

[9]   H. Sobotta, H. Neumann, R. Franzheld, W. Seifert [*Phys. Status Solidi B (Germany)* vol.174 (1992) p.K57-K60]

8.5   Raman and IR spectra of InN and ternary group III nitrides

L.E. McNeil

August 1993

A   INDIUM NITRIDE AND GALLIUM INDIUM NITRIDE

There appear to have been no Raman or IR measurements of pure InN to date, but Osamura et al [1] extrapolated their Kramers-Kronig analysis of the ternary compound $Ga_{1-x}In_xN$ to x = 1, using a Brout sum rule, and thereby obtained values of 478 cm^{-1} (TO) and 694 cm^{-1} (LO). The symmetry of these modes was not identified and the sample was polycrystalline but, since the value they obtained for the GaN (x = 0) TO mode was 563 cm^{-1}, the modes may be presumed to be of E_1 symmetry. The other infrared-active mode (presumably the A_1-symmetry vibration) was not observed. In the same study, the ternary compound $Ga_{1-x}In_xN$ was found to exhibit one-mode behaviour, in which the single observed mode shifts monotonically with concentration between the values observed for the pure compounds.

B   GALLIUM ALUMINIUM NITRIDE

The Raman spectrum of the ternary nitride $Ga_{1-x}Al_xN$ (0 < x < 0.15) has recently been reported [2] for the first time. Four peaks were observed in measurements on thin films, with frequencies at x = 0 of 534 cm^{-1} (A_1 (TO)), 560 cm^{-1} (E_1 (TO)), 569 cm^{-1} (E_2) and 741 cm^{-1} (E_1 (LO)). These values agree well with those reported elsewhere for GaN [3-7]. The peaks shifted to higher frequency as the concentration of Al was increased, displaying 'one-mode' behaviour with no evidence of an AlN localized mode at the highest values of x studied. However, the frequencies do not follow the expected linear interpolation between the values of GaN and AlN, with the TO and E_2 phonons shifting less than expected and the LO shifting more.

REFERENCES

[1]   K. Osamura, S. Naka, Y. Murakami [*J. Appl. Phys. (USA)* vol.46 (1975) p.3432-7]

[2]   K. Hayashi, K. Itoh, N. Sawaki, I. Akasaki [*Solid State Commun. (USA)* vol.77 (1991) p.115-8]

[3]   D.D. Manchon Jr., A.S. Barker Jr., P.J. Dean, R.B. Zetterstrom [*Solid State Commun. (USA)* vol.8 (1970) p.1227-31]

[4]   G. Burns, F. Dacol, J.C. Marinace, B.A. Scott, E. Burstein [*Appl. Phys. Lett. (USA)* vol.22 (1973) p.356-7]

[5]   A. Cingolani, M. Ferrara, M. Lugarà, G. Scamarcio [*Solid State Commun. (USA)* vol.58 (1986) p.823-4]

[6]   M.I. Eremets, V.V. Struzhkin, A.M. Shirkov, J. Jun, I. Grzegory, P. Perlin [*Acta Phys. Pol. A (Poland)* vol.75 (1989) p.875-7]

[7]   P. Perlin, C. Jauberthie-Carillon, J.P. Itie, A. San Miguel, I. Grzegory, A. Polian [*Phys. Rev. B (USA)* vol.45 (1992) p.83-9]

CHAPTER 9

DEFECTS AND IMPURITIES

9.1 General remarks on defects and impurities in group III nitride films

S.C. Strite

January 1994

The group III nitride semiconductor family is a highly viable material system for short wavelength optical device applications. Largely due to the difficulty of incorporating stoichiometric quantities of N and the lack of a suitable substrate material, the crystal quality of these materials has been difficult to improve. Because most of the efforts of the past twenty years have been directed towards developing better crystal growth techniques, studies of the defect and impurity levels of these materials are in comparatively short supply. Nevertheless, a reasonably coherent picture has emerged in many areas.

Extended defects in these materials are the result of lattice mismatched heteroepitaxy. Both dislocations and stacking faults have been observed in large quantities in GaN and are certainly present in the other group III nitride films as well. These defects have been shown to cooperate, encouraging the formation of polytype domains in GaN and InN. The best results to date have been obtained on sapphire substrates using various buffer layers. Further improvements may be possible through more elaborate buffer layer schemes, such as graded buffer layers or strained layer superlattices. In addition, serious attention should be paid to the emergence of better lattice and thermally matched substrates such as SiC, MgO and ZnO.

As researchers have suspected for years, native defects, notably the N vacancy, play an enormous role in the electronic and optical properties of the group III nitride materials grown currently. Optical and electrical measurements show good agreement with theoretical predictions that the N vacancy forms a shallow donor in InN and GaN and a deep electron trap in AlN. Improved crystal growth has reduced the electron background in GaN to as low as the mid 10^{15} cm^{-3} level and further improvement is being aggressively pursued in many laboratories. The properties of InN and AlN thin films remain dominated by the native defect background.

Great strides have also been made in doping GaN. In the past several years, both p- and n-type impurities have been found leading to the demonstration of high quality p-n junction LEDs and raising hopes that a GaN-based laser will be realized in the near future. A zinc blende BN p-n junction LED emitted ultraviolet light at the shortest wavelength ever produced by a semiconductor. However, nitride device technology will be seriously hindered if conductive AlGaN cannot be developed. With the ever increasing pace of nitride research, we can look forward to the answers to these issues and more.

9.2 Extended structural defects in heteroepitaxial group III nitride films

S.C. Strite

January 1994

A   INTRODUCTION

One of the major difficulties in producing high quality group III nitride semiconductor samples is the lack of an ideal substrate material. Their small lattice constants and thermal expansion coefficients cause them to be lattice and thermally mismatched with all of the common semiconductor substrate materials. The recent commercialization of SiC substrates has begun to alleviate some of the problems for GaN and AlN but, until bulk single crystals of the various nitrides can be grown, the issues of heteroepitaxy will remain linked with nitride research.

B   ORIENTATION AND POLARITY OF HETEROEPITAXIAL GaN AND AlN

Sapphire remains the most common substrate for nitride heteroepitaxy. Its crystal structure, though hexagonal, is not wurtzite. GaN grown on (0001) basal plane sapphire has its (0001) plane parallel to the substrate, but its $[2\bar{1}\bar{1}0]$ direction is parallel to the sapphire $[1\bar{1}00]$ direction. Nitrides have been grown on many different orientations of sapphire, but the (0001) basal plane remains the most popular and generally yields the best material. The orientational relationships between various sapphire planes and nitride epitaxial films have been studied by many workers [1-3].

Wurtzite GaN has also been grown on SiC substrates [4] while zincblende material has been deposited on GaAs [5], Si [6] and MgO [7]. In these cases the films are observed to be oriented exactly to the substrate.

Another issue which is of importance to nitride epitaxy is the polarity of the epitaxial film. Sasaki et al [8] have used X-ray photoemission to determine the polarity of GaN grown on both sapphire and SiC. Their data indicate that GaN grown on the (0001) sapphire plane is Ga-terminated. Recent Auger measurements [9] on GaN grown on AlN buffer layers yielded contradictory results, deducing that the GaN was N-terminated. In the case of (0001) SiC, Ga-terminated GaN was observed to grow on the C face while N-terminated GaN grows on the Si face indicating that N-C and Ga-Si bonds are preferred. Better GaN quality was achieved on $(0001)_{Si}$ SiC substrates, although an amorphous SiN_x interlayer can form if care is not taken to initiate growth with a Ga prelayer.

C   EXTENDED STRUCTURAL DEFECTS IN HETEROEPITAXIAL GaN

Due to the non-ideal substrate choices for nitride heteroepitaxy, large numbers of dislocations are inevitably formed in the epitaxial material to alleviate the lattice mismatch and the strain of post-growth cooling. Powell et al [3] have reported in detail the structure of threading dislocations in GaN on sapphire. They observed a concentration of 2×10^{10} cm^{-2} threading dislocations having Burgers vectors of $(a_0/3)<11\overline{2}0>$. Crystal growers, with considerable success, have tried to minimize these problems by initiating heteroepitaxy with an optimized buffer layer. Yoshida et al [10] were the first to observe improved GaN on sapphire when growth was initiated with a thin AlN buffer layer. Many researchers have since used this approach to substantially improve the quality of their GaN thin films. Nakamura et al [11] demonstrated that a low temperature GaN buffer layer is at least equally effective for heteroepitaxy on sapphire. Other groups have developed optimal techniques of their own for heteroepitaxy on SiC [4], Si [6] and GaAs [12].

Besides dislocations, the most prevalent structural defect observed in group III nitride thin films is the stacking fault. FIGURE 1 is a low resolution transmission electron micrograph of zinc blende GaN grown on a (001) GaAs substrate [5] which shows typical (111) stacking faults. Similar planar defect structures have been observed in zinc blende GaN grown on 3C SiC [13] and MgO [3,7] substrates. Lei et al [14], using X-ray diffraction (XRD), have shown the existence of stacking faults in wurtzite GaN grown on (111) Si. In lattice mismatched heteroepitaxy of the group III nitrides, stacking faults serve as a form of strain relief in the manner illustrated in FIGURE 2. In Datareview 9.3 we will see that the role of the stacking fault defect in the group III nitrides is even more complicated as a result of the bistability of the nitride semiconductor crystal structures.

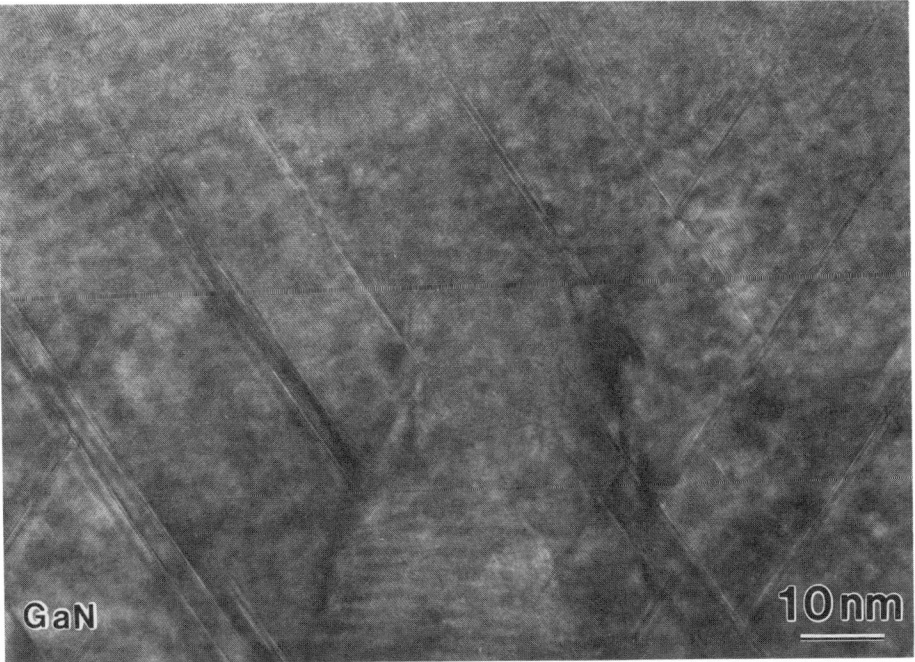

FIGURE 1 Electron micrograph of bulk zinc blende GaN grown on (001) GaAs. Numerous stacking faults are apparent as linear streaks propagating diagonally along <111> directions. From [5].

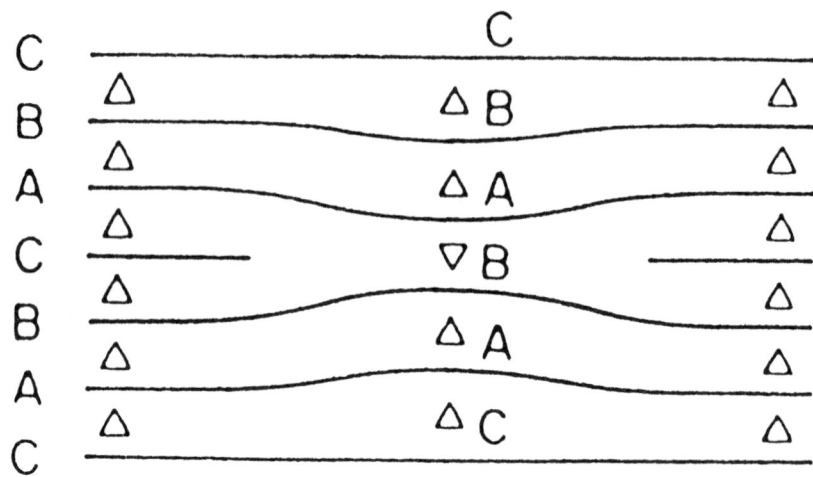

FIGURE 2 Schematic of a typical stacking fault defect. When growth is not along the stacking direction, the formation of a stacking fault requires the termination of an atomic plane at a dislocation.

D CRYSTAL GROWTH AND STRUCTURAL DEFECTS IN BN

Crystal growth methods for cubic BN, the polytype of most interest for semiconductor applications, lag behind those for GaN and AlN. Early research was mostly directed towards coating applications [15] and only recently have semiconductor devices [16,17] and heteroepitaxy been gaining interest. BN has been grown homoepitaxially at high pressure [16] and heteroepitaxially on Si substrates by laser ablation [18]. Little characterization of the crystal defects has been reported by either group. The metastability of cubic BN makes the growth of bulk material difficult. Ren et al [19] reported that cubic BN films grown on Si can revert to amorphous material under certain conditions. BN grown on Si by Friedmann et al [20] was observed to be multiphase, consisting largely of the hexagonal polytype with small (200 Å) zinc blende domains. These reports are typical of many present efforts at heteroepitaxial BN.

REFERENCES

[1]   R. Madar, D. Michel, G. Jacob, M. Boulou [*J. Cryst. Growth (Netherlands)* vol.40 (1977) p.239-52]

[2]   A.V. Kuznetsov, S.A. Semiletov, G.V. Chaplygin [in *Growth of Crystals* Eds E.I. Givargizov, S.A. Grinberg (Consultants Bureau, 1988) p.13-22]

[3]   R.C. Powell, N.-E. Lee, Y.-W. Kim, J.E. Greene [*J. Appl. Phys (USA)* vol.73 (1993) p.189-203]

[4]   M.E. Lin et al [*Appl. Phys. Lett. (USA)* vol.62 (1993) p.702-4]

[5]   S. Strite et al [*J. Vac. Sci. Technol. B (USA)* vol.9 (1991) p.1924-9]

[6]   T. Lei, M. Fanciulli, R.J. Molnar, T.D. Moustakas, R.J. Graham, J. Scanlon [*Appl. Phys. Lett. (USA)* vol.59 (1991) p.944-6]

[7]   R.C. Powell, G.A. Tomasch, Y.-W.Kim, J.A. Thornton, J.E. Greene [*Mater. Res. Soc. Symp. Proc. (USA)* vol.162 (1990) p.525-30]

[8]   T. Sasaki, T. Matsuoka, A. Katsui [*Appl. Surf. Sci. (Netherlands)* vol.41/42 (1989) p.504-8]

[9]   M.A. Khan, J.N. Kuznia, D.T. Olson, R. Kaplan [*J. Appl. Phys (USA)* vol.73 (1993) p.3108-10]

[10]   S. Yoshida, S. Misawa, S. Gonda [*J. Vac. Sci. Technol. B (USA)* vol.1 (1983) p.250-3]

[11]   S. Nakamura [*Jpn. J. Appl. Phys. (Japan)* vol.30 (1991) p.L1705-L1707]

[12]   S. Strite, D.S.L. Mui, G. Martin, Z. Li, D.J. Smith, H. Morkoc [*Inst. Phys. Conf. Ser. (UK)* vol.120 (1992) p.89-93]

[13]   M.J. Paisley, Z. Sitar, J.B. Posthill, R.F. Davis [*J. Vac. Sci. Technol. A (USA)* vol.7 (1989) p.701-5]

[14]   T. Lei, K.F. Ludwig Jr., T.D. Moustakas [*J. Appl. Phys (USA)* vol.74 (1993) p.4430-7]

[15]   S.P.S. Arya, A. D'Amico [*Thin Solid Films (Switzerland)* vol.157 (1988) p.267-82]

[16]   O. Mishima, J. Tanaka, S. Yamaoka, O. Fukunaga [*Science (USA)* vol.238 (1987) p.181-3]

[17]   O. Mishima, K. Era, J. Tanaka, S. Yamaoka [*Appl. Phys. Lett. (USA)* vol.53 (1988) p.962-4]

[18]   G.L. Doll, J.A. Sell, C.A. Taylor II, R. Clarke [*Phys. Rev. B (USA)* vol.43 (1991) p.6816-9]

[19]   S.L. Ren, A.M. Rao, P.C. Eklund, G.L. Doll [*Appl. Phys. Lett. (USA)* vol.62 (1993) p.1760-2]

[20]   T. A. Friedmann et al [*Appl. Phys. Lett. (USA)* vol.61 (1992) p.2406-8]

9.3 Polytypism and stacking faults in the group III nitrides

S.C. Strite

January 1994

A INTRODUCTION

The group III nitrides, like many wide bandgap semiconductors (most notably carbon and SiC), have been observed to crystallize in more than one structure. These different phases, which have identical chemical formulas, are referred to as polytypes. Zinc blende and hexagonal BN are similar to carbon so far as the cubic polytype is diamond-like with tetrahedral sp^3 bonds while the hexagonal phase has graphitic sp^2 bonding. Conversely, the cubic and hexagonal GaN, AlN and InN polytypes all have similar tetrahedral bonding structures and comparable bond lengths, being distinguished by the stacking sequence of the group III and N bilayers. The group III nitrides are most commonly observed to have hexagonal crystal structures, although each has also been stabilized in the zinc blende structure, generally through the use of a cubic substrate. Below, we briefly review the competing crystal structures which are important for the understanding of nitride polytypism, and then we move on to the experimental data.

B WURTZITE AND ZINC BLENDE CRYSTAL STRUCTURES

The difference between the wurtzite and zinc blende crystal structures, shown in FIGURE 1, lies in the stacking sequence of bilayers of the group III and N atoms. The wurtzite structure has an ABABAB... sequence along its [0001] direction which gives rise to the commonly used shorthand of 2H for this polytype. The number 2 denotes the periodicity of the stacking while H refers to the overall hexagonal symmetry of the crystal. Similarly, the zinc blende structure with its ABCABC... stacking sequence along the [111] direction is often referred to as the 3C phase, due to its three plane periodicity and overall cubic symmetry. A common defect in these types of crystal is the stacking fault in which the ideal stacking sequence is broken. In both structures, the atoms are tetrahedrally coordinated with almost identical nearest neighbour spacings. In the 3C structure, one atom's bonds are rotated 60° with respect to its nearest neighbours, while in the 2H structure there is no rotation between adjacent atoms (FIGURE 2).

C X-RAY AND TEM INVESTIGATIONS OF POLYTYPISM IN InN AND GaN

The issue of polytypism in the nitrides harkens back in the literature to the first mention of cubic GaN by Seifert and Tempel [1], who observed it as a minority phase in a bulk wurtzite crystal. Since then, other workers, notably Powell et al [2], have observed small polytype domains in bulk thin films. An extensive understanding of the formation mechanism of polytype domains required the epitaxial growth of 3C nitride thin films. Due to the higher total bulk energies of the zinc blende GaN and InN polytypes, 3C films generally have much higher polytype domain densities than 2H layers, providing an excellent medium for their study by XRD and TEM.

Wurtzite GaN

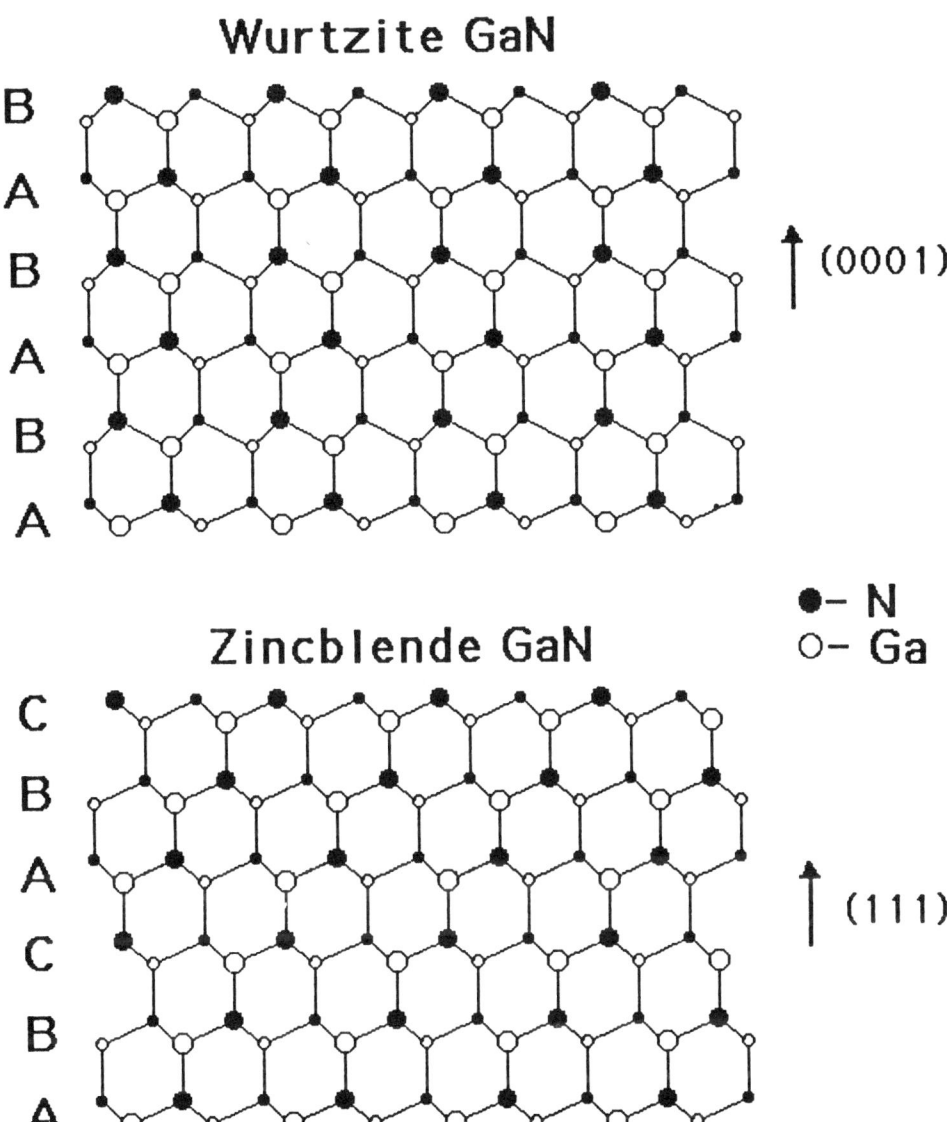

FIGURE 1 Crystal structures of the two GaN polytypes. Wurtzite or 2H (top) has the hcp unit cell with an ABABAB... stacking sequence normal to its (0001) face. Zinc blende or 3C (bottom) has an fcc unit cell with an ABCABC... stacking sequence normal to its (111) face.

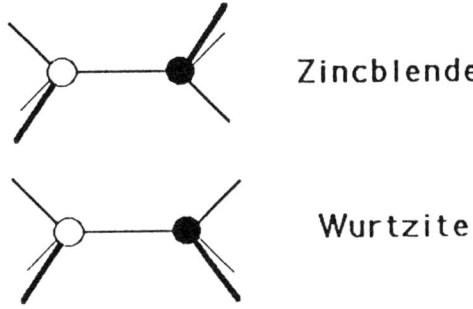

FIGURE 2 The local environment of atoms in the two crystal structures differs by a 60° rotation of adjacent tetrahedral bonds in the zinc blende case.

Lei et al [3] first elucidated the nature of the stacking fault defect in the group III nitrides by using XRD directly to observe the presence of 3C domains in 2H GaN grown on (111) Si as well as 2H domains in 3C GaN grown on (001) Si. The polytype domains were oriented with the [111] zinc blende direction parallel to the [0001] wurtzite direction. That is, both polytypes had identical stacking directions, and therefore the interface between the polytype domains was deduced to be a stacking fault in the (111) zinc blende plane, precisely the defect that was observed in such abundance in TEM measurements of 3C GaN [2,4,5].

This work demonstrated that there were a significant number of polytype domains present in GaN samples of both polytypes, and reminded the research community that to detect polytype domains by XRD successfully, scans in more than one orientation are required. FIGURE 3 shows a typical XRD theta scan of an epitaxial InN thin film grown on (001) GaAs [6]. The (002) and (004) zinc blende InN peaks are readily apparent and one might be led to conclude from the lack of other diffraction peaks that the film is single phase epitaxial 3C InN. However, if one looks for a 2H component stacked parallel to the 3C InN, a phi scan around the (110) wurtzite axis peak reveals the four <110> peaks characteristic of 2H InN (FIGURE 4).

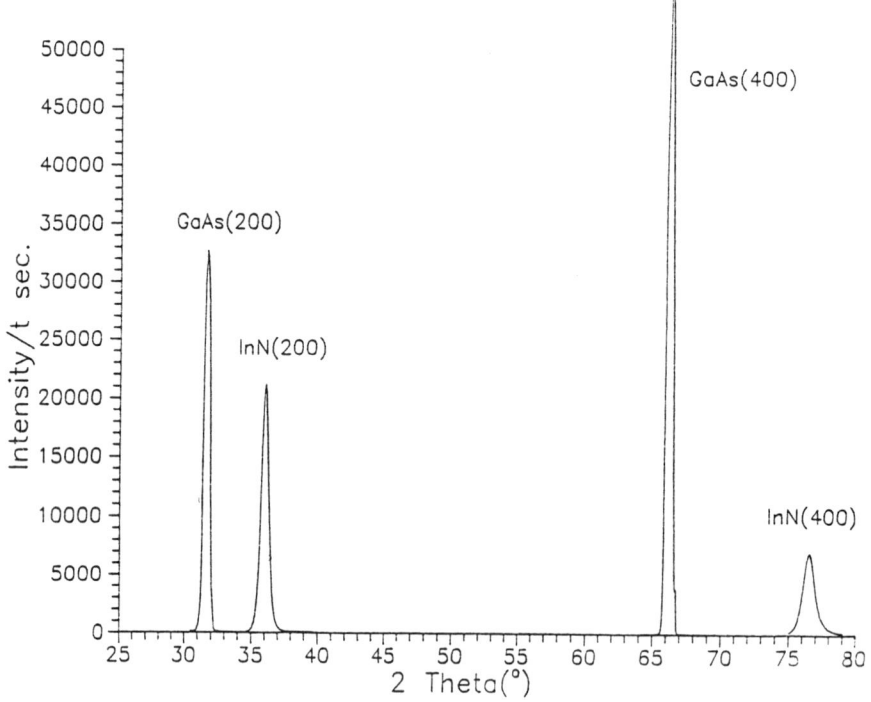

FIGURE 3 Theta - 2 x theta X-ray scan of an InN/(001) GaAs film showing the (200) and (400) peaks of the epitaxial zinc blende InN material and the GaAs substrate. From [6].

High resolution TEM performed by Smith and Chandrasekhar [6] has provided a direct image of a polytype domain boundary. FIGURE 5 is a micrograph of the InN film from FIGURE 4. The InN in the upper left is 2H and the InN in the lower right is 3C. The wurtzite phase has its [0001] stacking direction parallel to the zinc blende [111] stacking direction. The interface between the two polytype domains is a stacking fault, illustrated schematically in Figure 2 of Datareview 9.2.

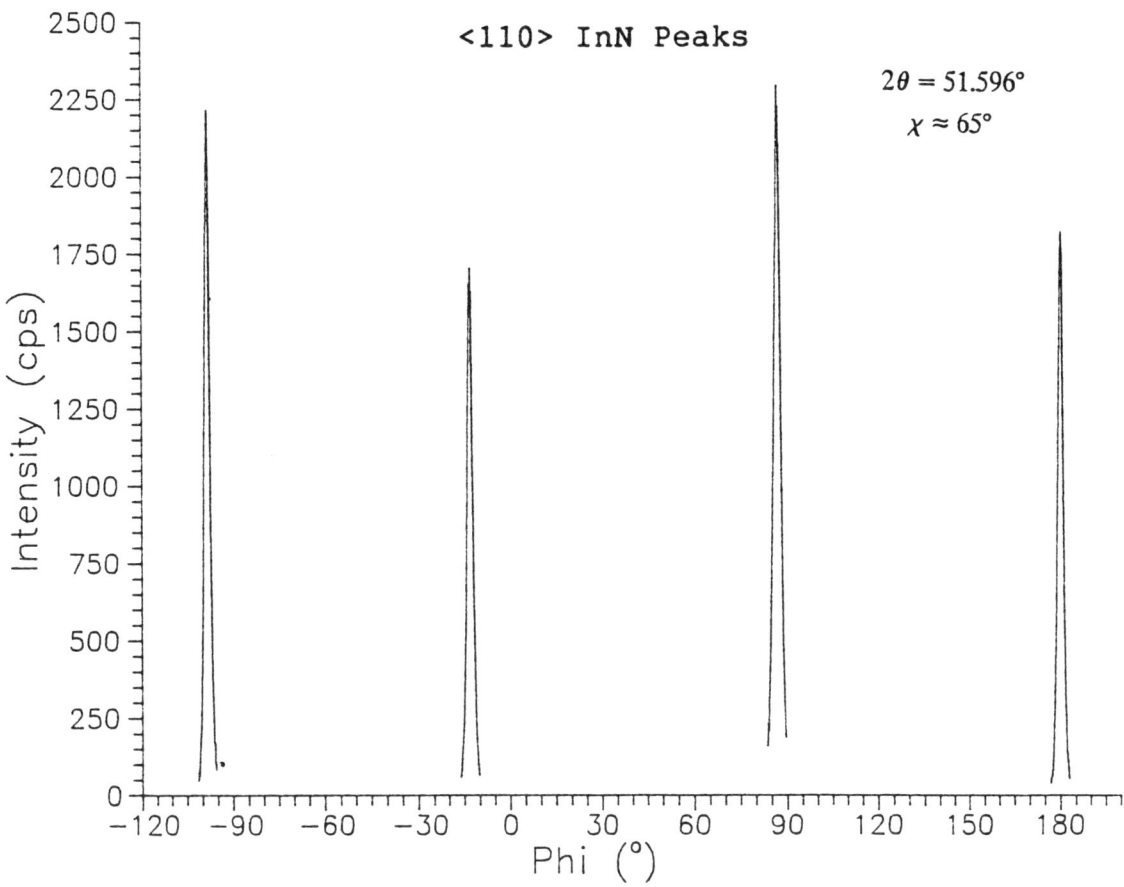

FIGURE 4 X-ray scan around the (110) wurtzite axis reveals the four <110> peaks at roughly 90° intervals confirming the presence of 2H InN stacked normal to the (111) stacking fault planes and parallel to the 3C InN stacking direction. From [6].

D   RELATIONSHIP OF STACKING FAULTS TO POLYTYPE DOMAINS

To understand the mechanisms governing polytype domain formation in heteroepitaxial GaN and InN thin films, one must consider the relative formation energies of polytype domains. It is generally accepted that the 2H phase is the energetically preferred polytype while the 3C phase is metastable. This conclusion follows the experimental observation that epitaxial GaN and InN grown on polytype neutral (111) zinc blende or basal plane hexagonal substrates, as well as polycrystalline material, is nearly always wurtzite. In order to stabilize the zinc blende phase, it is generally necessary to use (001) cubic substrates which provide a template for epitaxial 3C growth. Due to the large number of dislocations always present in nitride heteroepitaxy, there are numerous nucleation sites for stacking faults. The dislocation/stacking fault mechanism not only lowers the total crystal energy by relieving strain but it allows the formation of 2H domains which have a lower bulk energy than the metastable 3C polytype. Davis et al [7] have observed by TEM that 3C AlN grown on 3C SiC faults heavily into the 2H phase. InN also appears to strongly prefer the 2H phase. In GaN, the relative energies of the two polytypes appear to be closer in energy, as evidenced by one report of epitaxial 3C GaN grown on basal plane sapphire [8]. By changing their initial growth parameters somewhat, Lei and Moustakas [9] were able to reduce the 3C

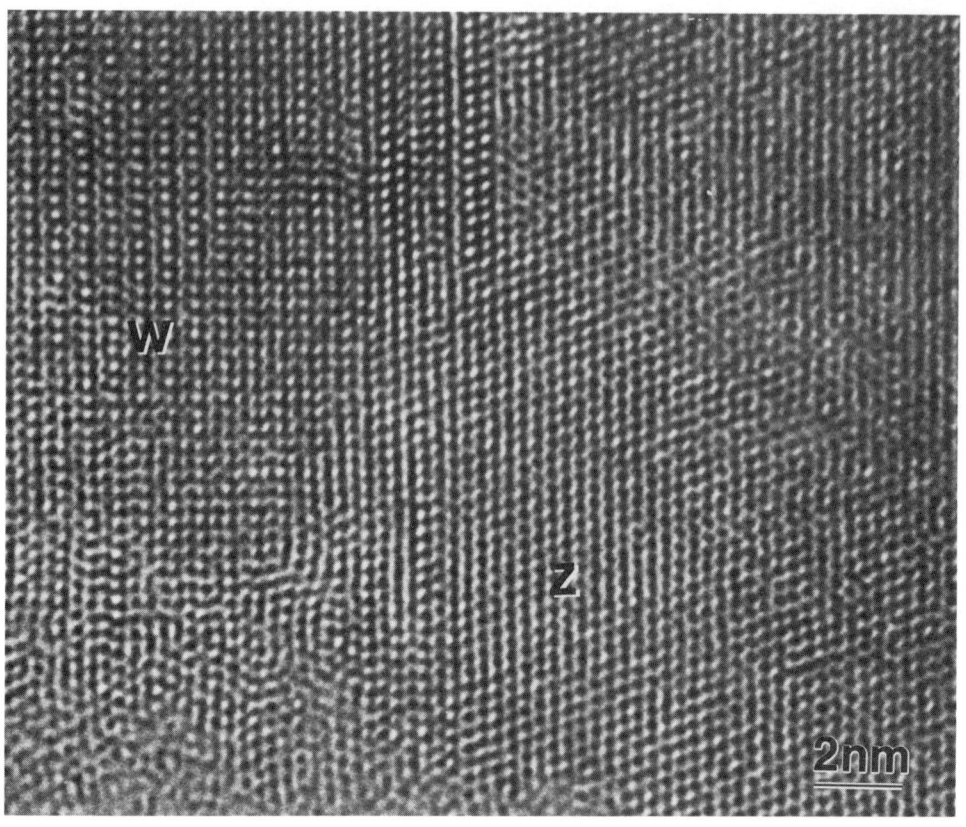

FIGURE 5 High resolution TEM micrograph of the InN film from FIGURES 3 and 4 showing the (111) stacking fault dividing the wurtzite (upper left) and zinc blende (lower right) domains. From [6].

content in their 2H GaN films grown on (111) Si to below the XRD detection limit. Thus there is some evidence that single domain GaN thin films can be achieved on lattice mismatched substrates.

The polytypism of the nitrides also introduces potential opportunities. If the substrate orientation is chosen so that the growth direction is parallel to the stacking direction, a condition satisfied by both (111) zinc blende and (0001) hexagonal substrates, the polytype domain boundary can be coherent as illustrated in FIGURE 6. The 2H and 3C polytypes of GaN have room temperature bandgaps of 3.4 eV and 3.2 eV, respectively, and the bandgap difference between the AlN polytypes is predicted [10] to be 1.1 eV. These types of band offsets may be exploited in future devices based on polytype (homo-hetero)junctions similar to those already demonstrated in the SiC polytype system.

B
A
B
A
C
B
A
C
B
A

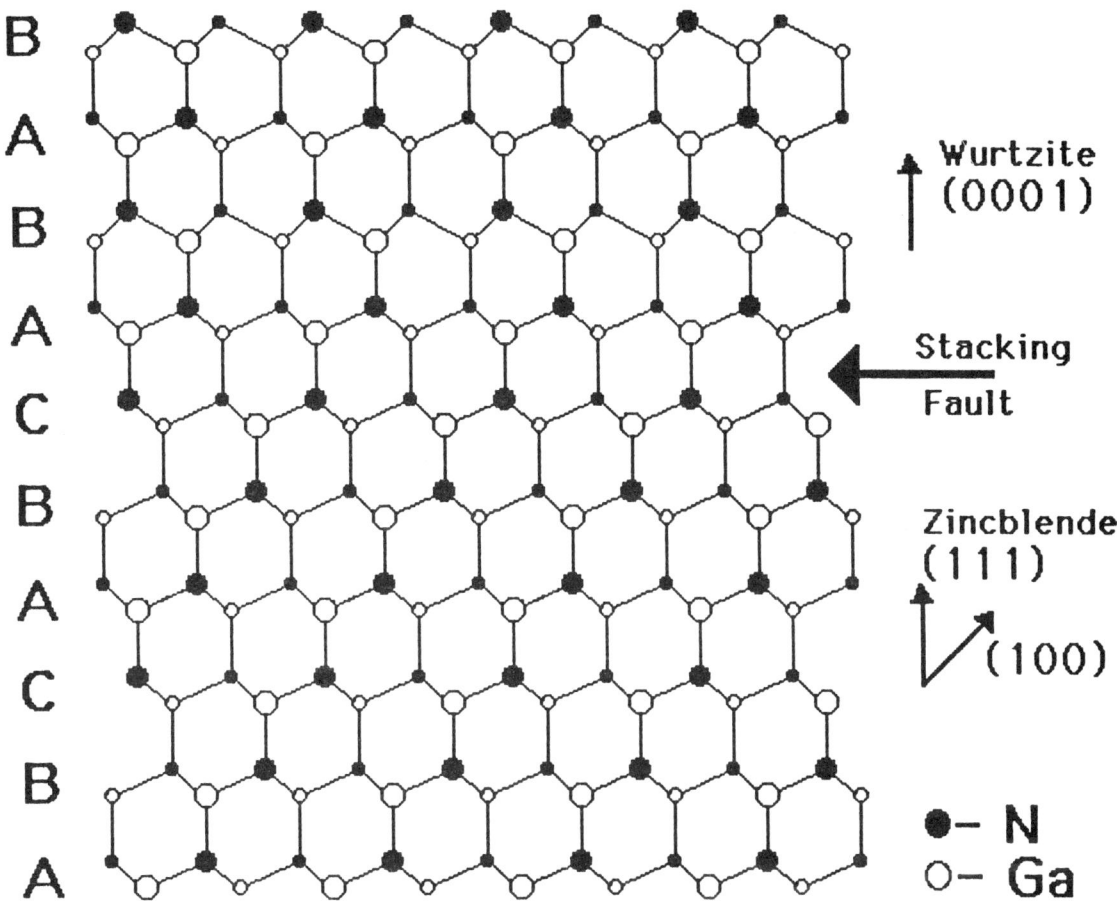

Wurtzite
(0001)

Stacking
Fault

Zincblende
(111)
(100)

●– N
○– Ga

FIGURE 6 Coherent GaN polytype heterojunction formed by the presence of a stacking fault at the interface of (111) 3C and (0001) 2H GaN.

REFERENCES

[1]   W. Seifert, A. Tempel [*Phys. Status Solidi A (Germany)* vol.23 (1974) p.K39-K40]

[2]   R.C. Powell, G.A. Tomasch, Y.-W. Kim, J.A. Thornton, J.E. Greene [*Mater. Res. Soc. Symp. Proc. (USA)* vol.162 (1990) p.525-30]

[3]   T. Lei, K.F. Ludwig Jr., T.D. Moustakas [*J. Appl. Phys (USA)* vol.74 (1993) p.4430-7]

[4]   S. Strite et al [*J. Vac. Sci. Technol. B (USA)* vol.9 (1991) p.1924-9]

[5]   M.J. Paisley, Z. Sitar, J.B. Posthill, R.F. Davis [*J. Vac. Sci. Technol. A (USA)* vol.7 (1989) p.701-5]

[6]   S. Strite et al [*J. Cryst. Growth (Netherlands)* vol.127 (1993) p.204-8]

[7]   R.F. Davis [Workshop on the Widegap Nitrides, St. Louis, USA, April 1992]

[8]   T.P. Humphreys et al [*Mater. Res. Soc. Symp. Proc. (USA)* vol.162 (1990) p.531-4]

[9]   T. Lei, T.D. Moustakas [private communication]

[10]   W.R.L. Lambrecht, B. Segall [*Phys. Rev. B (USA)* vol.43 (1991) p.7070-85]

9.4 Native defects in the group III nitrides

S.C. Strite

January 1994

A   INTRODUCTION

The study of native defects in GaN, AlN and InN has been an active one as a result of their profound influence on the electronic and optical properties of these materials. Native defects are believed to be responsible for many of the optical transitions reported in the nitrides as well as the background n-type conductivity in GaN and InN. In AlN, their presence as deep levels could be responsible for the inability to obtain conductive material. Tansley and Egan [1] recently correlated the existing experimental data on GaN, AlN and InN with the theoretical predictions of Jenkins and Dow [2] to produce a reasonably comprehensive view of the point defect energies from which we will borrow.

B   NATIVE DEFECTS IN GaN

Historically, one of the major challenges facing nitride researchers was bringing the carrier concentration of GaN under control. Until the late 1980s, nearly all of the GaN and InN material reported had n-type carrier concentrations of the order of $10^{19}\,\mathrm{cm^{-3}}$. No impurity has been shown to be present in sufficient quantity to justify the observed carrier concentrations, so native defect level(s) of an unknown nature, widely speculated to be N vacancies, generally were cited as the cause. Recently, modern crystal growth techniques have brought the background electron concentration down to the mid $10^{16}\,\mathrm{cm^{-3}}$ level in many laboratories, but all unintentionally doped GaN grown today remains n-type.

Theoretical studies support the notion that N vacancies act as shallow donors in GaN. Jenkins et al [2,3] calculated the energies of the N and Ga vacancies and antisite defects in GaN. Nitrogen vacancies were found to lie roughly 40 meV below the conduction band, while Ga vacancies formed shallow acceptors. Both types of antisite defect were predicted to lie deep within the forbidden band.

Several groups [4-7], using both optical and electrical techniques, have reported the existence of a shallow donor level having energies between 10 and 40 meV. Electron spin resonance measurements concluded that the n-type conductivity arises from a band of delocalized donors [8]. Optically detected magnetic resonance [9] was used to study the deep 2.2 eV photoluminescence (PL) band present in even the best material grown today. A sharp signal characteristic of shallow donors was clearly detected indicating that recombining conduction electrons contribute to the 2.2 eV PL band. A second weaker signal was also observed which was attributed to N vacancies. The coexistence of both shallow donors and deep levels is consistent with the triplet of N vacancy states predicted by Jenkins and Dow [2].

Given the experimental and theoretical evidence described above, it seems probable that N vacancies, as has long been suspected, are responsible for the electron background in GaN. Less experimental data exists concerning the deeper antisite levels and the Ga vacancy. Based on the available experimental and theoretical data, the assignments of Tansley and Egan [1], shown in FIGURE 1, seem reasonable.

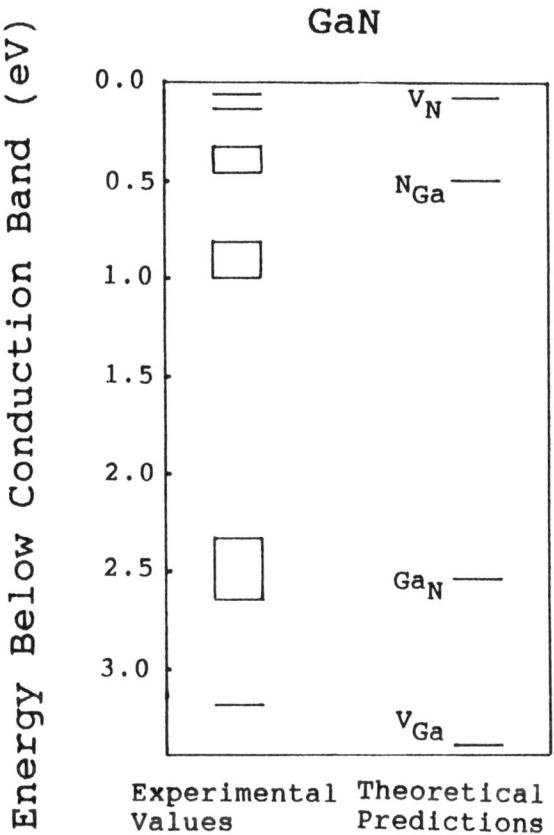

FIGURE 1 Comparison of experimental data with theoretical predictions for point defects in wurtzite GaN (after [1]).

C   NATIVE DEFECTS IN InN

InN, like GaN, is commonly characterized by a large n-type background concentration. Typical electron concentrations in InN are in the 10^{19}-10^{20} cm^{-3} range, higher than generally observed in GaN. Because InN is quite difficult to grow due to its thermal instability, research has not yet progressed to the point where the relatively higher concentrations can be fully attributed to the material properties or the crystal growth. InN has attracted less experimental attention than the other three nitride semiconductors, partly because its bandgap lies in the orange optical wavelengths, a region at which alternative semiconductor technology is already available. As with the case for GaN, it has long been assumed that N vacancies are responsible for the large background electron concentration in unintentionally doped samples.

Jenkins and Dow [2] have calculated the point defect energies in InN, obtaining results similar to the GaN case. N and In vacancies are predicted to be shallow donors and acceptors respectively, while the antisite defects were found to be deep levels.

Due to the high carrier concentrations, which generally obscure the band edge, it has been difficult to directly observe the N vacancy level which is predicted to be nearly resonant with the conduction band. Tansley et al [10] have observed donor levels at roughly 40 meV and 200 meV. No carrier freeze out was observed down to 25 K, so a more shallow donor level, probably the N vacancy, was inferred. Several deeper levels were also detected in that study, although their nature is not fully understood.

D   NATIVE DEFECTS IN AlN

In stark contrast to GaN and InN, unintentionally doped AlN thin films are highly resistive, generally in the $10^{11} - 10^{13}$ ohm-cm range, indicating that any native defects present are sufficiently deep that they are not ionized at room temperature. Experimental investigations [11-13] have observed states lying roughly 170, 500, and 800 - 1000 meV below the conduction band edge which are reasonably well matched to the expected donor triplet of the N vacancy in AlN. Other workers [13,14] have detected deeper levels lying between 1.4 and 1.85 eV, speculated to be due to the N antisite defect.

Jenkins and Dow [2] have also calculated the point defect energies of AlN. They predicted the N vacancy to lie roughly 1 eV below the conduction band forming a deep electron trap. The Al vacancy was calculated to be slightly more than 1 eV above the valence band forming a deep hole trap. Both types of antisite defect were also predicted to form deep traps in the middle of the forbidden band. These findings are consistent with the insulating nature of AlN.

E   NATIVE DEFECTS IN GROUP III NITRIDE ALLOYS

Alloy properties will become increasingly important as nitride thin films continue to approach device quality. Optimized designs rely on heterojunctions to upgrade the performance of semiconductor devices. The insulating nature of AlN may eventually limit the range of applications of the nitride semiconductors. AlGaN, InGaN, AlInN and even AlN-SiC solid solutions are all thought to be promising alloy systems for device applications. The AlGaN alloy system is the most thoroughly studied to date. Yoshida et al [15] reported on the entire range of Al content. Unintentionally doped AlGaN films were observed to become insulating at Al mole fractions above 30% probably reflecting the movement of the N vacancy into the forbidden gap with increasing mole fraction. A theoretical study [3] of the N vacancy in AlGaN confirms that the level moves deeper into the forbidden band as Al mole fraction is increased. AlN-SiC solid solutions have also been observed to conduct only at low AlN mole fraction. The presence of deep trap levels in Al-containing alloys will make it extremely difficult to dope these materials. Applications such as lasers, in which low resistivity is a key operational parameter, will be limited in their use of wide bandgap AlN alloys until deep level densities can be reduced and suitable dopants are found.

REFERENCES

[1]   T.L. Tansley, R.J. Egan [*Phys. Rev. B (USA)* vol.45 (1992) p.10942-50]

[2]   D.W. Jenkins, J.D. Dow [*Phys. Rev. B (USA)* vol.39 (1989) p.3317-29]

[3]   D.W. Jenkins, J.D. Dow, M.-H. Tsai [*J. Appl. Phys. (USA)* vol.72 (1992) p.4130-3]

[4]   R.J. Molnar, T. Lei, T.D. Moustakas [*Appl. Phys. Lett. (USA)* vol.62 (1993) p.72-4]

[5]   R. Dingle, M. Ilegems [*Solid State Commun. (USA)* vol.9 (1970) p.175-9]

[6]   J.I. Pankove, S. Bloom, G. Harbeke [*RCA Rev. (USA)* vol.36 (1975) p.163-71]

[7]   O. Lagerstedt, B. Monemar [*J. Appl. Phys. (USA)* vol.45 (1974) p.2266-72]

[8]   M.A. Khan, D.T. Olson, J.N. Kuznia, W.E. Carlos, J.A. Freitas [*J. Appl. Phys. (USA)* vol.74 (1993) p.5901-3]

[9]   E.R. Glaser et al [*Appl. Phys. Lett. (USA)* vol.63 (1993) p.2673-5]

[10]   T.L. Tansley, C.P. Foley [*J. Appl. Phys. (USA)* vol.60 (1986) p.2092-5]; T.L. Tansley, R.J. Egan, E.C. Horrigan [*Thin Solid Films (Switzerland)* vol.164 (1988) p.441-8]

[11]   J. Edwards, K. Kawabe, G. Stevens, R.J. Tredgold [*Solid State Commun. (USA)* vol.3 (1965) p.99-100]

[12]   M. Morita, K. Tsubouchi, N. Mikoshiba [*Jpn. J. Appl. Phys. (Japan)* vol.21 (1982) p.728-30]

[13]   G.A. Cox, D.O. Cummings, K. Kawabe, R.H. Tredgold [*J. Phys. Chem. Solids (UK)* vol.28 (1968) p.543-8]

[14]   R.W. Francis, W.L. Worrell [*J. Electrochem. Soc. (USA)* vol. 123 (1976) p.430-3]

[15]   S. Yoshida, S. Misawa, S. Gonda [*J. Appl. Phys. (USA)* vol.53 (1982) p.6844-8]

9.5   Impurities and dopants in the group III nitrides

S.C. Strite

January 1994

A   INTRODUCTION

As the crystal quality of nitride samples continues to improve, the search for well-behaved dopants has moved to the forefront.   In the past few years, the background carrier concentrations of GaN films have been reduced to the point that donor impurities have become necessary for the first time [1].   In that same time period, a reliable procedure for p-type doping of GaN with Mg has been discovered.   The literature is extensive and well documented regarding the many impurities which have been introduced into the nitride semiconductors [2].   In this Datareview, we choose to focus primarily on shallow impurities which are potentially useful as dopants in GaN-based device structures.

B   IMPURITY LEVELS AND DOPANTS IN GaN

Most of the work cataloguing the energy levels of impurities in GaN was undertaken in an unsuccessful effort to discover a suitable p-type dopant.   Early efforts at p-type doping normally resulted in highly resistive compensated material.   Pankove and Hutchby [3], in the most comprehensive early study of impurities in GaN, implanted 35 elements into GaN. FIGURE 1 summarizes the estimated binding energies of most of the obvious potential acceptor impurities as measured by various workers [2].

Akasaki et al [4] reported the first successful technique for p-type doping of GaN.   A Mg concentration of roughly $10^{20}\,cm^{-3}$ was incorporated into GaN resulting in compensated material.   When the sample was treated with low energy electron beam irradiation (LEEBI), the resistivity was drastically reduced and the sample exhibited p-type conductivity. Nakamura et al [5] have improved upon these initial results and have reported in detail on the Mg doping mechanism in GaN.   As grown Mg-doped layers had a hole concentration of $p = 2 \times 10^{15}\,cm^{-3}$ and a mobility $\mu_p = 9\,cm^2\,V^{-1}\,s^{-1}$.   LEEBI treatment improved the electrical characteristics to $p = 3 \times 10^{18}\,cm^{-3}$ with no degradation of the mobility and overall hole concentrations as large as $p = 8 \times 10^{18}\,cm^{-3}$ were achieved.   Further investigation showed that by thermally annealing under an N_2 ambient, a similar transformation could be induced. Annealing also permitted the nature of the conversion to be studied.   The effects of N_2 annealing were shown to be reversible when the sample was annealed under NH_3 which identified a Mg-H neutral complex as the GaN compensating agent.

Recent theoretical insights have provided a viable explanation for the success of the Mg acceptor in GaN compared to the other group II impurities.   Due to the strong binding of the N anion, the group III nitrides are more ionic than typical III-V semiconductors.   Their band structures resemble those of II-VI semiconductors in many ways, including the presence of a large ionicity gap in the valence band density of states which pushes the lower valence bands (LVB) deeper beneath the valence band edge.   In GaN, the shallow Ga 3d electron core level energies are predicted to overlap with the N 2s-like LVB states as a result of them lying

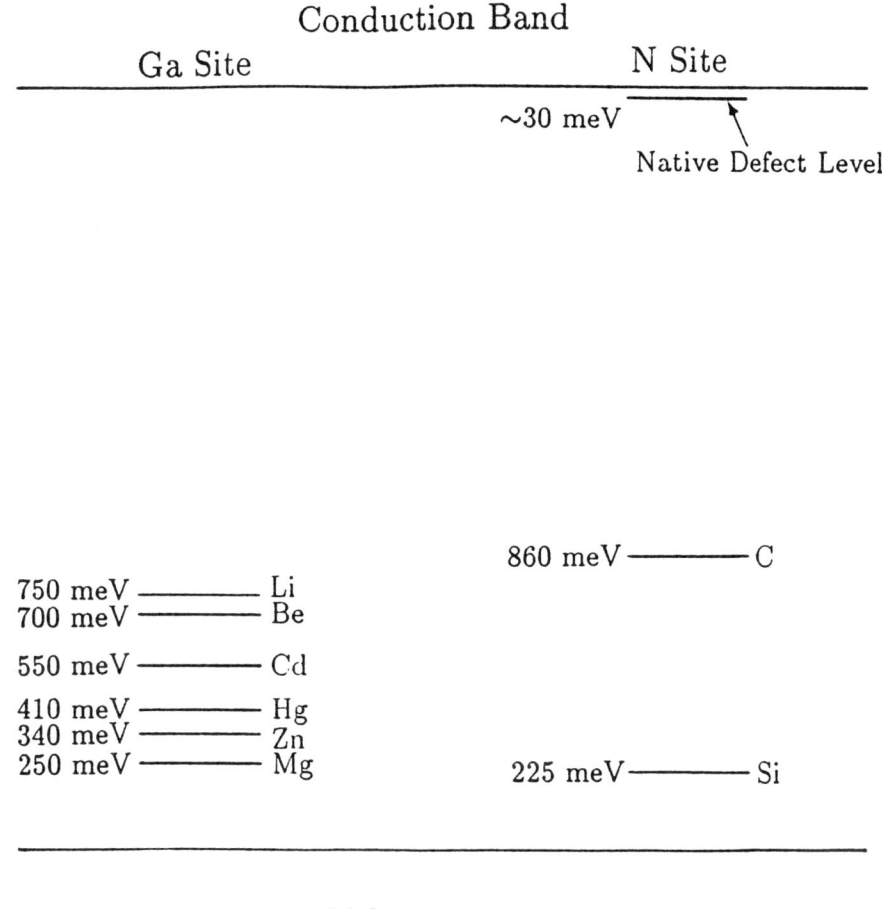

FIGURE 1 Estimated binding energies of some substitutional acceptors in GaN (after [2]).

deeper beneath the valence band edge [6]. The resulting energy resonance causes the Ga 3d electrons to strongly hybridize with the valence band s- and p-levels having a profound influence on the properties of GaN, including such quantities as the bandgap, lattice constant, acceptor levels and valence band heterojunction offsets [7]. It is known in the cases of ZnS and ZnSe, that potential acceptors such as Cu, whose d-electrons are resonant with the LVB, are repelled by the d-hybridized upper valence band resulting in a deep level, while impurities without d-electron resonances form shallow acceptors [7]. Mg is the final group II element which does not have d-electrons and, possibly as a result, is sufficiently shallow in GaN to be ionized at room temperature. On the other hand, Zn, Cd and Hg, which have d-electrons, form deep levels in GaN. That is, because of the greater ionicity of the group III nitrides, the LVB states lie near in energy to the shallow d electron states, which considerably complicates the energy levels of potential acceptors which have shallow d electrons.

C   IMPURITIES IN AlN AND InN

Historically, the most common impurity in AlN has been unintentional oxygen contamination. Early measurements on oxygen-contaminated material led to a number of the physical properties of AlN, such as the bandgap and the lattice constant, being mis-reported. However, once grown, AlN is relatively impervious to further atmospheric oxygen contamination in contrast to other Al-bearing semiconductors.

Little is known about the spectrum of impurity levels in AlN and InN. Researchers are generally interested in the properties of AlN as a buffer layer for GaN growth or as an extremely hard, thermally conductive material for coating and packaging applications. The electrical properties are dominated by deep level compensation as discussed in Datareview 9.4, and no effort at doping has yet produced high quality conductive material. In the case of InN, the background electron concentration has not been brought under control and impurity studies would be premature.

D   IMPURITIES IN BN

Unlike its smaller bandgap counterparts GaN and AlN, BN has proved relatively easy to dope, although ohmic contacts remain a problem. Wentorf [8] first reported n- and p-BN. Recently, Mishima and co-workers [9,10] used Si and Be, as n- and p-type dopants respectively, to form a p-n junction light emitting diode (LED) from zinc blende BN. The same group [11] reported, from electrical measurements, that the Be acceptor level had an activation energy of 0.3 eV. Detailed studies of the impurity levels in BN probably will not become available until better samples can be grown.

E   IMPURITIES IN ALLOYS

As discussed in Datareview 9.4, many of the device applications envisioned for the group III nitrides will require AlGaN/GaN heterojunctions in which the AlGaN is doped. It is reasonable to expect, since AlN has not yet been successfully doped, that AlGaN and AlInN alloys may also prove resistant to doping at larger AlN mole fractions.

Jenkins and Dow [12] have considered the problem of impurity levels theoretically in both InAlN and InGaN. Their calculations predict that many impurity levels which are deep levels in InN move up into the conduction band or down into the valence band as the Al or Ga mole fraction is increased. Therefore, at specific alloy concentrations, it would then be possible to dope these alloys both n- and p-type. There has not yet been an experimental verification of this intriguing prediction.

Nakamura and co-workers have studied Si- [13] and Cd- [14] doped InGaN. Si doping did not shift the main band edge emission peak, but did increase its intensity by 20 - 36 times. In contrast, Cd impurities reduced the InGaN PL peak a constant 0.5 eV at all InN mole fractions up to 20 %, while yielding intensity enhancements comparable to Si doping. Reducing the InGaN emission wavelength is important for increasing the apparent brightness of blue LEDs because the human eye is more sensitive at longer wavelengths. No electrical measurements were reported for Si- or Cd-doped InGaN.

REFERENCES

[1]   S. Nakamura, T. Mukai, M. Senoh [*Jpn. J. Appl. Phys. (Japan)* vol.31 (1992) p.195-200]

[2]   S.Strite, H. Morkoc [*J. Vac. Sci. Technol. B (USA)* vol.10 (1992) p.1237-66]; S.Strite, M.E. Lin, H. Morkoc [*Thin Solid Films (Switzerland)* vol.231 (1993) p.197-210]

[3]   J.I. Pankove, J.A. Hutchby [*J. Appl. Phys (USA)* vol.47 (1976) p.5387-90]

[4]   I. Akasaki, H. Amano, M. Kito, K. Hiramatsu [*J. Lumin. (Netherlands)* vol.48/49 (1991) p.666-70]; H. Amano, M. Kito, K. Hiramatsu, I. Akasaki [*Jpn. J. Appl. Phys. (Japan)* vol.28 (1989) p.L2112-4]

[5]   S. Nakamura, N. Iwasa, M. Senoh, T. Mukai [*Jpn. J. Appl. Phys. (Japan)* vol.31 (1992) p.1258-66]; S. Nakamura, T. Mukai, M. Senoh, N. Iwasa [*Jpn. J. Appl. Phys. (Japan)* vol.31 (1992) p.L139-42]; S. Nakamura, M. Senoh, T. Mukai [*Jpn. J. Appl. Phys. (Japan)* vol.30 (1991) p.L1708-11]

[6]   V. Fiorentini, M. Methfessel, M. Scheffler [*Phys. Rev. B (USA)* vol.47 (1993) p.13353-62]

[7]   S.H. Wei, A. Zunger [*Phys. Rev. B (USA)* vol.37 (1988) p.8958-81]

[8]   R.H. Wentorf [*J. Chem. Phys. (USA)* vol.36 (1962) p.1990-6]

[9]   O. Mishima, J. Tanaka, S.Yamaoka, O. Fukunaga [*Science (USA)* vol.238 (1987) p.181-3]

[10]   O. Mishima, K. Era, J. Tanaka, S.Yamaoka [*Appl. Phys. Lett. (USA)* vol.53 (1988) p.962-4]

[11]   T. Taniguchi, J. Tanaka, O. Mishima, T. Ohsawa, S.Yamaoka [*Appl. Phys. Lett. (USA)* vol.62 (1993) p.576-8]

[12]   D.W. Jenkins, J.D. Dow [*Phys. Rev. B (USA)* vol.39 (1989) p.3317-29]

[13]   S. Nakamura, T. Mukai, M. Senoh [*Jpn. J. Appl. Phys. (Japan)* vol.32 (1993) p.L16-9]

[14]   S. Nakamura, N. Iwasa, S. Nagahama [*Jpn. J. Appl. Phys. (Japan)* vol.32 (1993) p.L338-41]

CHAPTER 10

MATERIAL INTERFACES WITH GROUP III NITRIDES

10.1 Material interfaces with BN

D.R. Gilbert and R.K. Singh

February 1994

A   INTRODUCTION

Cubic boron nitride (c-BN) has the largest bandgap energy of all the III-V compound semiconductors, making it a leading candidate for high temperature electronic operation. Its physical properties, such as hardness and thermal conductivity, are comparable to those of diamond. Unlike diamond, c-BN is believed to be the thermodynamically stable phase of boron nitride at room temperature and pressure [1]. However, c-BN also requires special fabrication techniques for its production. To make practical use of c-BN's unique properties, it is necessary to form electrically active interfaces with other materials.

B   OHMIC CONTACTS

Formation of ohmic contacts with metal conductors is an essential operation in the production of semiconductor devices. The exceptional properties of c-BN make this process very difficult. To date, very little work has been published regarding the formation of ohmic contacts to c-BN. Yoshida and Tsuji [2] currently hold a patent for the formation of ohmic contacts to n-type c-BN. Their process involves the use of an intermediate metal(Au)-semiconductor(Si, Ge) alloy film between the metal contact and the c-BN semiconductor.

C   SCHOTTKY CONTACTS

Due to c-BN's wide bandgap, direct metal contacts to c-BN are Schottky-type contacts. However, information regarding the characterization of these contacts is not readily available. Non-ohmic contacts have been formed in work involving c-BN p-n junction diodes using silver painted electrodes [3]. Andreyev [4] has investigated the wetting behaviour of various metals on c-BN surfaces. A high pressure, high temperature experimental apparatus was used to provide contact between the metal melts and the polycrystalline c-BN surface. Although direct observation of the liquid behaviour was not possible, measurements made using the

TABLE 1   Andreyev experimental data on the wettability of c-BN by metal melts.

System	Wetting angle (°)	Adhesion energy W_{ad} (J m^{-2})	Interfacial energy γ_{sl} (J m^{-2})
Fe - c-BN	120	0.812	5.513
Co - c-BN	154	0.178	6.278
Ni - c-BN	135	0.465	5.823
Cu - c-BN	153	0.127	5.831
Ag - c-BN	142	0.180	5.367
Au - c-BN	145	0.189	5.558
Al - c-BN	78	0.943	4.538

equilibrium solidified droplets were subsequently corrected. TABLE 1 shows the results of these experiments for a variety of metals, including values for the adhesion energy and interfacial energy of these systems.

D   HETEROJUNCTIONS

The similarities between c-BN and diamond make the two materials uniquely suited for the formation of heterojunctions between the two. This system represents the most extensively published research involving c-BN interfaces, driven largely by the difficulties involved in obtaining heteroepitaxy for both materials. According to strain energy calculations by Braun et al [5] common low index planes ({100} and {111}) constitute the preferred epitaxial systems for diamond/c-BN structures with the diamond {100} ∥ c-BN {221} system predicted as the next most likely configuration. Experiment has shown that the matching diamond/c-BN {100} and {111} systems do occur [6,7]. Using Raman spectroscopy, Yoshikawa et al [7] have shown that diamond grown epitaxially on (100) c-BN experiences a tensile stress due to lattice mismatch. In forming bonds across the interface there is a charge transfer to compensate for the difference in valences of the two materials [8,9]. Bond strength at the interface is predicted to be only 10% less than bonding in the bulk materials [9]. A type-II band alignment is anticipated between the two semiconductors, with a valence band offset of - 1.4 eV and a conduction band offset of approximately - 0.5 eV for c-BN relative to diamond [8,9].

E   CONCLUSION

This Datareview has presented an overview of issues involved in the formation of interfaces with cubic boron nitride. The small number of references concerning this topic shows it to be an area open for further investigation. At present, c-BN is a material in need of much research and development if it is to fulfil its potential.

REFERENCES

[1]   V. Solozhenko, V. Leonidov [*Russ. J. Phys. Chem. (UK)* vol.62 (1988) p.1646-7]
[2]   K. Yoshida, K. Tsuji [Sumitomo Electric Industries Ltd, US Patent No.5187560 (February 16 1993)]
[3]   O. Mishima, K. Era, J. Tanaka, S. Yamaoka [*Appl. Phys. Lett. (USA)* vol.53 (1988) p.962-4]
[4]   A.V. Andreyev [*Proc. 1st Int. Conf. on the Applications of Diamond Films and Related Materials*, Auburn, Alabama, USA, 17-22 Aug 1991, Eds Y. Tzeng, M. Yoshikawa, M. Murakawa, A. Feldman (Elsevier Science Publishers, 1991) p.143-8]
[5]   M.W.H. Braun, H.S. Kong, J.T. Glass, R.F. Davis [*J. Appl. Phys. (USA)* vol.69 (1991) p.2679-81]
[6]   S. Koizumi, T. Murakami, T. Inuzuka, K. Suzuki [*Appl. Phys. Lett. (USA)* vol.57 (1990) p.563]
[7]   M. Yoshikawa, H. Ishida, A. Ishitani, S. Koizumi, T. Inuzuka [*Appl. Phys. Lett. (USA)* vol.58 (1991) p.1387-8]

[8]   W.E. Pickett [*Phys. Rev. B (USA)* vol.38 (1988) p.1316]

[9]   W.R.L. Lambrecht, B. Segall [*Phys. Rev. B (USA)* vol.40 (1989) p.9909-19]

10.2 Ohmic contacts to AlN, GaN and InN

L.L. Smith and R.F. Davis

March 1994

A   INTRODUCTION

The formation of ohmic contacts with semiconductor materials and devices is a fundamental component of solid state device architecture. As device size has diminished and the scale of integration has increased, the quality of these interfaces has become an increasingly important concern. In addition, the presence of parasitic resistances and capacitances, such as those existing at contact interfaces, becomes more detrimental at higher operating powers and higher oscillation frequencies. The development of adequate and reliable ohmic contacts to the compound semiconductors - particularly those with wider bandgaps - has met a number of challenges. The subject of ohmic contacts to p- and n-type III-V compounds - mostly GaAs, AlGaAs and InP - has received a great deal of attention over the past decade, and significant advances have been made [1-11]. By comparison, the group III nitrides have received little attention in this regard. However, interest in these materials has been renewed in recent years as thin film growth techniques have improved, p-type doping in GaN and AlGaN solid solutions has been achieved, and p-n junctions have been fabricated.

The majority of successful ohmic contact systems that have so far been implemented with the more conventional compound semiconductors have relied upon alloying (liquid-phase reaction) or sintering (solid-phase reaction) via post-deposition annealing treatments, and/or the presence of high carrier concentrations near the interface [1,2,6]. It has frequently been observed that the lowest resistivity contacts tend to form easily on semiconductors with high carrier concentrations, usually much higher than the levels needed for device operation. For this reason many semiconductor contact areas are implanted with an appropriate dopant species prior to metallization. In general, it has been difficult to achieve low-resistivity contacts directly on lightly-doped material, regardless of the choice of metal. In addition, many otherwise successful ohmic contact systems have only limited thermal stability and are subject to degradation under subsequent thermal processing steps. Such contact degradation usually takes the form of extensive interdiffusion, interfacial reaction, and interphase growth, accompanied by an increase in contact resistivity. It is reasonable to suppose that the cleanliness and preparation of the semiconductor surface prior to contact deposition plays a significant role in the behaviour of the interface, and there are indications in the recent literature that support this view [11,45]. Thorough oxide removal is especially important, though it may well prove to be a persistent challenge with Al-containing compounds in particular.

B   AlN

Most of the AlN single crystals and films grown to date have exhibited very high resistivities, and the impurity doping behaviour of this material is not yet well understood. This has so far limited the use of AlN in device structures, and there has been little study of the electronic properties of AlN/metal interfaces. However, there is a rapidly growing interest in the application of AlN as an IC packaging substrate, primarily due to its superior thermal properties. This effort has already resulted in a growing number of reports describing the formation, reactions and microstructures of AlN/metal interfaces [46-48]. Of particular interest for contact studies is the potential for forming metal nitrides at group III nitride interfaces, since many metal nitrides are semiconducting or conducting and exhibit good chemical and thermal stability. Carim and Loehman [48] noted the formation of continuous layers of TiN at the AlN interface with an Ag-Cu-Ti braze alloy, as well as a continuous $(Ti,Cu,Al)_6N$ phase; there was evidence that the TiN phase preferentially wet the AlN grain boundaries. Norton et al [46] found that coating the AlN surface with TiH_2 or ZrH_2 allowed an Ag-Cu-Zn braze alloy to flow readily on the coated AlN surface. Reaction products were formed at the AlN interfaces: Ti_2N and Ti_3Al in the presence of TiH_2, and Zr_3Al and ZrN from ZrH_2. Asai et al investigated a different metallization approach for AlN, involving a paste containing TiN and Mo powders [47].

Metallization studies have revealed important factors in the interfacial reaction behaviour of AlN with some metals; the electronic behaviour of AlN/metal interfaces has not yet been extensively characterized, but some references can be found among the earlier investigations of the group III nitrides. Edwards et al [12] studied the electrical behaviour of small AlN single crystals. They acquired current-voltage measurements by evaporating In, Ni and Au contacts onto the crystals, and found significant variations in the log j-log V characteristics to be dependent on electrode material and heat treatment. Gold electrodes, annealed at 160°C and slow-cooled, gave the lowest resistivity measurements. Rutz [13] reported the fabrication of an AlN-based blue and UV electroluminescent diode that used Al contacts for both electrodes.

C   GaN

In comparison to AlN and InN, GaN has received significantly more attention. Its electronic and optical properties have been characterized more thoroughly, and over the years a number of devices have been fabricated [14-23,50,52]. Most of the GaN films reported to date have been grown via chemical vapour deposition (CVD) or molecular beam epitaxy (MBE) and have exhibited high n-type carrier concentrations (10^{17} - 10^{18} cm^{-3}), generally attributed to nitrogen vacancies, and low mobilities (below $100\,cm^2\,V^{-1}\,s^{-1}$) even in the undoped state. At the time of writing, significant progress is beginning to be made in control over both p- and n-type doping and the reduction of background carrier concentrations. The reported ohmic contact metals for both p- and n-type GaN have been Al [14-16,19,21,22,27,49,51] or Au [17-19,21,22,51]. Khan and co-workers have fabricated several GaN-based devices, including a MESFET [20], a HEMT [50], and a Schottky barrier photodetector [52]. The ohmic source and drain contacts for the MESFET and the HEMT were formed from annealed Ti/Au layers. For the p-type Mg-doped GaN used in the photodetector, Cr/Au layers were deposited and annealed as the ohmic contact. Abernathy et al [28] have proposed the deposition of low-resistivity InN as an ohmic contact layer for GaN. The thermal treatment schedules and resulting electrical data for these GaN/metal contact systems are summarized in TABLE 1.

TABLE 1 Reported ohmic contact systems for GaN.

Contact	GaN type*	Carrier concentration (cm^{-3})	Thermal treatment	Contact resistivity ρ_c (Ω cm^2)	Ref
Al	n-u:GaN	$1 - 2 \times 10^{17}$	NR	NR	[16,19,44]
	n-u:GaN	NR	NR	NR	[13,14,49]
	n-u:GaN	1×10^{17}	500 °C/20 s	$> 10^{-2}$	[51]
	n-u:GaN	3×10^{18}	575 °C/10 min	$0.12 - 4.4 \times 10^{-3}$	[21,22]
	n-Si:GaN	5×10^{18}	NR	NR	[18]
	n$^+$-Si:GaN	NR	NR	NR	[27]
	i-Zn:GaN	NR	NR	NR	[27]
	p-Mg:GaN	2×10^{16}	NR	NR	[16]
Au	n-u:GaN	3×10^{18}	575 °C/10 min	$1.6 - 3.1 \times 10^{-3}$	[21,22]
	n-u:GaN	1×10^{17}	500 °C/20 s	$> 10^{-2}$	[51]
	p-Mg:GaN	4×10^{16}	NR	NR	[19]
	p-Mg:GaN	8×10^{18}	NR	NR	[18]
In	n-u:GaN	3×10^{18}	575 °C/10 min	8.1×10^{-3}	[21]
Ti/Au	n-u:GaN	1×10^{17}	250 °C/30 s	7.8×10^{-4}	[20]
	n-u:GaN	NR	250 °C/60 s	NR	[50]
Ti/Al	n-u:GaN	1×10^{17}	900 °C/30 s	8×10^{-6}	[51]
Cr/Au	p-Mg:GaN	(7×10^{17})†	NR	NR	[52]
InN (proposed)	-	-	NR	NR	[28]

NR = not reported.
* 'u:' designation signifies undoped or 'unintentionally doped' material.
† Intended doping level.

The majority of contact systems listed in TABLE 1 were reported in connection with simple device structures rather than as studies of contact behaviour as such. It should also be pointed out that the quality and growth parameters of the GaN material varied from case to case; more rigorous characterization of the relationships between contact performance and the Fermi level, defect structures, dopant behaviour etc. awaits further study.   The only reported systematic investigations of GaN ohmic contact materials and heat treatment have been the work of Foresi and Moustakas [21,22] and Lin et al [51]. In both cases transfer-length measurements (TLM) were performed to measure the specific contact resistivity of various metals on undoped n-GaN before and after heat treatments. Both of these studies employed GaN films grown via ECR-activated MBE, on R-plane (1120) and basal-plane (0001) sapphire substrates respectively; the GaN films used by Lin et al had lower carrier concentrations (1×10^{17} cm^{-3}, compared with 3×10^{18} cm^{-3}) and higher mobilities (100 cm^2 V^{-1} s^{-1}, compared with ~ 20 cm^2 V^{-1} s^{-1}). Aluminium and Au contacts were studied in both cases; Lin et al also characterized Ti/Au and Ti/Al contacts, in which 20 nm layers of Ti were deposited directly on GaN and followed by 100 nm of either Au or Al. After a sequence of rapid annealing ending with 900 °C, the Ti/Al contacts were found to be completely alloyed and to have resulted in a very low specific contact resistivity of 8×10^{-6} Ω cm^2.

D   InN

While a number of studies have been published describing InN deposition and electronic and optical property characterization [29-42], little attention has been paid to metallization and contact interfaces. Of all the group III nitrides, InN remains the least documented and least understood. Most of the InN films grown to date have been n-type with very high carrier concentrations (10^{18}-10^{20} cm^{-3}) and low mobilities. These characteristics have been attributed to native nitrogen vacancies. Tansley and Foley [37] reported the deposition of InN with much lower carrier concentrations ($\sim 10^{16}$ cm^{-3}) and high mobilities (~ 5000 cm^2 V^{-1}s^{-1}). In its more conductive form InN has been proposed as an ohmic contact material for GaN [28], as noted above. It is reportedly relatively easy to make ohmic contacts to InN using In metal or silver paste [43].

E   CONCLUSION

This Datareview discussed ohmic contact formation to AlN, GaN and InN which has been reported to date. As yet, GaN has received the most attention and a number of devices have been made, while the AlN and InN systems await better characterization and control of semiconductor properties. Many new developments coupled with extensive characterization of metal interfaces with the group III arsenides and phosphides have been made in the past several years, and these will probably be very helpful in the effort to develop corresponding nitride technology. Interest in and study of the group III nitrides are now progressing rapidly, and many new advances can be expected in the near future.

REFERENCES

[1]   T.C. Shen, G.B. Gao, H. Morkoç [*J. Vac. Sci. Technol. B (USA)* vol.10 (1992) p.2113-32]

[2]   R. Williams [in *Modern GaAs Processing Techniques* (Artech House, Norwood, MA, 1990) p.218-27]

[3]   M. Murakami [*Mater. Sci. Rep. (Netherlands)* vol.5 (1990) p.273-317]

[4]   A. Piotrowska, E. Kaminska [*Thin Solid Films (Switzerland)* vol.193/194 (1990) p.511-27]

[5]   A. Piotrowska, A. Guivarc'h, G. Pelous [*Solid-State Electron. (UK)* vol.26 (1983) p.179-97]

[6]   V.L. Rideout [*Solid-State Electron. (UK)* vol.18 (1975) p.541-50]

[7]   K. Tanahashi, H.J. Takata, A. Otsuki, M. Murakami [*J. Appl. Phys. (USA)* vol.72 (1992) p.4183-90]

[8]   H.J. Takata, K. Tanahashi, A. Otsuki, H. Inui, M. Murakami [*J. Appl. Phys. (USA)* vol.72 (1992) p.4191-6]

[9]   M.C. Hugon, B. Agius, F. Varniere, M. Froment, F. Pillier [*J. Appl. Phys. (USA)* vol.72 (1992) p.3570-7]

[10]   W.O. Barnard, G. Myburg, F.D. Auret [*Appl. Phys. Lett. (USA)* vol.61 (1992) p.1933-5]

[11]   G. Stareev [*Appl. Phys. Lett. (USA)* vol.62 (1993) p.2801-3]

[12]   J. Edwards, K. Kawabe, G. Stevens, R.H. Tredgold [*Solid State Commun. (USA)* vol.3 (1965) p.99-100]

[13]   R.F.Rutz [*IBM Tech. Discl. Bull. (USA)* vol.17 (1975) p.2800-1]

[14]   J.I. Pankove, P.E. Norris [*RCA Rev. (USA)* vol.33 (1972) p.377-82]

[15]   J.I. Pankove [*RCA Rev. (USA)* vol.34 (1973) p.336-43]

[16]   H. Amano, M. Kito, K. Hiramatsu, I. Akasaki [*Jpn. J. Appl. Phys. (Japan)* vol.28 (1989) p.L2112-L2114]

[17]   S. Nakamura [*Jpn. J. Appl. Phys. (Japan)* vol.30 (1991) p.L1705-L1707]

[18]   S. Nakamura, T. Mukai, M. Senoh [*Jpn. J. Appl. Phys. (Japan)* vol.30 (1991) p.L1998-L2001]

[19]   I. Akasaki, H. Amano, M. Kito, K. Hiramatsu [*J. Lumin. (Netherlands)* vol.48&49 (1991) p.666-70]

[20]   M.A. Khan, J.N. Kuznia, A.R. Bhattarai, D.T. Olson [*Appl. Phys. Lett. (USA)* vol.62 (1993) p.1786-7]

[21]   J.S. Foresi [Ohmic Contacts and Schottky Barriers on Gallium Nitride (MS thesis, Boston University, 1992)]

[22]   J.S. Foresi, T.D. Moustakas [*Appl. Phys. Lett. (USA)* vol.62 (1993) p.2859-61]

[23]   I. Akasaki, H. Amano, N. Koide, M. Kotaki, K. Manabe [*Physica B (Netherlands)* vol.185 (1993) p.428-32]

[24]   H.P. Maruska, D.A. Stevenson, J.I. Pankove [*Appl. Phys. Lett. (USA)* vol.22 (1973) p.303]

[25]   H.P. Maruska, D.A. Stevenson [*Solid-State Electron. (UK)* vol.17 (1974) p.1171]

[26]   Y. Morimoto, S. Ushio [*Jpn. J. Appl. Phys. (Japan)* vol.13 (1974) p.365-6]

[27]   N. Koide et al [*J. Cryst. Growth (Netherlands)* vol.115 (1991) p.639-42]

[28]   C.R. Abernathy, S.J. Pearton, F. Ren, P.W. Wisk [*J. Vac. Sci. Technol. B (USA)* vol.11 (1993) p.179-82]

[29]   Q. Guo, O. Kato, A. Yoshida [*J. Appl. Phys. (USA)* vol.73 (1993) p.7969-71]

[30]   A. Wakahara, T. Tsuchiya, A. Yoshida [*J. Cryst. Growth (Netherlands)* vol.99 (1990) p.385-9]

[31]   B.T. Sullivan, R.R. Parsons [*J. Appl. Phys. (USA)* vol.64 (1988) p.4144-9]

[32]   C.P. Foley, J. Lyngdal [*J. Vac. Sci. Technol. A (USA)* vol.5 (1987) p.1708-12]

[33]   T.L. Tansley, C.P. Foley [*J. Appl. Phys. (USA)* vol.60 (1986) p.2092-5]

[34]   T.L. Tansley, C.P. Foley [*J. Appl. Phys. (USA)* vol.59 (1986) p.3241-4]

[35]   K. Kubota, Y. Kobayashi, K. Fujimoto [*J. Appl. Phys. (USA)* vol.66 (1989) p.2984-8]

[36]   C.P. Foley, T.L. Tansley [*Phys. Rev. B (USA)* vol.33 (1986) p.1430-3]

[37]   T.L. Tansley, C.P. Foley [*Electron. Lett. (UK)* vol.20 (1984) p.1066-8]

[38]   T.L. Tansley, C.P. Foley [in *Semi-Insulating III-V Materials* Eds D.L. Look, J.S. Blakemore (Shiva, London, 1984)]

[39]   B.R. Natarajan, A.H. Eltoukhy, J.E. Greene [*Thin Solid Films (Switzerland)* vol.69 (1980) p.217-27]

[40]   B.R. Natarajan, A.H. Eltoukhy, J.E. Greene [*Thin Solid Films (Switzerland)* vol.69 (1980) p.201-16]

[41]   H.J. Hovel, J.J. Cuomo [*Appl. Phys. Lett. (USA)* vol.20 (1972) p.71-3]

[42]   J.B. MacChesney, P.M. Bridenaugh, P.B. O'Connor [*Mater. Res. Bull. (USA)* vol.5 (1970) p.783-92]

[43]   W.A. Bryden [private communication]

[44]   M.R.H. Khan, I. Akasaki, H. Amano, N. Okazaki, K. Manabe [*Physica B (Netherlands)* vol.185 (1993) p.480-4]

[45]   F.W. Ragay, M.R. Leys, J.H. Wolter [*Appl. Phys. Lett. (USA)* vol.63 (1993) p.1234-6]

[46]   M.G. Norton, J.M. Kajda, B.C.H. Steele [*J. Mater. Res. (USA)* vol.5 (1990) p.2172-6]

[47]   H. Asai et al [*IEEE Trans. Compon. Hybrids Manuf. Technol. (USA)* vol.13 (1990) p.457-61]

[48]   A.H. Carim, R.E. Loehman [*J. Mater. Res. (USA)* vol.5 (1990) p.1520-9]

[49]   P. Hacke, T. Detchprohm, K. Hiramatsu, N. Sawaki [*Appl. Phys. Lett. (USA)* vol.63 (1993) p.2676-8]

[50]   M.A. Khan, A. Bhattarai, J.N. Kuznia, D.T. Olson [*Appl. Phys. Lett. (USA)* vol.63 (1993) p.1214-5]

[51]   M.E. Lin, Z. Ma, F.Y. Huang, Z.F. Fan, L.H. Allen, H. Morkoç [*Appl. Phys. Lett. (USA)* vol.64 (1994) p.1003-5]

[52]   M.A. Khan, J.N. Kuznia, D.T. Olson, M. Blasingame, A.R. Bhattarai [*Appl. Phys. Lett. (USA)* vol.63 (1993) p.2455-6]

10.3 Schottky barriers and band offsets in AlN, GaN and InN

L.L. Smith and R.F. Davis

March 1994

A   INTRODUCTION

The formation of Schottky barriers is important for a variety of semiconductor device applications, and the properties of semiconductor/semiconductor and semiconductor/insulator interfaces are essential aspects of nearly all microelectronic architecture. In comparison to the arsenides and phosphides the group III nitrides have received only limited study of Schottky contacts or the electronic properties of interfaces with other semiconductors, insulators and substrates. Such studies of the arsenide and phosphide systems have progressed rapidly in recent years [1-6]; the nitride materials are experiencing a revival of interest as a result of recent improvements in film quality, doping, and p-n junction fabrication.

The group III nitride compounds are more ionically bonded than their phosphide and arsenide counterparts as a result of larger electronegativity differences between the component elements. According to the observations of Kurtin et al [7] this fact indicates that the nitrides should experience less Fermi level stabilization or 'pinning' at the surface than do the more covalent compounds. Thus, the barrier heights of contacts to the nitrides should be more dependent on the contact material than is the case with the more conventional and more covalent semiconductors such as Si, GaAs, InP, SiC, etc. With the work of Foresi and Moustakas [8,9], and more recently Hacke et al [31], this concept is beginning to be investigated. Their studies and others have revealed significant differences in the as-deposited I-V character of a variety of metals on undoped n-type GaN; thus there is some evidence that barrier heights on GaN vary with the choice of contact metal. However, a more thorough understanding of Fermi level pinning and interface states in GaN and its fellows awaits more rigorous investigation using material of lower defect density and lower background carrier concentration, and material having activated extrinsic n-type and p-type behaviour.

B   AlN

Although a number of studies of the band structure and electronic properties of AlN have been conducted over the years there has been little direct investigation of its interfacial electronic structure and properties with metals. The rapidly growing interest in the application of AlN as an IC packaging substrate has resulted in several studies that have characterized the chemical and physical structure of AlN/metal interfaces [10,11,31]; these studies are described in more detail elsewhere in this volume [36]. The electronic properties of AlN/metal interfaces have yet to be reported in detail, but as improvements are made in film growth and doping techniques, successful contact strategies will become important for device development.

Interfaces of AlN with substrate crystals and in heterostructures have been better documented. Within the past three years, Strite and Morkoç [12] and Davis [13] have published comprehensive reviews of group III nitride developments and included discussions of heterostructure characteristics. Layered structures consisting of AlN/GaN [14,17,18], AlGaN/GaN [16,19,20] and AlN/SiC [15,29] have been grown. However, for engineered device applications it is very important that the heterojunction band offsets be known. Significant progress in this direction has recently been reported in a paper by Martin et al [33], which describes measurements of the valence band discontinuity of a wurtzite (0001) AlN/GaN heterostructure. The valence band discontinuity was measured directly from the core-level binding energies obtained from X-ray photoelectron spectra. From these measurements the valence-band maximum was found to be 0.8 ± 0.3 eV.

All of the high-purity AlN films grown to date have been highly resistive. The lack of suitable doping techniques has so far limited the use of AlN as a semiconductor, but there has been some exploration of its use as an insulating layer. Alexandre et al [21] have described the preparation of an AlN/GaAs metal-insulator-semiconductor (MIS) diode, in which the AlN layer showed good insulating properties and a low interface trap density.

C   GaN

The semiconductor properties of GaN have been more thoroughly characterized than those of AlN or InN, but as yet there are still only a few examples of rectifying metal/GaN interfaces described in the literature. An early report by Kopeliovich et al [30] indicated that metals with a work function greater than ~4 eV yield finite barrier heights on GaN. Khan et al [22] reported forming a Cr/Au Schottky barrier on n-type AlGaN that had been compensated by N^+ implantation. More recently, Khan and co-workers have fabricated several GaN-based devices, including a MESFET [23], a HEMT [34], and a Schottky barrier photodetector [35]. An Ag Schottky barrier was used for the gate contact in the n-GaN-based MESFET; the HEMT gate contact metal was reported to be TiW and a non-annealed Ti/Au contact was used for the p-GaN photodetector. Khan et al [24] evaporated an Al contact as a Schottky electrode on Zn-compensated i-GaN for a MIS-type LED. The authors have deposited Pt contacts on both p- and n-type MBE-grown GaN; both systems exhibited rectifying behaviour in current-voltage (I-V) measurements. From I-V measurements the ideality factors of the Pt/n-GaN contacts were in the range 1.7-2.0, and the barrier height was calculated to be 1.0 eV.

Foresi and Moustakas [8,9] described the rectifying character of as-deposited Au contacts on undoped n-type films grown on sapphire via MBE. More recently, Hacke et al [31] have reported a study of Au in Schottky contact formation on undoped n-type GaN grown on sapphire by hydride vapour phase epitaxy. In this study temperature-dependent I-V measurements and C-V measurements were employed. The barrier height was found to be 0.84 eV and 0.94 eV from I-V and C-V measurements, respectively, with a forward current ideality factor of about 1.03 and a reverse bias leakage current below 10^{-10} A at -10 V.

The environmental stability of contact structures and properties is of considerable importance in the production and performance of microelectronic devices. Bermudez et al [37] have investigated the growth and behaviour of thin Ni films on GaN. A detailed analysis of surface and interfacial reactions was described. Chemical reaction at the interface was found to occur even near room temperature. Annealing at $\geq 600\,°C$ resulted in extensive intermixing of Ga and Ni, and loss of N. Observations such as these point out important considerations to be included in evaluating contact strategies.

Like AlN, high-resistivity GaN has been used as an insulating layer in GaAs-based device structures. Martin et al [25] and Strite et al [26] reported the formation of a GaAs/cubic GaN/GaAs semiconductor-insulator-semiconductor (SIS) diode structure, from which they calculated an effective conduction band offset of 0.9 eV. In addition, nitridation has been used as a means of improved chemical and electronic passivation of GaAs surfaces. According to Pankove et al [27], whereas AlN coatings on GaAs served merely as chemical encapsulants, GaN coatings effectively 'sealed in' the mobile charge carriers in the GaAs surface region and resulted in a fourfold improvement in luminescence efficiency.

D   InN

Schottky barriers and band offsets in pure InN have received little attention, in comparison with AlN and GaN. Most of the InN materials grown to date have had very high carrier concentrations, low mobilities, and poor thermal stability above 500 °C. However, some properties of alloys and heterostructures with the other group III nitrides have been described. Kubota et al [28] grew a layered GaN/InN structure and found a large degree of InGaN alloy formation at the interface. The same workers also describe the deposition of InAlN alloy and estimate the bandgap and lattice parameter of $In_{0.17}Al_{0.83}N$ to be very closely matched to GaN. According to these results, such a heterostructure would have no barrier.

E   CONCLUSION

This Datareview focused on examples of and issues relating to Schottky barriers and band offsets to AlN, GaN and InN. As yet, only a few examples of rectifying contacts and interfaces with other semiconductors have been described. A small number of devices and layered heterostructures have been formed, from which information on band structure discontinuities have been obtained, and several metal systems have been employed as Schottky barriers on GaN. There has been a renewed interest in the group III nitrides in recent years, and significant progress has been achieved. Additional research, including basic and applied studies of rectifying contacts to these materials, must be conducted to realize their full potential.

REFERENCES

[1]   S.X. Jin et al [*Appl. Phys. Lett. (USA)* vol.62 (1993) p.2719-21]
[2]   F.D. Auret, G. Myburg, H.W. Kunert, W.O. Barnard [*J. Vac. Sci. Technol. B (USA)* vol.10 (1992) p.591-5]
[3]   G. Myburg, F.D. Auret [*Appl. Phys. Lett. (USA)* vol.60 (1992) p.604-6]
[4]   J. Ding, Z. Lilienthal-Weber, E.R. Weber, J. Washburn, R.M. Fourkas, N.W. Cheung [*Appl. Phys. Lett. (USA)* vol.52 (1988) p.2160-2]
[5]   H.J. Chae et al [*J. Appl. Phys. (USA)* vol.72 (1992) p.3589-92]
[6]   D.T. Quan, H. Hbib [*Solid-State Electron. (UK)* vol.36 (1993) p.339-44]

[7]   S. Kurtin, T.C. McGill, C.A. Mead [*Phys. Rev. Lett. (USA)* vol.22 (1969) p.1433-6]

[8]   J.S. Foresi [Ohmic Contacts and Schottky Barriers on Gallium Nitride (MS thesis, Boston University, 1992)]

[9]   J.S. Foresi, T.D. Moustakas [*Appl. Phys. Lett. (USA)* vol.62 (1993) p.2859-61]

[10]   M.G. Norton, J.M. Kajda, B.C.H. Steele [*J. Mater. Res. (USA)* vol.5 (1990) p.2172-6]

[11]   H.Asai et al [*IEEE Trans. Compon. Hybrids Manuf. Technol. (USA)* vol.13 (1990) p.457-61]

[12]   S. Strite, H. Morkoç [*J. Vac. Sci. Technol. B (USA)* vol.10 (1992) p.1237-66]

[13]   R.F. Davis [*Proc. IEEE (USA)* vol.79 (1991) p.702-12]

[14]   Z. Sitar, M.J. Paisley, B. Yan, J. Ruan, W.J. Choyke, R.F. Davis [*J. Vac. Sci. Technol. B (USA)* vol.8 (1990) p.316-22]

[15]   L.B. Rowland, R.S. Kern, S. Tanaka, R.F. Davis [*Appl. Phys. Lett. (USA)* vol.62 (1993) p.3333-5]

[16]   M.A. Khan, R.A. Skogman, J.M. Van Hove, S. Krishnankutty, R.M. Kolbas [*Appl. Phys. Lett. (USA)* vol.56 (1990) p.1257-9]

[17]   S. Yoshida, S. Misawa, S. Gonda [*Appl. Phys. Lett. (USA)* vol.42 (1983) p.427-9]

[18]   S. Yoshida, S. Misawa, S. Gonda [*J. Vac. Sci. Technol. B (USA)* vol.1 (1983) p.250-3]

[19]   M.A. Khan, J.M. Van Hove, J.N. Kuznia, D.T. Olson [*Appl. Phys. Lett. (USA)* vol.58 (1991) p.2408-10]

[20]   M.A. Khan, J.N. Kuznia, J.M. Van Hove, D.T. Olson [*Appl. Phys. Lett. (USA)* vol.59 (1991) p.1449-51]

[21]   F. Alexandre, J.M. Masson, G. Post, A. Scavennec [*Thin Solid Films (Switzerland)* vol.98 (1982) p.75-80]

[22]   M.A. Khan, R.A. Skogman, R.G. Schulze, M. Gershenzon [*Appl. Phys. Lett. (USA)* vol.43 (1983) p.492-4]

[23]   M.A. Khan, J.N. Kuznia, A.R. Bhattarai, D.T. Olson [*Appl. Phys. Lett. (USA)* vol.62 (1993) p.1786-7]

[24]   M.R.H. Khan, I. Akasaki, H. Amano, N. Okazaki, K. Manabe [*Physica B (Netherlands)* vol.185 (1993) p.480-4]

[25]   G. Martin, S. Strite, J. Thornton, H. Morkoç [*Appl. Phys. Lett. (USA)* vol.58 (1991) p.2375-7]

[26]   S. Strite, D.S.L. Mui, G. Martin, Z. Li, D.J. Smith, H. Morkoç [in *GaAs and Related Compounds* Ed G.B.Stringfellow (Institute of Physics, Bristol, 1991) p.89-93]

[27]   J.I. Pankove, J.E. Berkeyheiser, S.J. Kilpatrick, C.W. Magee [*J. Electron. Mater. (USA)* vol.12 (1983) p.359-70]

[28]   K. Kubota, Y. Kobayashi, K. Fujimoto [*J. Appl. Phys. (USA)* vol.66 (1989) p.2984-8]

[29]   V.A. Dmitriev [*Physica B (Netherlands)* vol.185 (1993) p.440-52]

[30]   É.S. Kopeliovich et al [*Sov. Phys.-Semicond. (USA)* vol.9 (1975) p.125]

[31]   P. Hacke, T. Detchprohm, K. Hiramatsu, N. Sawaki [*Appl. Phys. Lett. (USA)* vol.63 (1993) p.2676-8]

[32]   A.H. Carim, R.E. Loehman [*J. Mater. Res. (USA)* vol.5 (1990) p.1520-9]

[33]   G.A. Martin et al [submitted to *Appl. Phys. Lett. (USA)* February 1993]

[34]   M.A. Khan, A. Bhattarai, J.N. Kuznia, D.T. Olson [*Appl. Phys. Lett. (USA)* vol.63 (1993) p.1214-5]

[35]   M.A. Khan, J.N. Kuznia, D.T. Olson, M. Blasingame, A.R. Bhattarai [*Appl. Phys. Lett. (USA)* vol.63 (1993) p.2455-6]

[36]   L.L. Smith, R.F. Davis [Datareview in this book: 10.2 Ohmic contacts to AlN, GaN and InN]

[37]   V.M. Bermudez, R. Kaplan, M.A. Khan, J.N. Kuznia [*Phys. Rev. B (USA)* vol.48 (1993) p.2436-44]

SUBJECT INDEX

SUBJECT INDEX